Use of Hydrocolloids to Control Food Appearance, Flavor, Texture, and Nutrition

Use of Hydrocolloids to Control Food Appearance, Flavor, Texture, and Nutrition

Amos Nussinovitch
The Robert H. Smith Faculty of Agriculture, Food and Environment
The Hebrew University of Jerusalem
Israel

Madoka Hirashima
Faculty of Education
Mie University
Japan

This edition first published 2023
© 2023 John Wiley & Sons Ltd

All rights reserved. No part of this publication may be reproduced, stored in a retrieval system, or transmitted, in any form or by any means, electronic, mechanical, photocopying, recording or otherwise, except as permitted by law. Advice on how to obtain permission to reuse material from this title is available at http://www.wiley.com/go/permissions.

The right of Amos Nussinovitch and Madoka Hirashima to be identified as the authors of this work has been asserted in accordance with law.

Registered Offices
John Wiley & Sons, Inc., 111 River Street, Hoboken, NJ 07030, USA
John Wiley & Sons Ltd, The Atrium, Southern Gate, Chichester, West Sussex, PO19 8SQ, UK

For details of our global editorial offices, customer services, and more information about Wiley products visit us at www.wiley.com.

Wiley also publishes its books in a variety of electronic formats and by print-on-demand. Some content that appears in standard print versions of this book may not be available in other formats.

Trademarks: Wiley and the Wiley logo are trademarks or registered trademarks of John Wiley & Sons, Inc. and/or its affiliates in the United States and other countries and may not be used without written permission. All other trademarks are the property of their respective owners. John Wiley & Sons, Inc. is not associated with any product or vendor mentioned in this book.

Limit of Liability/Disclaimer of Warranty
While the publisher and authors have used their best efforts in preparing this work, they make no representations or warranties with respect to the accuracy or completeness of the contents of this work and specifically disclaim all warranties, including without limitation any implied warranties of merchantability or fitness for a particular purpose. No warranty may be created or extended by sales representatives, written sales materials or promotional statements for this work. The fact that an organization, website, or product is referred to in this work as a citation and/or potential source of further information does not mean that the publisher and authors endorse the information or services the organization, website, or product may provide or recommendations it may make. This work is sold with the understanding that the publisher is not engaged in rendering professional services. The advice and strategies contained herein may not be suitable for your situation. You should consult with a specialist where appropriate. Further, readers should be aware that websites listed in this work may have changed or disappeared between when this work was written and when it is read. Neither the publisher nor authors shall be liable for any loss of profit or any other commercial damages, including but not limited to special, incidental, consequential, or other damages.

Library of Congress Cataloging-in-Publication Data
Names: Nussinovitch, A., author. | Hirashima, Madoka, author.
Title: Use of hydrocolloids to control food appearance, flavor, texture, and nutrition / Amos Nussinovitch, The Robert H. Smith Faculty of Agriculture, Food and Environment, The Hebrew University of Jerusalem, Israel, Madoka Hirashima, Faculty of Education, Mie University, Japan.
Description: Hoboken, NJ, USA : Wiley, 2023. | Includes bibliographical references and index.
Identifiers: LCCN 2022048786 (print) | LCCN 2022048787 (ebook) | ISBN 9781119700821 (cloth) | ISBN 9781119700838 (adobe pdf) | ISBN 9781119700906 (epub)
Subjects: LCSH: Food additives. | Food–Sensory evaluation. | Hydrocolloids–Industrial applications.
Classification: LCC TX553.A3 N665 2023 (print) | LCC TX553.A3 (ebook) | DDC 664/.06–dc23/eng/20221207
LC record available at https://lccn.loc.gov/2022048786
LC ebook record available at https://lccn.loc.gov/2022048787

Cover Design: Wiley
Cover Image: Courtesy of Amos Nussinovitch & Madoka Hirashima

Set in 9.5/12.5pt STIXTwoText by Straive, Pondicherry, India
Printed and bound by CPI Group (UK) Ltd, Croydon CR0 4YY

C9781119700821_201222

"Imagination is more important than knowledge. For knowledge is limited, whereas imagination embraces the entire world, stimulating progress, giving birth to evolution. It is, strictly speaking, a real factor in scientific research."

— **Albert Einstein**
In: Cosmic Religion: With Other Opinions and Aphorisms.
New York, Covici-Friede; 1st Edition (January 1, 1931)

Contents

Preface *xiii*
Acknowledgments *xxi*
About the Authors *xxiii*

1 Use of Hydrocolloids to Control Food Size and Shape *1*
1.1 Introduction *1*
1.2 The Attractive Shape of Foods *1*
1.2.1 Triangular and Prism-Shaped Foods *1*
1.2.2 Rectangular and Cube-Shaped Foods *4*
1.2.3 Circular and Spherical-Shaped Foods *4*
1.3 Selected Geometrical Properties of Foods *6*
1.3.1 Size *6*
1.3.2 Characterization of Size *6*
1.3.3 Size Reduction *7*
1.3.4 Energy Requirements for Size Reduction of Solid Materials *8*
1.4 Size Enlargement and Reduction Processes *10*
1.4.1 Definition of Forming and Its Aims *10*
1.4.2 Confectionery Molders *10*
1.4.3 Pie-Casing Formers *10*
1.4.4 Hydrocolloids in Food Fillings *11*
1.4.5 Cutting and Shaping Spherical Edible Products *12*
1.5 Shape – Definition and Implications *12*
1.5.1 Shape of a Food Commodity *12*
1.5.2 Roundness and Sphericity *12*
1.5.3 Average Projected Area and Sphericity of Hydrocolloid Beads *14*
1.5.4 How Are Gels Shaped? *15*
1.5.5 Silicone Molds to Modify Gel Shapes and Sizes *16*
1.6 Miscellaneous Shapes and Sizes of Edible Hydrocolloid Products *17*
1.6.1 Edible Hydrocolloid Gel Beads *17*
1.6.2 Parameters to Be Considered Upon Formation of Beads Through Capillary Jet Breakage *18*
1.6.3 Bead Shape and Its Improvement *20*
1.6.4 Shape and Size of Hydrocolloid Beads and Their Estimation *23*

1.7 Assorted Specially Shaped and Sized Hydrocolloid Foods *23*
1.7.1 Ham Consommé with Alginate Melon Beads *23*
1.7.2 Extruded Gel Noodles *24*
1.7.3 Cold Gels *24*
1.7.4 Knot Foie *24*
1.7.5 Shapes of Gummy Worms *25*
1.7.6 Gel Films *25*
1.8 Foods for the Elderly *26*
1.8.1 Effects of Hydrocolloid Addition on the Mastication of Minced Foods *27*
1.8.2 Hydrocolloids for the Design of Food for the Elderly *27*
1.9 Demonstrating the Use of Hydrocolloids in Controlling Food Size and Shape *28*
1.9.1 Agar Spaghetti *31*
1.9.2 Commercial Experimental Set to Produce Artificial Salmon Roe *32*
References *32*

2 Use of Hydrocolloids to Modulate Food Color and Gloss *40*
2.1 Introduction *40*
2.2 Appearance of Objects *40*
2.3 Optical Properties *41*
2.4 Color *42*
2.4.1 Color of Food Commodities *42*
2.4.2 Expressing Color Numerically *42*
2.4.3 The Kubelka–Munk Concept *47*
2.5 Gloss *48*
2.5.1 General Approach *48*
2.5.2 What Is Gloss and Why Is It Measured? *48*
2.5.3 Gloss Units and What Differences in Gloss Can Be Detected by Humans *49*
2.5.4 How Gloss Is Measured and Glossmeter Types *50*
2.6 On the Psychological Impact of Food Color and Gloss *51*
2.7 Where and When Are Hydrocolloids Utilized to Modulate Food Color and Gloss? *51*
2.7.1 Color of Fruit Leathers and Bars *51*
2.7.2 Gloss and Transparency of Edible Films *54*
2.7.3 High-Gloss Edible Coating *55*
2.7.4 Gloss and Transparency of HPMC Films Containing Surfactants as Affected by Their Microstructure *55*
2.7.5 Hydrocolloids in Forming Properties of Cocoa Syrups *56*
2.7.6 Color of Deep-Fat-Fried Products *56*
2.7.7 Spray-Dried Products *58*
2.7.8 Interaction of Anthocyanins with Food Hydrocolloids *59*
2.8 Demonstrating the Use of Hydrocolloids to Prepare Colored and Glossy Products/Recipes *60*
2.8.1 *Teriyaki* Fish with Pullulan *63*
2.8.2 Neutral Mirror Glaze (nappage neutre) *64*
References *65*

3 Use of Hydrocolloids to Modify Food Taste and Odor 74
3.1 Introduction 74
3.2 Flavor Perception: Aroma, Taste, and Volatile Compounds 74
3.3 Flavor of Hydrocolloid-Supplemented Value-Added Foods 78
3.3.1 Low-Fat Cheddar Cheese 78
3.3.2 Wholegrain Sorghum Bread 79
3.3.3 Fish Fingers 80
3.3.4 Meat Analogs 80
3.3.5 Spreads 81
3.3.6 Protein Beverages 81
3.4 Interactions of Flavor Compounds with Different Food Ingredients 82
3.4.1 Interactions Between Proteins and Flavor Compounds 82
3.4.2 Interactions Between Starch and Flavor Compounds 83
3.4.3 Interactions Between Hydrocolloids and Flavor Compounds 84
3.5 Effect of Hydrocolloids on Sensory Properties of Selected Model Systems and Beverages 86
3.6 Influence of Hydrocolloids on the Release of Volatile Flavor Compounds 88
3.7 Gels and Flavor 90
3.7.1 Hydrocolloid Gels and Flavor Release 90
3.7.2 Phase-Separated Gels and Aroma Release 91
3.8 The Influence of Flavor Molecules on the Behavior of Hydrocolloids 92
3.9 Demonstrating the Use of Hydrocolloids in Modifying Food Taste/Odor 92
3.9.1 Fried Chicken with Methylcellulose (MC) 95
3.9.2 Gluten-Free Bread with Hydroxypropyl Methylcellulose (HPMC) 96
References 97

4 Use of Hydrocolloids to Control Food Viscosity 107
4.1 Viscosity of Fluids 107
4.1.1 The Field of Flow and Viscosity 107
4.1.2 Laminar Flow and Turbulent Flow 108
4.2 Important and Useful Definitions 111
4.2.1 Dynamic Viscosity and Fluidity 111
4.2.2 Kinematic Viscosity 111
4.2.3 Relative Viscosity 111
4.3 Flow Equations 112
4.3.1 Definitions of Apparent Viscosity, Shear Stress, and Shear Rate 112
4.3.2 The General Equation for Viscosity 112
4.3.3 The Power Equation 113
4.3.4 The Herschel-Bulkley Model 113
4.3.5 Casson Equation 114
4.4 Thickening and Viscosity-Forming Abilities of Hydrocolloids – A General Approach 115
4.5 Hydrocolloids as Viscosity Formers in Foods 116
4.6 Time Dependence of Hydrocolloid Solutions 121
4.7 Fluid Gels 124

4.8	Demonstrating the Use of Hydrocolloids to Control Viscosity in Foods	126
4.8.1	Creamy Italian Dressing	128
4.8.2	French Dressing	128
	References	129

5 Use of Hydrocolloids to Improve the Texture of Crispy, Crunchy, and Crackly Foods 136

5.1	Introduction	136
5.2	Definitions of Crispness and Crunchiness	136
5.3	Dependence of Crunchiness and Crispness on Moisture and Oil Content	137
5.4	Mechanical, Acoustical, and Temporal Aspects of Crunchiness and Crispness	140
5.5	Crackly Foods	142
5.6	Methods for Improving the Texture of Crispy and Crunchy Foods Using Hydrocolloids	146
5.6.1	Vacuum Frying	146
5.6.2	Coating and Batter	148
5.7	Enhancement of Food Acoustic Properties Using Various Hydrocolloids	149
5.7.1	Contribution of Inulin to Crispness of Biscuits, Pizza, and Wafers	149
5.7.2	Crispness of Banana Chips	149
5.7.3	Specialty Starches as Functional Ingredients	150
5.7.4	Specialty Starches in Snack Foods	151
5.7.5	Protein-Rich Extruded Snack	152
5.8	Demonstrating the Preparation of Crunchy Products	154
5.8.1	Baked Tortilla Chips	156
5.8.2	Commercial Fabricated Potato Chips	157
5.8.3	Commercial Fabricated Fried Potato	157
	References	157

6 Use of Hydrocolloids to Improve the Texture of Hard and Chewy Foods 166

6.1	Texture Definitions	166
6.1.1	Hardness	166
6.1.2	Chewiness	167
6.1.3	Juiciness	167
6.2	Use of Hydrocolloids to Improve Bread Texture	168
6.3	Dairy Products	171
6.3.1	Dairy Foods	171
6.3.2	Cheeses	172
6.3.3	Functionality of Selected Hydrocolloids on Texture of Ice Cream	172
6.4	Fish Products	174
6.5	Further Contributions of Hydrocolloids to Textural Improvement	175
6.6	Other Miscellaneous Applications	176
6.6.1	Rice Starch Pastes	176
6.6.2	Rice Starch–Polysaccharide and Other Mixed Gels	177
6.6.3	Hydrocolloid Effects on Pea Starch	178

6.7 Demonstrating the Use of Hydrocolloids in Creating/Controlling Food Hardness and Chewiness *179*
6.7.1 Agar Jelly, *Seiryu* *182*
6.7.2 Low-Concentration Carrageenan Jelly, *mizu-Shingen mochi* *183*
References *183*

7 Use of Hydrocolloids to Control the Texture of Multilayered Food Products *192*
7.1 Introduction *192*
7.2 Multilayered Hydrocolloid-Based Foodstuffs *192*
7.2.1 Confectionery Products *192*
7.2.2 Cream-Filled Multilayered Food Products *194*
7.2.3 Gelled Multilayered Food Products *195*
7.2.4 Multilayered Films *197*
7.2.5 Nano-Multilayer Coatings *198*
7.2.6 Multilayered Liposomes and Capsules *199*
7.2.7 Multilayered Particles *199*
7.3 Methods to Estimate Properties of Multilayered Products *200*
7.3.1 Assessment of Stiffness and Compressive Deformability of Multilayered Texturized Fruit and Gels *200*
7.3.2 Calculating the Stress–Strain Relationships of a Layered Array of Cellular Solids *202*
7.3.3 Other Techniques to Assess Multilayered Products *205*
7.4 Current Systems and Methods to Prepare Multilayered Products *205*
7.4.1 Extrusion and Coextrusion *205*
7.4.2 Injection Molding *207*
7.4.3 3D-Printing and Layered Products *208*
7.4.4 Multilayered Emulsions *208*
7.5 Further Matters Related to Multilayered Products *209*
7.5.1 Natural Food-Grade Emulsifiers and Interfacial Layers *209*
7.5.2 Multilayer Adsorption *210*
7.5.3 Gelled Double-Layered Emulsions *210*
7.6 Complications Related to Multilayered and Colored Products *211*
7.7 Future Potential Biotechnological Uses of Multilayered Gels *215*
7.8 Demonstrating the Use of Hydrocolloids to Prepare Multilayered Products/Recipes *216*
7.8.1 Multilayered Gelatin Jelly *219*
7.8.2 Beer-Like Jelly *219*
References *220*

8 Hydrocolloids to Control the Texture of Three-Dimensional (3D)-Printed Foods *230*
8.1 Introduction *230*
8.2 A Brief History of 3D Printing *230*
8.3 3D Printing of Foods *231*
8.3.1 3D Options in Foods *231*

8.3.2 Special Personalized Foods for the Elderly 233
8.4 3D-Printed Food Products 234
8.4.1 Printed Sugar Products 234
8.4.2 Chocolate 235
8.4.3 Pastes, Pizza, Cookies, and Meat 236
8.5 Production of Snacks 237
8.5.1 Cereal-Based 3D Snacks 237
8.5.2 Fruit Snacks 238
8.6 Printability of Food Additives 238
8.6.1 Issues Related to 3D Food Printing 238
8.6.2 Printability of Hydrocolloids 238
8.6.3 Protein Products Applicable for 3D Printing 239
8.6.4 The Effect of 3D Printing on Lipids 240
8.7 Infill Percentage and Pattern 241
8.8 Modifying Food Texture to Suit Personal and Other Requirements by 3D Printing Technology 242
8.9 Hydrocolloids in 3D Printing 243
8.10 3D Printing of Hydrocolloid Foods Served in Restaurants 244
8.11 3D Printing and Laser Cooking 248
8.12 Novel Application for 4D Food Printing 248
References 249

9 Use of Hydrocolloids to Control Food Nutrition 255
9.1 Nutritional Applications of Natural Hydrocolloids 255
9.2 Types of Dietary Fibers 256
9.3 Dietary Fiber as a Versatile Food Component 257
9.4 Food Enriched in β-Glucans 258
9.5 Cereal Polysaccharides as the Foundation for Useful Ingredients in the Reformulation of Meat Products 259
9.6 Health Claims of Hydrocolloids 261
9.7 Miscellaneous Cases of Nutritional and Health Benefits 262
9.7.1 Health Benefits of Lactic Acid Bacteria (LAB) Exopolysaccharides (EPSs) 262
9.7.2 Fat Replacers 264
9.7.3 Benefits of Dietary Fermentable Fibers for Chronic Kidney Disease (CKD) 265
9.8 Demonstrating the Use of Hydrocolloids in Controlling Nutrition 266
9.8.1 Keto Bread Rolls with Inulin 269
References 270

Index 279

Preface

Hydrocolloids are among the most commonly used and versatile ingredients in the food industry. They function as thickeners, gelling agents, texturizers, stabilizers, and emulsifiers; they also have applications in the areas of edible coatings and flavor release. Manufactured foods that are reformulated for reduced fat rely primarily on hydrocolloids to provide suitable sensory quality. Furthermore, hydrocolloids are currently finding increasing applications in the health arena; they provide low-calorie dietary fiber, among many other uses.

Many books have been devoted to descriptions of the different water-soluble polymers (hydrocolloids) and their uses. In 1965, a monograph by M. Glicksman, *Gum Technology in the Food Industry* (Academic Press), presented a technical compilation of information in the area of hydrocolloid technology as it pertains to the food industry. The need for such a book was apparent to most food technologists and scientists, particularly those engaged in the development of convenience foods. Glicksman followed that book with three more volumes (1982–1984) entitled *Food Hydrocolloids, volumes I, II, and III* (CRC Press). Volume I was composed of two parts, the first dealing with comparative properties of hydrocolloids and the second with biosynthetic gums. Volume II dealt with natural food exudates and seaweed extracts, and Volume III described cellulose gums, plant seed gums, and plant extracts. Those books were much more comprehensive than Glicksman's first monograph and were very useful for both food technologists and academics.

In 1982, an excellent book entitled *Handbook of Water-Soluble Gums and Resins* (McGraw Hill Company) was edited by R. L. Davidson. The book comprised 23 chapters written by advisors and contributors from universities and the industry. It contained information on where water-soluble gums and resins come from, how they are used, how they work, and their individual uses to obtain specific properties and performance. It gave an encyclopedic description of the major commercial varieties of both natural and synthetic gums and resins, each listing beginning with a concise overview, followed by full details on the chemistry, properties, handling uses, and other pertinent factors.

In 1997, a monograph by one of us (A. Nussinovitch) entitled *Hydrocolloid Applications: Gum Technology in the Food and Other Industries* (Blackie Academic & Professional) was published, comprised of two parts. The first dealt briefly with describing the known hydrocolloids. The second was devoted to information, which is more difficult to locate, namely uses of hydrocolloids in ceramics, cosmetics, and explosives, for glues, for immobilization and encapsulation, in inks and paper, and for the creation of spongy matrices, textiles, and different texturized products. Another monograph by A. Nussinovitch entitled *Water-Soluble Polymer Application in Foods* (Blackwell Science) from 2003 was devoted to the uses of hydrocolloids in foods and in biotechnology and discussed topics such as hydrocolloid adhesives, hydrocolloid coatings, dry macro- and liquid-core capsules, multilayered products, flavor encapsulation, texturization, cellular solids, and hydrocolloids in the production of special textures. Yet another monograph by A. Nussinovitch from 2010, *Plant Gum Exudates of the World: Sources, Distribution, Properties and Applications* (CRC), provided a description of the most extensive collection of plant gum exudates in print. The book included a chapter specifically devoted to food uses of plant exudates, including confectionery, salad dressings and sauces, frozen products, spray-dried products, wine, adhesives, baked products, and beverages, among many other industrial products and animal foods.

In 2021, the 3rd edition of *Handbook of Hydrocolloids*, edited by G. O. Phillips and P. A. Williams (Woodhead Publishing, an imprint of Elsevier), was published. This excellent book reviewed hydrocolloids obtained from plants and trees, seaweed, bacteria, and fungi together with chapters on chitin derived from crustaceans, collagen and gelatin from animals, flaxseed gum and mustard gum, as well as dendronan and beta glucans from fungal sources. The introductory chapter gives an overview of the main hydrocolloids, their regulatory status, a comparison of their thickening and gelation properties, their ability to control the stability of emulsions and dispersions and their function as dietary fiber. The book also includes chapters dealing with techniques for the chemical and physicochemical characterization of hydrocolloids, oral processing and tribology, mixed hydrocolloid systems, hydrocolloids for the encapsulation and delivery of active compounds and their application as edible films, coatings, and food packaging, as well as a chapter on the health aspects of hydrocolloids, reviewing their impact on gut health, digestion, and the absorption of nutrients and postprandial plasma constituents.

The practical guide book: *Food Stabilizers, Thickeners and Gelling Agents* (Wiley-Blackwell, 2010), edited by A. Imeson, reviewed the incorporation of hydrocolloids into foods to give them structure, flow, stability, and the eating qualities desired by consumers.

These are just a few examples of the wealth of material covering this field of science. Note that the inclusion of a book in this short list does not imply that it is any better than other published books on hydrocolloids or their widespread applications.

Although food recipes could be found in a few of these many books, there had been no scientific books fully devoted to the fascinating topic of hydrocolloids and their unique applications in the kitchen. A kitchen can be regarded as an experimental laboratory, with food preparation and cookery involving processes that are well described by the chemical or physical sciences. It is well established that an understanding of the chemistry and physics of cooking, and of the involvement of different ingredients (such as hydrocolloids) in these processes, will lead to improved performance and increased innovation in this realm. Since the use of hydrocolloids is on the rise in many fields, the writing of a book that covers both past and future uses of hydrocolloids in the kitchen was both timely and of great interest. In 2013, we co-authored the book *Cooking Innovations: Using Hydrocolloids for Thickening, Gelling and Emulsification* (CRC, Taylor & Francis Group). That very successful book did not include several very important hydrocolloids, among them chitin and chitosan, gum karaya, gum tragacanth, and milk proteins. These were included in our book (CRC, Taylor & Francis Group, 2018) *More Cooking Innovations: Novel Hydrocolloids for Special Dishes*, which completed the work of the first volume. In this latter volume, we added chapters on unique hydrocolloids that, in our opinion, will not only be used in future cooking but can pave the way to new and fascinating recipes and cooking techniques. These books were much more than just traditional cookbooks. The useful, albeit not always purely theoretical background provided for the different hydrocolloids was extensive and to the point. We made an attempt to include not only recipes that everyone can follow within the confines of their own kitchen – be they professional chefs or amateur cooks, but also products and scientific experiments that can be conducted and studied in university laboratories, such as those currently being performed by one of the authors (M. Hirashima) in Japan. As such, these books can be used as textbooks for cooking science and food-processing classes and provide recipes that can be scaled up for industrial use, making them ideal for food technologists and engineers. They bridge the gap between scientist and chef, or in fact anyone who is interested in novel applications and textures in the kitchen. In addition, they were designed to serve as a guide for all those who want to introduce the fascinating world of hydrocolloids to the public. These cooking books serve to advance both the evolution and revolution of hydrocolloid uses in today's kitchen. If in 1965, Glicksman's first book about hydrocolloids paved the way for those who were trying to use hydrocolloids in food technology, cooking, and the industry, these two books on cooking innovations shed light on some hydrocolloids that were not well known 50 years ago and that can produce fantastic recipes and unique results.

The format of most hydrocolloid textbooks, including our scientific cooking books, consists of a chapter for each hydrocolloid that provides data on its chemical and physical properties, then more or less detailed information on its possible inclusion in a particular food formulation(s), followed every now and then by a brief explanation of how the hydrocolloid contributes to the food's texture, eliminating or decreasing problems such as syneresis, improving stability, etc.

The current book, *Use of Hydrocolloids to Control Food Appearance, Flavor, Texture and Nutrition*, aims to present the whole field of hydrocolloid utilization in foods from a totally different angle. In accordance with Professor Bourne's classical book on *Food Texture and Viscosity: Concept and Measurement* (Academic Press, 2002), the four principal quality factors in foods are appearance, flavor, texture, and nutrition. Appearance covers color, shape, size, gloss, appealing to the visual sense. Flavor comprises taste and odor. Texture is primarily the response of the tactile senses to physical stimuli, and nutrition includes major nutrients (carbohydrates, fats, and proteins) and minor nutrients (fiber, minerals, and vitamins). Each of these four quality factors can be controlled, improved, modified, and/or changed by hydrocolloids and in this unique book, we will show how this is achieved. Although this is not a cookbook, most chapters will include figures of relevant commercial or homemade food products and except for Chapter 8, at least one, and usually a few recipes demonstrating a particular hydrocolloid's unique abilities for achieving the abovementioned food quality factors, and those abilities will be elaborated upon. In this book, several formulations have been chosen specifically for the food technologist, who will be able to manipulate them for large-scale use or practice them as a starting point for novel industrial formulations.

Chapter 1. Use of Hydrocolloids to Control Food Size and Shape

Appearance – one of the four quality factors of foods – comprises color, shape, size, gloss, and visual appeal. In general, foods' physical properties include size and shape, and number, nature, and conformation of constituent structural elements. During mastication, the size and shape of food particles and their surface roughness are sensed, making them inseparable attributes of the overall textural sensation. The ability to use hydrocolloids to control food shape and size is common knowledge. This chapter begins by defining size and shape. Round and spherical shapes are emphasized, along with the means to estimate/calculate roundness and sphericity. Other shapes, geometrical and irregular, are also illustrated. Explanations of how to obtain different food shapes using different molds for setting and other techniques, such as flow of a preset gel through silicone tubes and simultaneous cooling of an agar–agar or other gelling agent solution to obtain a noodle (spaghetti) shape, are described. The chapter also includes a description of cutting techniques for gels or foods that are created with the help of hydrocolloids, as well as spherification to produce spherical and ellipsoid shapes (e.g. artificial salmon eggs), and different illustrations of these foods' appearance.

Chapter 2. Use of Hydrocolloids to Modulate Food Color and Gloss

Many foods are difficult to identify when their texture or color are concealed, leaving flavor as the only recognizable attribute. Both color and gloss can be formulated and applied to compound or processed foods. Control of food gloss and color through coating with sauces that include hydrocolloids as stabilizers can be easily achieved using the suitable hydrocolloid(s) in combination with other food ingredients. Colors of baked and fried foods can also be changed through the inclusion of hydrocolloids in their composition. The controlled introduction of changes in foods is explained and exemplified for special food coatings containing hydrocolloid films, for example cooking fish with the addition of pullulan to produce a neutral mirror glaze (*nappage neutre*) with pectin.

Chapter 3. Use of Hydrocolloids to Modify Food Taste and Odor

Flavor, comprised of taste (perceived by the tongue) and odor (perceived by the olfactory center in the nose), results from the response of receptors in the oral and nasal cavities to chemical stimuli. These are called "the chemical senses." In this chapter, flavor, aroma, taste, and volatile compounds are discussed. Examples of foods whose flavor is governed, enhanced, or changed by the addition of hydrocolloids, such as low-fat cheddar cheese, fish and meat analog products, specialty breads, protein beverages and spreads, are provided. A section is devoted to the detailed interactions of flavor and different food ingredients, including interactions between proteins and flavor compounds, hydrocolloids and flavor compounds, and a special discussion devoted to starch. Other distinct topics include the effect of hydrocolloids on sensory properties of selected model systems and beverages; the influence of hydrocolloids on the release of volatile flavor compounds; and the relationship between gels and flavor release. Most of the available literature emphasizes the effect of hydrocolloids on flavor release and perception. Nevertheless, from a practical perspective, it is just as important to understand the influence of flavor molecules on the behavior of hydrocolloids, and therefore, a section is devoted to this issue. Finally, the use of hydrocolloids to modify the taste/odor of foods is exemplified by two recipes. In the first, the batter for fried chicken includes methylcellulose to improve taste and flavor; in the second, hydroxypropyl methylcellulose is used to create gluten-free bread, with good mouthfeel, satisfactory volume, and retained softness over time.

Chapter 4. Use of Hydrocolloids to Control Food Viscosity

In contrast to aroma, color, and taste, which can be formulated and introduced into compound or processed foods, texture cannot be added "from a bottle." The viscosity of a medium is enhanced through immobilization of water by macromolecules with some potential intermolecular bonding. Viscosity is defined as the internal friction of a fluid or its tendency to resist flow. The International Handbook of Food Additives mentions numerous substances that can be added to foods. Over 10% of these constituents are described as texturizers, thickeners, viscosity modifiers, and bodying agents. They give the food technologist a large selection of aids to develop desired textures. This chapter includes sections on flow and viscosity, including definitions, flow equations and models, a description of the thickening abilities of hydrocolloids and their use as viscosity formers in foods, time dependency of hydrocolloid solutions and fluid gels, and a demonstration of the use and performance of hydrocolloids to control viscosity of foods in recipes for Italian and French dressings that include typical hydrocolloids.

Chapter 5. Use of Hydrocolloids to Improve the Texture of Crispy, Crunchy, and Crackly Foods

Food texture is primarily judged by the tactile senses' detection of physical stimuli resulting from contact between some part of the body and the food. Touch is the main means of sensing texture but kinesthetics (sense of movement and position), and sometimes sight (degree of slump, rate of flow) and sound (associated with crisp, crunchy, and crackly textures) also come into play. This chapter starts with definitions of crispness and crunchiness and their dependence on a food's moisture and oil content. The mechanical, acoustical, and temporal aspects of crunchiness and crispness are described and analyzed. An entire section is devoted to the definition of crackly foods and how to influence their properties. Methodologies for improving the texture of crispy and crunchy foods using hydrocolloids are described in detail, with sections on coatings and batters for vacuum frying. A special section is devoted to foods that are treated with various hydrocolloids to enhance their acoustical properties. Cases such as the contribution of inulin to the crispness of biscuits, pizza, and wafers; crispness of banana chips; specialty starches as functional ingredients and in snack foods; and protein-rich extruded snacks are also described. A recipe for baked tortilla chips is provided, as well as photographs of typical crisp and crunchy products.

Chapter 6. Use of Hydrocolloids to Improve the Texture of Hard and Chewy Foods

Among the many textural properties of foods are hardness/softness, juiciness, and chewability, which may be variously favored by different cultures. In this chapter, we first define some textural terms and suggest their importance and similarities/differences among different cultures. Objective and subjective methods for estimating these properties are very briefly mentioned and several representative food families and models are discussed – including bread, dairy products, fish products – for improvement of textural features and special applications. The chapter also includes recipes that demonstrate how hydrocolloids serve to create/improve textural and sensory effects.

Chapter 7. Use of Hydrocolloids to Control the Texture of Multilayered Food Products

A simple technique for achieving different textures and tastes in the same bite is to make a food product consisting of different layers. This chapter starts with a description of a handful of such multilayered hydrocolloid products. It explains layer adhesion and how some of the mechanical properties of the layered array can be estimated from the properties of the individual layers. Different adhesion methods for multilayered hydrocolloid gels might lead to innovative products in the food industry and to the development of novel foods and dishes. Multilayered confectionery compositions made up of hydrocolloids and formed using extrusion or co-extrusion processes with no less than two different confectionery compositions which differ visually or sensorially are described. This chapter considers multilayered confectionery products, cream-filled products, multilayered gel products, and multilayered films, coatings, particles, and liposomes. Problems related to multilayered and colored products are described. Recipes for multilayered gelatin jelly and beer jelly, and photographs of multilayered Japanese sweets are also included to demonstrate how simple, and sometimes complicated, it is to construct such products, as well as the usefulness and beauty of these foods.

Chapter 8. Hydrocolloids to Control the Texture of Three-Dimensional (3D)-Printed Foods

In recent years, the potential applications of 3D food printing have revealed this technique's prospects. In general, printable materials have different tastes, nutritional values, and textures. Some of them are sufficiently stable to hold their shape after deposition/extrusion, whereas others may require a post-cooking process. Non-printable, traditional foods require the addition of hydrocolloids to be processed. This chapter includes a brief history of 3D printing and a description of 3D printing options for food. Other topics include cereal-based snacks, and the printability of proteins, carbohydrates and lipids; infill percentage and pattern; and modification of food texture to suit individual and other requirements for 3D printing technology. This chapter focuses on hydrocolloids in the 3D printing process, with a mention of their use to control the texture of 3D-printed foods. The chapter concludes with some futuristic options, such as 3D printing and laser cooking and a novel application for four-dimensional (4D) food printing.

Chapter 9. Use of Hydrocolloids to Control Food Nutrition

Hydrocolloids have numerous and growing applications in the health realm; e.g., they provide low-calorie dietary fiber, among many other uses. Food hydrocolloid research has recently been launched into, for example, dietary fiber with physiological effects for an elderly society, where the number of people with mastication and swallowing difficulties, and of patients with lifestyle-related diseases is on the rise; hydrocolloids that have conventionally been used only to provide texture but are now being used for their nutritional value; and processing of food fibers to enable their use as a food ingredient with nutritional value and a textural role. Furthermore, novel functional food fibers with textural and nutritional benefits are emerging. Hydrocolloids themselves can be exploited for their inherent nutritional value, as they can contain a large proportion of dietary fiber. One such hydrocolloid is the commercially available partially hydrolyzed guar gum, whose low viscosity enables its use at sufficiently high concentrations to obtain the benefits of its high dietary fiber content. The use of functional fibers in baked goods and meat applications, where water binding is key, but also as constituents of beverages, is discussed.

Each chapter in this book addresses one or more particular quality factor. The chapter starts with a description of the nature of the factor. It is important to note that this book is not intended as a replacement for the already published books on hydrocolloid properties (some of which are mentioned above); our aim is not to compete with or repeat any of the information found in those books. Our book may serve to inspire students and scientists, and to introduce food technologists, professional chefs and amateur cooks alike to the myriad uses of hydrocolloids – how they are used and their specific purposes. In summary, the volume is written such that chefs, food engineers, food science students, and other professionals will be able to cull ideas from the contents and recipes and be initiated into the what, where, and why of specific hydrocolloids' use.

We believe that this is going to be an extremely useful book and an essential purchase for personnel involved in food formulation, food science, and food technology, in particular food scientists (chemists/microbiologists/technologists) working in product development, food engineers whose job typically involves figuring out how to make the products envisioned by developers, research chefs, e.g. members of the Research Chefs Association, professional restaurant chefs who like to experiment with new creations, members of the American Culinary Federation and culinary education programs, and members of equivalent organizations in foreign countries, as well as amateur cooks. It will also serve those who are interested in developing novel foods for emerging areas, e.g. the growing elderly market and 3D food production, and will be a welcome addition to the traditional libraries of universities and research institutes where food science, chemistry, agriculture, and other theoretical and practical industrial food issues are taught or studied. In that sense, the book is unique and we are confident that it will be a great success.

Amos Nussinovitch and Madoka Hirashima

Acknowledgments

This book was written over the course of two years, following the success of our previous two books: *Cooking Innovations: Using Hydrocolloids for Thickening, Gelling, and Emulsification* and *More Cooking Innovations: Novel Hydrocolloids for Special Dishes*, both with CRC Press, Taylor & Francis Group. The current title: *Use of Hydrocolloids to Control Food Appearance, Flavor, Texture, and Nutrition* (Wiley) deals with four principal quality factors in foods: appearance, flavor, texture, and nutrition. Each of these factors can be controlled, improved, modified, and/or changed by hydrocolloids. The chapters include demonstrations of how particular hydrocolloids' unique abilities are used to achieve desired food quality factors. In addition, the book includes several formulations that have been chosen specifically for the food technologist, who will be able to manipulate them for large-scale use or practice them as a starting point for novel industrial formulations. The book was written with the belief that an understanding of the chemistry and physics of the involvement of hydrocolloids in food science and technology will lead to improved performance and increased innovation in this realm. The use of hydrocolloids is on the rise in numerous fields. The time is therefore ripe for this book, which covers both past and future uses of hydrocolloids in the market, industry, and kitchen. Our hope is that this book will assist readers who are in search of comprehensive knowledge about the fascinating field of hydrocolloids, as well as those seeking up-to-date information on the very different past, current, and future uses and applications of hydrocolloids in foods. The volume is written such that food engineers, food science students, chefs, amateur cooks, and other professionals will be able to cull ideas from the text, photographs, and included recipes and be initiated into the what, where, and why of specific hydrocolloids' uses.

We wish to thank the publishers for giving us the opportunity to write this book. Special thanks to Rebecca Ralf, Commissioning Editor, Life Sciences at John Wiley & Sons Limited Chichester, West Sussex, United Kingdom, for her efficient handling of the project, from its conception to the moment that we got the green light to start writing, photographing, and cooking. Rebecca's genuine interest, enthusiasm, and encouragement during the process were phenomenal and deeply appreciated. We wish to thank Rosie Hayden, the very efficient Senior Managing Editor of Wiley, for maintaining a very good connection with us throughout the project and responding very quickly with full understanding to all of our numerous queries. We wish to thank our editor, Camille Vainstein, for working shoulder-to-shoulder with us when time was getting short. We are grateful to the illustrator Netta Kasher for drawing the wonderful cover art for this book. We wish to thank Prof. Hiroshi

Wada, PhD, from the Plant Biophysics/Biochemistry Research Laboratory Graduate School of Agriculture, Ehime University, Japan, for giving us the permission to use a photograph related to the hidden mechanisms of apple watercore formation. We wish to thank Akinori Yusa from the Ricoh Institute of Sustainability and Business and Dr. Masaru Kawakami from Yamagata University for providing us with the 3D food printer photographs. We also wish to thank Sugar Lab (https://sugarlab3d.com/) for giving us the permission to use the 3D romantic red ombré photograph. We are also grateful to OPENMEALS (https://www.open-meals.com/) for offering us their files of 3D-printed images.

Last, but not least, we wish to thank the Hebrew University of Jerusalem and Mie University for being our home and refuge for many years of very extensive research and teaching.

June 2022

Amos Nussinovitch & Madoka Hirashima
Israel & Japan

About the Authors

Professor Amos Nussinovitch was born in Kibbutz Megiddo, Israel. He studied Chemistry at Tel Aviv University, and Food Engineering and Biotechnology at the Technion – Israel Institute of Technology. He has worked as an engineer at several companies and has been involved in a number of R&D projects in both the United States and Israel, focusing on the mechanical properties of liquids, semisolids, solids, and powders. He is currently at the Biochemistry and Food Science Department of the Robert H. Smith Faculty of Agriculture, Food and Environment of the Hebrew University of Jerusalem, where he leads a large group of researchers working on theoretical and practical aspects of hydrocolloids. Professor Nussinovitch is the sole author of the books: *Hydrocolloid Applications: Gum Technology in the Food and Other Industries*; *Water-Soluble Polymer Applications in Foods*; *Plant Gum Exudates of the World: Sources, Distribution, Properties, and Applications*; *Polymer Macro- and Micro-Gel Beads: Fundamentals and Applications*; *Adhesion in Foods: Fundamental Principles and Applications*. He and his present co-author Professor Madoka Hirashima (see below) recently co-authored the books: *Cooking Innovations: Using Hydrocolloids for Thickening, Gelling, and Emulsification* and *More cooking innovations: Novel Hydrocolloids for Special Dishes*. Professor Nussinovitch is the author or co-author of numerous papers on hydrocolloids and on the physical properties of foods, and he has many patents. Several years ago, Professor Nussinovitch received a lifetime award from the Manufacturers Association of Israel for his unique and considerable contributions to both academia and the food industry in Israel. In 2020, Professor Nussinovitch received the prestigious Kaye prize for his studies on developing edible protective films to extend postharvest shelf life of fresh and processed fruit and vegetables. In parallel, he founded the successful startup company "Sufresca" (coating of fruit and vegetables for shelf-life extension) and serves as its CSO. He has also served as the representative of Israeli food chemists in the European Food Chemistry Division for many years now.

Professor Madoka Hirashima was born in Kyoto, Japan. She studied the rheological properties of curdlan and cornstarch at the Graduate School of Human Life Sciences, Osaka City University. She worked at a food company as a new food developer and then as a lecturer at several colleges. She is currently in Home Economics Education at the Faculty of Education, Mie University, where she teaches cooking as well as cooking science. Professor Hirashima is co-author of several books on cooking and cooking science for

Japanese students and several books reviewing hydrocolloids. She and her present co-author Professor Nussinovitch recently co-authored the books: *Cooking Innovations: Using Hydrocolloids for Thickening, Gelling, and Emulsification* and *More cooking innovations: Novel Hydrocolloids for Special Dishes*. She continues to study the rheological properties of polysaccharides, with a focus on the textures of starch and konjac products. She was nominated as "Finalist in Best Refereed Paper for the IFHE Congress 2016" for the paper: Creating a new texture by controlling the bubble content in konjac.

1

Use of Hydrocolloids to Control Food Size and Shape

1.1 Introduction

Appearance – one of the four quality factors of foods – comprises color, shape, size, gloss, and visual appeal. In general, foods' physical properties include size and shape, and number, nature, and conformation of constituent structural elements. During mastication, the size and shape of food particles and their surface roughness are sensed, becoming inseparable attributes of the overall textural sensation. The ability to use hydrocolloids to control food shape and size is common knowledge. This chapter begins by defining size and shape. Round and spherical shapes are emphasized, along with the means to estimate/calculate roundness and sphericity. Other shapes, geometrical and irregular, are also illustrated. Explanations of how to obtain different food shapes using different molds for setting and other techniques, such as flow of a preset gel through silicone tubes and simultaneous cooling of an agar–agar or other gelling agent solution to obtain a noodle (spaghetti) shape, are described. This chapter also includes a description of cutting techniques for gels or foods that are created with the help of hydrocolloids, as well as spherification to produce spherical and ellipsoid shapes (e.g. artificial salmon eggs), and different illustrations of these foods' appearance (https://www.amazingfoodmadeeasy.com/info/modernist-techniques/more/gelification-technique).

1.2 The Attractive Shape of Foods

1.2.1 Triangular and Prism-Shaped Foods

Consumers consistently associate a diversity of tastes and flavors to food shape. The shape of a product/food might affect the consumer's sensory and hedonic responses to it. Many foods are triangular in shape (i.e. a polygon with three edges and three vertices, one of the basic shapes in geometry). Samosa is a fried or baked pastry with a savory filling such as spiced potatoes, onions, peas, chicken and other meats, or lentils (Figure 1.1). Aside from its triangular shape, it might appear as a cone or half-moon shape, depending on the region

Use of Hydrocolloids to Control Food Appearance, Flavor, Texture, and Nutrition, First Edition.
Amos Nussinovitch and Madoka Hirashima.
© 2023 John Wiley & Sons Ltd. Published 2023 by John Wiley & Sons Ltd.

Figure 1.1 Medieval Indian Persian manuscript Nimatnama-i-Nasiruddin-Shahi [The Book of Recipes], c. 16th century, showing samosas being served. Date: 1495–1505. *Source:* The British Library.

from which it originates (Mudireddy 2019). The outer deep-fried crust remains pretty much the same, whereas the filling changes (Reza 2015; Duggal 2021). Nachos are a regional Mexican dish consisting of heated tortilla chips or *totopos* (i.e. a flat, round, or triangular corn product similar to a tortilla) covered with melted cheese or a cheese-based sauce, often served as a snack or appetizer (Figure 1.2) (Mudireddy 2019). Borek, or burek, is a filled pastry made of a thin flaky dough such as filo with a variety of fillings, for instance, cheese, meat, potatoes, or spinach (https://en.wikipedia.org/wiki/B%C3%B6rek). In Albania, it is called *byrek*, and is traditionally prepared from a number of dough layers rolled out by hand. The final form can be small, individual triangles (https://en.wikipedia.org/wiki/B%C3%B6rek). In Armenia, *byorek* consists of dough folded into triangles and stuffed with different spiced fillings. In Tunisia, the local deep-fried variant, called *brik*, is a triangular-shaped pastry pocket with chopped onion, tuna, harissa (i.e. hot chili pepper paste) and parsley (Field and Field 1970). Pizza, perhaps the most popular fast food in the

Figure 1.2 Nachos with beef and beans. *Source:* Renee Comet.

Figure 1.3 A single slice of pizza with sausage and cheese. *Source:* Renee Comet.

world, was originally a food for the poor in 18th century Naples. It is a source of national and regional pride in Italy, as well as of cultural identity (Mudireddy 2019). It consists of a flat base of leavened dough topped with tomatoes, cheese, and various other ingredients and baked at high temperatures, traditionally in a wood-fired oven (Helstosky 2008). Pizza is generally round and usually cut into triangular slices (sectors) (Figure 1.3). A sector is a region bounded by two radii of a circle and the intercepted arc of the circle (Figure 1.4). The angle formed by the two radii is called the central angle. A sector with a central angle less than 180° is called a minor sector. Which Wich sells customizable submarine sandwiches called "Wiches" – there are 10 categories and 60 topping options (https://www.whichwich.com/menu/) – that can be cut into

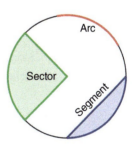

Figure 1.4 A circle showing the arc, segment and sector. *Source:* Limane/ Wikimedia Commons/ Public domain.

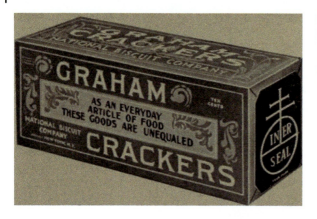

Figure 1.5 A box of National Biscuit Company graham crackers, c. 1915, which was priced at 10 cents. *Source:* Internet Archive Book Images/Wikimedia Commons/Public Domain.

little triangles. Toblerone is a chocolate bar produced in Bern, Switzerland, with a distinctive triangular shape and packaging, the bar being composed of a series of joined triangular prisms. The decision to space out the distinctive triangular chocolate chunks in two sizes of Toblerone bars sold in the UK upset its fans, who say that they mind the gap (https://www.bbc.com/news/uk-37904703). A waffle is a food item prepared from leavened batter or dough that is cooked by sandwiching it between two hot plates that are decorated to provide a distinct size, shape, and surface impression. Many places serve them in miniature triangular patterns with remarkable toppings (https://en.wikipedia.org/wiki/Waffle).

1.2.2 Rectangular and Cube-Shaped Foods

A rectangle is a special case of a parallelogram in which each pair of adjacent sides is perpendicular (https://en.wikipedia.org/wiki/Rectangle). De Villiers defined a rectangle more generally as any quadrilateral with axes of symmetry through each pair of opposite sides (De Villiers 1996). Many foods have a rectangular shape (https://thingsthatarerectangles.wordpress.com/category/food/). A few examples include toast, which is a rectangular breakfast treat, and graham crackers (Figure 1.5), a classical rectangular snack (https://thingsthatarerectangles.wordpress.com/category/food/) used to sandwich roasted marshmallows and a piece of chocolate in the traditional campfire treat *s'more*. A saltine or soda cracker is a thin, usually square cracker made from white flour, yeast, and baking soda, with most varieties lightly sprinkled with coarse salt (https://en.wikipedia.org/wiki/Saltine_cracker). Brisket is a cut of meat from the breast or lower breast of beef or veal which can have a rectangular shape. Rice provides an example of a food that can be shaped into cubes.

1.2.3 Circular and Spherical-Shaped Foods

A circle is a shape consisting of all points in a plane that are at a given distance from a given point, the center; accordingly, it is the curve traced out by a point that moves in a plane such that its distance from the center point is constant (https://en.wikipedia.org/wiki/Circle). A sphere is a geometrical object in a three-dimensional space likened to the surface of a ball. Similar to the circle in a two-dimensional space, a sphere is defined mathematically as the set of points that are all at the same distance r from a given point (center) in a three-dimensional space (https://en.wikipedia.org/wiki/Sphere). The circle is a highly symmetrical shape: every line through the center forms a line of reflection symmetry, and it has rotational symmetry around the center for every angle. As a result of this beautiful geometry, many foods have a circular shape. A few examples are: cut hard-boiled eggs, pizza, burgers, crepes, savory and sweet pies, pancakes, doughnuts, okonomiyaki, scones, tacos, meatballs, pork buns, cupcakes, cob loaf dip, tart, cookies, sushi, wheels of cheese, bagels, and cakes (https://www.taste.com.au/articles/proof-circular-foods-are-the-best/ecugi02b). Other web sources provide more examples of spherical foods. For example, buffalo chicken cheese balls, spinach balls, fried macaroni and cheese balls, kicked-up cheese balls, salmon croquets, Oreo balls, cheesy fried-chicken farm balls, puff balls, jello popcorn balls, sausage balls, cookie dough balls, pickled green cherry tomatoes, Italian meatballs, sauerkraut balls, peanut butter balls, Mexican rice balls, no-bake granola balls, ham and cheese arancini, cake balls, popcorn balls, rum balls, shrimp balls, Braunschweiger balls, sun-dried tomato and mozzarella rice balls, potato rice balls, strawberry balls, and cheesy artichoke chicken balls, to name a few (https://www.food.com/ideas/on-a-roll-ball-shaped-snacks-sweets-6350#c-379847). It is essential to note that the terms "sphere" and "ball" can be used interchangeably, although in mathematics, a distinction is made between a *sphere* – a two-dimensional closed surface set in a three-dimensional Euclidean space (Figure 1.6), and a *ball* – a three-dimensional shape that comprises the sphere and everything *inside* the sphere (a *closed ball*) or, more often, just the points inside the sphere (an *open ball*) (https://en.wikipedia.org/wiki/Sphere).

Figure 1.6 Detail from Raphael's *The School of Athens* featuring a Greek mathematician – perhaps representing Euclid or Archimedes – using a compass to draw a geometric construction. *Source:* Raffaello Sanzio/Wikipedia Commons/Public Domain.

1.3 Selected Geometrical Properties of Foods

1.3.1 Size

The textural characteristics of foods include mechanical, geometrical, and other primary parameters. Under the heading of geometrical characteristics, we include particle size and shape (e.g. gritty, grainy, coarse); and particle shape and orientation (e.g. fibrous, cellular, crystalline) (Szczesniak 1963). Size and shape are always part of a physical object, and are essential to adequately defining that object. In defined uses, both size and shape affect the process (Mohsenin 1970). This connection can be revealed using a two-dimensional system such as:

$$I = f(\text{sh}, s) \tag{1.1}$$

where I is the index influenced by both shape, sh, and size, s. In further uses, the index I may be a function of not only size and shape but also of other parameters, such as orientation, o, packing index, p, firmness, f, and so on.

$$I = f(\text{sh}, s, o, p, f, \ldots) \tag{1.2}$$

A good illustration of this relationship would be determining the number of specified objects to fill a container. If I is substituted by Y, with X_1 for shape, X_2 for size, X_3 for orientation, X_4 for packing and X_5 for firmness, the equation in multiple regression form will be:

$$Y = b_1 x_1 + b_2 x_2 + b_3 x_3 + b_4 x_4 + b_5 x_5 \tag{1.3}$$

This relationship can be evaluated by assessing a set of specimens, where the magnitude of the contribution of each x to the variation in y can be estimated by means of analysis of variance and multiple correlation (Quenoville 1952; Mohsenin 1970). This methodology was used to define the correlation between permeability of a reservoir and the petrographic properties of sediments, such as shape, size, mineral composition, arrangement, and orientation (Griffiths 1958).

The size and shape of an agricultural commodity, or of any item for consumption, not only affect the degree of consumer approval; in numerous cases, they also affect packaging, distribution of stresses when forces are applied, and processability (Szczesniak 1983). In the area of food powders, such as soluble coffee manufacturing, the existence of fines could clog extraction columns and in a similar manner fines might block filters in home brewing equipment. Particle size and, specifically, particle size distribution in porous media control permeability to water and impact extractability and solubility characteristics (Szczesniak 1983).

1.3.2 Characterization of Size

Size is frequently characterized by determining whether a whole or ground product can be passed through a sieve or screen of defined mesh size. This method is widely used for grading agricultural commodities; however, it has substantial disadvantages, such as the fact that sieving only divides the product in relation to its smallest dimension. It is, however, appropriate for products that are close to spherical in shape. Slit screens are used for oblong materials (i.e. the vertical diameter of the screen hole is greater than the horizontal diameter), such as rice (Szczesniak 1983). Sieve analysis refers to the practice of assessing a material's particle size distribution by passing it through a sequence of sieves of progressively smaller mesh size (i.e. the number of openings in 1 sq. inch of screen) and weighing the quantity of material that remains in each sieve as a portion of the full mass (https://fintec1.com/wp-content/uploads/2020/07/FF-24.0101-Determing-Screen-Mesh-Size.pdf).

Additional applied methods for defining the size of fruit and vegetables are length or diameter measurements, and count per volume or weight. For fruit, larger size is generally crucial for the fresh market, whereas for vegetables, which are preferred at immaturity, a smaller size is desirable (Szczesniak 1983). The size and shape of food powders affect aerodynamic and hydrodynamic features, as well as mixing, segregation, and other characteristics that influence both processing and quality. Average size and particle size distribution are very important parameters for powders. The size and shape of dry particles may be determined by microscopic examination. A particle size analyzer uses the resultant photographs to estimate the size and distribution of the particles. The size of particles suspended in a fluid food product can be determined microscopically, by wet-sieving technique or by Coulter Counter particle size analyzer (Szczesniak 1983). When particles are distributed in a liquid or semi-liquid food, their size and shape influence the rheological parameters of the system. In fruit juices, particles of increasing size increase viscosity, whereas with globules in an emulsion, decreasing size leads to increasing viscosity. In commercial tomato juice, particle shape and size (diameter) influence viscosity (Surak et al. 1979). Elongated particles in tomato juice contribute more to viscosity than spherical ones (Kattan et al. 1956).

1.3.3 Size Reduction

It is often necessary to reduce the size of foods, both in the industry and in the kitchen. Size control is frequently needed to speed up transfer of both heat and mass. Size is also controlled for mixing purposes and to obtain uniformity in cooking, during consumption and on the plate. Size reduction continues on the plate as the bite is portioned out, and then in the oral cavity during mastication (Chen 2009). Essential unit operations in nearly all food processes include size reduction and mixing. These are used to manufacture specific formulations, to assist in processing, or to adjust sensory features of foods to the required quality (Fellows 2000). Size-reduction operations expose the cut surfaces to air (i.e. oxygen), leading to further oxidation of the food. A reduction in size/increase in area also increases hot-air drying rates or the action of oxidative enzymes, such as lipoxygenase or peroxidase (Fellows 2000). In engineering, size reduction is one of many "unit operations" or actions that must be taken as part of the material's processing. Size reduction is divided into two major categories depending on whether the material is a solid (grinding and cutting) or a liquid (emulsification or atomization) (Earle 1983). Technically, forces acting on a body produce fracturing or shearing of the material. While engineers are mostly concerned with the energy involved in grinding, cooks appreciate the effects of size reduction in cooking (i.e. color, flavor, and texture).

1.3.4 Energy Requirements for Size Reduction of Solid Materials

Theoretical considerations propose that the energy dE necessary to yield a slight change dx in the size of a unit mass of material can be expressed as a power function of the size of the material (Brennan et al. 1990):

$$\frac{dE}{dx} = -\frac{K}{x^n} \tag{1.4}$$

A number of researchers have used this equation. Peter von Rittinger was an Austrian pioneer of mineral processing (Figure 1.7), whose work included the early scientific study of comminution, and his theory on the relations of work and power in when breaking rocks, sometimes referred to as Rittinger's law (Lynch and Rowland 2005). Rittinger's inventions and mathematical descriptions of unit operations established mineral processing as a modern technology (Steiner 1991). He was the first to provide a sound mathematical basis for the physics of wet-classification and gravity-separation processes, and the first to discuss the quantitative relationship between energy consumed and size reduction achieved in crushing and grinding (Steiner 1991).

For the grinding of solids, Rittinger considered that the required energy should be proportional to the newly produced surface, with a factor of 2. Then

$$\frac{dE}{dx} = -\frac{K}{x^2} \tag{1.5}$$

Figure 1.7 Peter von Rittinger, lithograph by Josef Kriehuber, 1856. *Source:* Peter Geymayer/Wikimedia Commons/Public Domain.

or, integrating

$$E = K\left[\frac{1}{x_2} - \frac{1}{x_1}\right] \tag{1.6}$$

where x_1 is the average initial feed size, x_2 is the average final product size, E is the energy per unit mass required to create this new surface and is generally measured in horsepower hour per ton, and K is Rittinger's constant for a specific machine and material. Rittinger's law is applicable mostly to that part of the process during which the new surface is being created and holds most accurately for fine grinding, where the increase in surface per unit mass of material is large (Richardson et al. 2002). Due to the inaccuracy of the assumptions on which it is based, and the difficulty in determining its parameters, Rittinger's law is only approximate (Berk 2009).

Friedrich Kick was a professor at the Technical University in Vienna from 1892 to 1910. He became well-known for his experiments on material properties and his development of test equipment to measure those properties. Kick's experiments led him to hypothesize a link between energy and size reduction (Lynch and Rowland 2005): "For any unit weight of ore particles the energy required to produce any given reduction ratio in the volumes is constant no matter what may be the original size of the particles" (Richards and Locke 1940). Kick considered that the energy required for a given size reduction was proportional to the size-reduction ratio, which requires $n = 1$ (Kick 1883). Then

$$\frac{dE}{dx} = -\frac{K}{x} \tag{1.7}$$

or

$$E = K \ln \frac{x_1}{x_2} \tag{1.8}$$

with $\frac{x_1}{x_2}$ being the size-reduction ratio.

Kick's law has been found to hold more accurately for coarser crushing, where most of the energy is used to create fractures along existing fissures. It gives the energy required to deform particles within the elastic limit. For numerous crushing operations, the energy prerequisite suggested by Kick's law appears to be too low, whereas that required by Rittinger's equation appears to be excessive (Brennan et al. 1990).

Fred Bond was the engineer who did much to define the relationship between ore hardness, tonnage processed, size reduction achieved, and power required (Lynch and Rowland 2005). Throughout the late 1930s and 1940s, Bond dedicated much of his time to understanding the energy–size reduction correlation. He advanced the concept that the energy needed for grinding was the total energy needed to make the grinding mill product minus the energy needed to make the feed (Lynch and Rowland 2005). In Bond's work, n takes the value 3/2 giving

$$\frac{dE}{dx} = -\frac{K}{x^{3/2}} \tag{1.9}$$

or

$$E = 2K\left[\frac{1}{(x_2)^{1/2}} - \frac{1}{(x_1)^{1/2}}\right] \quad (1.10)$$

where x_1 and x_2 are measured in micrometers and E in kWh short per ton (907.16 kg),

$$K = 5Ei \quad (1.11)$$

where Ei is the bond work index – the energy required to reduce the unit mass of a material from an infinite particle size to a size at which 80% of it passes a 100-μm sieve. The bond work index is obtained from laboratory crushing tests on the feed material. The third theory holds reasonably well for a variety of materials undergoing coarse, intermediate, and fine grinding (Bond 1952; Brennan et al. 1990).

1.4 Size Enlargement and Reduction Processes

1.4.1 Definition of Forming and Its Aims

Forming is a size-enlargement process in which foods that have a dough-like consistency are molded into a variety of shapes and sizes, frequently following a mixing operation. Forming is used to intensify the range and accessibility of baked goods, confectionery, and snack foods (Fellows 2000). Controlling the size of a formed food piece is important to ensuring identical rates of heat transfer into the center of the baked product, to guarantee the weight of the food piece, and to confirm the standardization of smaller food items for the subsequent control of fill weights. Extrusion also has a forming function (Fellows 2000).

1.4.2 Confectionery Molders

Confectionery depositing–molding equipment is comprised of separate molds of the required shape and size for a particular product, associated with a specific conveyor. The depositor has a piston filler that places a precise specified volume of hot sugar mass into each mold. Depositors can place food of a single type, layered foods, or center-filled foods. In the case of chocolate paste, the food is then cooled in a cooling tunnel. When it has hardened satisfactorily, separate sweets are ejected and the molds restart the cycle (Fellows 2000).

1.4.3 Pie-Casing Formers

Pie casings are shaped by putting a portion of dough into aluminum foil containers or refillable pie molds and pressing it with a die. Then a filling is placed into the casing and a continuous layer of dough is positioned over the top. The lids are then cut by countering blades (Fellows 2000). Hayashi (1989) and Rheon (2016) described equipment for forming and encasing balls of dough with other materials. Throughout this process, the inner and outer materials are coextruded, distributed and shaped by two "encrusting discs"

(Fellows 2000). The comparative thickness of the outer layer and the diameter of the inner sphere are determined by the separate materials' flow rates (Fellows 2000). The comparative thicknesses of the inside and external layers can therefore be modified by merely fine-tuning the flow rates, allowing a high degree of flexibility for the manufacturer of various consumable items. Such equipment is used in Japan for manufacturing cakes with an outer layer of rice dough and filled with adzuki bean paste. It can also be applied to produce doughnuts, fish filled with vegetables, hamburgers filled with cheese, meat pies, and sweetbreads filled with jam (Fellows 2000).

1.4.4 Hydrocolloids in Food Fillings

Hydrocolloids are frequently used in fruit pie filling (Saha and Bhattacharya 2010). Fruit preparations for baked goods are used as toppings or fillings for biscuits, cakes, pies or rolls. Methylcellulose and hydroxypropyl methylcellulose (HPMC) are used to control boil-out of bakery fillings and of hot structures in potato croquettes, meat analogues, processed meat, or fish products. Boil-out occurs when during the baking or microwaving process, the viscosity of the filling decreases, causing it to boil and be expelled from the dough product. Carboxymethylcellulose (Figure 1.8) stabilizes and thickens waffle fillings with a high sugar content (Onyango et al. 2009; Correa et al. 2010; Gómez et al. 2010). Addition of 2–5% inulin (Figure 1.9) inhibits the phase separation of pasta fillings. Inulin creates and maintains a soft, velvety filling with a diminutive bite that does not disintegrate during ingestion. κ- and ι-carrageenan, and their mixtures, can be used in pie fillings and ready-to-eat desserts (Soukoulis et al. 2008). In addition, the synergistic gelling between high-guluronate alginate and highly esterified pectin may be used in fruit fillings (Toft et al. 1986). Inclusion of xanthan gum to fruit pie fillings increases consistency, stability, flavor release and syneresis control.

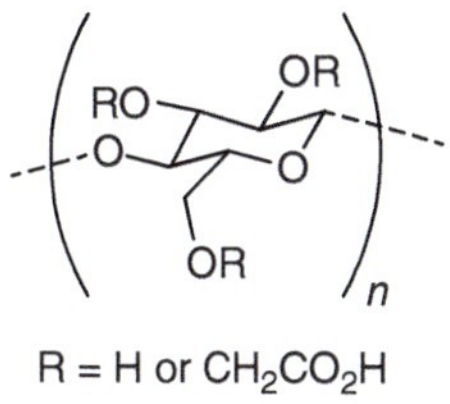

Figure 1.8 Chemical structure of carboxymethylcellulose. Source: Edgar181/Wikimedia Commons/Public domain.

Figure 1.9 Structural formula of inulin (fructan). Additional features (carbon numbers and repeated fructosyl unit) are highlighted with colors. Source: Florian Fisch/Wikimedia Commons/Public domain.

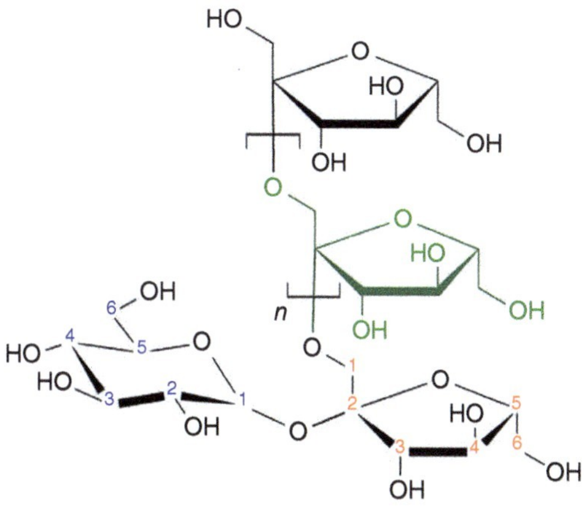

1.4.5 Cutting and Shaping Spherical Edible Products

For cutting and shaping spherical edible products, specialized companies and machines are essential. For instance, Rheon Automatic Machinery Co. manufactures unique food-processing machines. The company develops technologies to automate food-production systems in processing and baking. Rheon Automatic Machinery operates branches in the United States and Germany, with Yasunori Tashiro serving as its current chairperson (https://www.bloomberg.com/profile/person/3232015). One of their inventions relates to a method for shaping a spherical body consisting of a dough crust and filling. A continuously fed cylindrical body, consisting of dough crust and filling, is constricted by at least three slidable members, which form an opening, or close it. The members slide one along the other to constrict the cylindrical body. The contact area of the members with the surface of the cylindrical body gradually decreases as the opening is closed, thereby constricting the cylindrical body and shaping the spherical body (Tashiro 1989). Another patent deals with an improved apparatus for cutting and shaping a spherical body of sticky material. The apparatus is particularly useful for processing a substantially round food product consisting of soft, very sticky material while avoiding problems of dough adhesion or clogging, or of protruding filling (Tashiro 1993).

1.5 Shape – Definition and Implications

1.5.1 Shape of a Food Commodity

"To define the shape of a body fully, one must specify the location of all points on the external surface" (Medalia 1980). This is a time-consuming process. In addition, it can raise mathematical difficulties, particularly in the case of powders/agglomerated powders (Szczesniak 1983). As a result, qualitative shape descriptions of, for example, fruit and vegetables have been prepared (Mohsenin 1970).

1.5.2 Roundness and Sphericity

Roundness is defined as the degree of sharpness of the corners of a solid (Curray 1951; Mohsenin 1970).

$$\text{Roundness} = \frac{A_p}{A_c} \qquad (1.12)$$

where A_p is the largest projected area of an object in its natural resting position (m²) and A_c is the area of the smallest circumscribing circle (m²) (Mohsenin 1970).

Roundness can also be estimated by:

$$\text{Roundness} = \sum_i^n \frac{r}{NR} \qquad (1.13)$$

where r is the radius of the curvature of a corner of the particle surface, R is the radius of the maximum inscribed circle in the longitudinal section of the particle, and N the number of corners. If the corners are worn down, then r approaches R, and roundness = 1 (when $r \rightarrow R$) (Mohsenin 1970).

Furthermore:

$$\text{Roundness ratio} = \frac{r}{R} \tag{1.14}$$

where R in this case is the mean radius of the object and r is the radius of the curvature of the sharpest corner (Mohsenin 1970).

Sphericity is a measure of how spherical (round) an object is. Sphericity can be estimated by different methods. For instance:

$$\text{Sphericity} = \frac{d_e}{d_c} \tag{1.15}$$

where d_e is the diameter of a sphere of the same volume as the object, and d_c is the diameter of the smallest circumscribing sphere or, usually, the longest diameter of the object (Curray 1951; Mohsenin 1970). The advantages of using such an expression are that it relates the shape of the solid to that of a sphere of the same volume.

The volume of a solid can be equal to the volume of a *triaxial ellipsoid* which has diameters equivalent to those of the sample. This ellipsoid is a type of quadric surface that is a higher-dimensional analogue of an ellipse. The equation describing a standard axis-aligned ellipsoid body in an *xyz*-Cartesian coordinate system is:

$$\frac{x^2}{a^2} + \frac{y^2}{b^2} + \frac{z^2}{c^2} = 1 \tag{1.16}$$

where a and b are the equatorial radii (along the x and y axes) and c is the polar radius (along the z axis), all of which are fixed positive real numbers determining the shape of the ellipsoid. If all three radii are equal, the solid body is a sphere; if two radii are equal, the ellipsoid is a spheroid (Weisstein 2003). In other words: $a = b = c$ (sphere); $a = b > c$ (oblate spheroid, i.e. disk-shaped); $a = b < c$ (prolate spheroid, i.e. like a rugby ball) (Figure 1.10). Finally, if $a > b > c$, the solid is a scalene ellipsoid (i.e. it has three unequal sides).

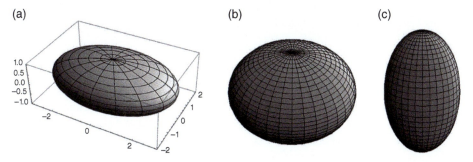

Figure 1.10 (a) Ellipsoid, (b) Oblate and (c) prolate spheroids. *Source:* Adapted from Wikimedia Foundation.

Scalene ellipsoids are frequently called "triaxial ellipsoids" the implication being that all three axes need to be specified to define the shape (Weisstein 2003).

As stated, if the volume of the solid is equal to the volume of a triaxial ellipsoid with intercepts a, b, c, and in addition, the diameter of the circumscribed sphere is the longest intercept a of the ellipsoid, then the degree of sphericity can be calculated as:

$$\text{Sphericity} = \sqrt[3]{\frac{\text{Volume of solid}}{\text{Volume of circumscribed sphere}}} = \frac{\sqrt[3]{abc}}{a} \qquad (1.17)$$

or, the geometric mean diameter divided by the major diameter, where a is the longest intercept, b is the longest intercept normal to a, and c is the longest intercept normal to a and b (Mohsenin 1970). It is important to note that the intercepts need not intersect at a common point. Another definition of *sphericity* is given by $\frac{d_i}{d_c}$, where d_i is the diameter of the largest inscribed circle and d_c is the diameter of the smallest circumscribed circle (Mohsenin 1970).

Sphericity can also be defined as the ratio of the surface area of a sphere having the same volume as the object to the actual surface area of the object (Sahin and Sumnu 2006):

$$\phi = \frac{\pi D_p^2}{S_p} = 6 \frac{V_p}{D_p S_p} \qquad (1.18)$$

where ϕ is the sphericity, D_p is the equivalent or nominal diameter of the object (m), S_p is the surface area of the object (m²), and V_p is the volume of the object (m³) (Sahin and Sumnu 2006). The equivalent diameter is regarded as the diameter of a sphere that has the same volume as the spherical object before (Sahin and Sumnu 2006).

Another approach to estimating sphericity is:

$$\phi = \frac{\Sigma(D_i - D_{ave})^2}{(D_{ave} N)^2} \qquad (1.19)$$

where D_i is any measured dimension (m), D_{ave} is the average dimension or equivalent diameter (m), and N is the number of measurements (an increase in N increases the accuracy of the result; Sahin and Sumnu 2006). In accordance with this formula, if the sample sphericity value is close to zero, it can be considered spherical (Sahin and Sumnu 2006).

1.5.3 Average Projected Area and Sphericity of Hydrocolloid Beads

Another method of approximating the degree of sphericity of a convex body was established nearly 65 years ago when sizing machines were developed for fruit (Houston 1957). A novel criterion for size, the average projected area, was suggested, based on the theories of convex bodies. For such bodies:

$$\frac{V^2}{S^3} \geq \frac{1}{36\pi} \qquad (1.20)$$

where V is the volume and S is the surface area of the convex body. A_c, the average projected area of the convex body, is one-fourth of the surface area. Substituting $4A$ for S in Eq. (1.20), and after some manipulation, we get:

$$A \leq KV^{2/3} \tag{1.21}$$

In the case of a sphere, $K = \dfrac{9\pi}{16} = 1.21$, when equality in Eq. (1.21) is achieved. In other words, this method enables evaluating the degree of sphericity of a bead and checking how K changes when the bead is not spherical. A few further conclusions can be derived: the dimensionless K constant varies with the characteristic dimensions of the object; the variation of A_c for unrestricted orientation is too great for a satisfactory measure of volume, but this variation can be reduced to an acceptable level (Houston 1957).

A small number of manuscripts have dealt with roundness and sphericity of hydrocolloid beads. Indeed, they are in most cases small items; consequently, axial dimension measurements can be derived from the outlined projections of the bead. Tracings need to be made of both the maximum and minimum projected areas, and the axis dimensions can be derived from those drawings (Nussinovitch and Gershon 1996). Identical results can be achieved by shadowgraph (Mohsenin 1970). An approximation of the deviations of hydrocolloid beads from sphericity is essential in studies of mass and heat transfer. The sphericity of alginate beads was estimated by two techniques (Nussinovitch and Gershon 1996): a method initially developed for quartz grains, which expresses the shape character of a bead relative to that of a sphere of the same volume, and a method formerly established for convex bodies, which determines a dimensionless constant that relates surface area to volume. Deviations from sphericity were easily estimated by the former method. The difference between the two measures of bead sphericity was not more than 5%. Deviations from perfect sphericity were ~7% for the tested alginate beads under the examined conditions (Nussinovitch and Gershon 1996).

1.5.4 How Are Gels Shaped?

Gelling agents are used to produce different food product shapes. Cubes, spheres and noodle shapes are very common, and add visual charm to the dishes. It is preferable to set soft gels directly in the platter in which they will be served, as they are prone to tearing upon unmolding. It is also possible to set harder gels in serving dishes which might provide much visual appeal, particularly if glass serving dishes are used. It is easy to form gels by setting them in a block and then cutting them into the desired shapes. For this purpose, the container can be filled with gel to the final height and then the gel is cut into cubes with a knife. If harder gel is preferred, a cookie cutter (i.e. a tool to cut cookie/biscuit dough into a particular shape) can be used for a wide range of shapes. Sheets of elastic gels can set in a thin layer of not more than 3 mm thickness on a smooth plastic surface. This creates flexible sheets that can be wrapped over or around foods. Another common way to set artistically shaped gels is to use silicone or plastic molds. Ice cube trays that can be purchased in a wide variety of shapes and sizes might be used. For instance, hemispherical and spherical molds of dissimilar sizes can be utilized to manufacture interesting shapes. An additional remarkable option is the preparation of gel noodles, usually used as a garnish (see Section 1.9.1 for a detailed description of such a dish). Another shape with marked visual appeal is rounded gel pearls, or beads. There are numerous procedures to prepare these. The methodologies and more details about such shapes are provided in this section and Section 1.6 (see also https://www.amazingfoodmadeeasy.com/info/modernist-techniques/more/gelification-technique).

1.5.5 Silicone Molds to Modify Gel Shapes and Sizes

Siloxanes, frequently called silicones, are molecules with an oxygen–silicon backbone (Si-O-Si) in which each Si atom carries two organic groups (Rücker and Kümmerer 2015). Fluid siloxanes are used to form silicone elastomers or silicone rubbers through vulcanization, with the addition of crosslinking agents to link the linear polymers (Lassen et al. 2010). Silicone elastomers can be used in consumer goods such as diving masks and other scuba equipment, protective masks, earplugs, baking tins, consumer product packaging, showerhead membranes, soft-touch products (e.g. pencils), spatulas, etc., and for food packaging (fruit labels, bakery papers, wrappers for candy, chewing gum, meat and frozen food) (Lassen et al. 2010). Silicones can safely contact foods, as is the case with kitchenware such as baking molds, dough scrapers, or food packaging (Gross 2015).

Linear and cyclic volatile methyl siloxanes (VMSs) are chemicals with no natural source. VMSs are ubiquitous anthropogenic pollutants that have recently come under suspicion for potential toxicity and environmental persistence (Alton and Browne 2020). They have been extensively used in cosmetics, personal care products and coatings, among many others. As a result of their wide use, VMSs can be found in different environmental media, as well as in the human body (Fromme et al. 2019). The topic of siloxane in baking molds, emission to indoor air and migration to food during baking in an electric oven was studied (Fromme et al. 2019). Fourteen new silicone baking molds and three commercial metallic molds were used in different baking experiments. VMSs were measured in the indoor air throughout the baking process and at the edge and center of the finished cakes using gas chromatography–mass spectrometry. In addition, particle number concentration (PNC) and particle size distribution were measured in the indoor air. The conclusions of that study demonstrated that the silicone molds with the highest concentrations of cyclic VMSs were associated with distinctly higher concentrations of the compounds in indoor air. Using a mold for more than one baking cycle decreased the indoor air concentrations markedly. Samples collected from the edge of the cake had higher concentrations than samples from the center. The overall impression was that as a general rule, silicone molds should be used only after precleaning while strictly following the producers' temperature guidelines (Fromme et al. 2019).

Used as bakeware, silicone cooking molds do not need to be coated with oil; in addition, their flexibility allows easy removal of the baked foods from the mold. Silicone molds are microwave-safe, freezer-safe, and oven-safe. In 1979, the FDA declared that silicone was safe for use in cooking and everyday use. Changes in temperature do not cause food-grade silicone to leach into foods, despite common consumer concern (https://www.mindbodygreen.com/articles/is-silicone-toxic). The non-stick, bendable molds are available in numerous shapes and sizes. It is recommended to first check the molds with water to determine how much volume they will hold. Using this technique, it is possible to mold gels with suspended solids. To accomplish this, only a portion of the gel is cast. Then it is left to set until a soft texture is achieved, and the solid is carefully laid upon the soft gel and the remaining gelling fluid is cast. Thick gel sheets can be cast using a mold formed from pastry bars (Myhrvold et al. 2011). Silicone molds can be used to achieve differently shaped products. For example, spaghetti is a thin, long, round type of pasta, whereas linguine is thin, long and flat. To make gel "linguine," a thin gel sheet is cast in a mold. Once the gel has set, a pasta cutter can be used to cut the sheet into strips. Although, perfect gel spheres can be challenging to manufacture, silicone molds can be used to create them in various different sizes. To obtain gel spheres, the hot gel fluid formulation is poured into the cavities of the chosen silicone mold and allowed to set. Some gels require refrigeration, others set at room temperature after just a few minutes. The set spheres are then popped out of the mold (Myhrvold et al. 2011).

1.6 Miscellaneous Shapes and Sizes of Edible Hydrocolloid Products

1.6.1 Edible Hydrocolloid Gel Beads

Agar is a mixture of at least two polysaccharides –agarose (Figure 1.11), a neutral polymer, and agaropectin, a sulfated polymer (Nussinovitch 1997, 2010). Firm gel beads can be prepared by dropping, for example, an agar/agarose solution from a syringe into a cold oil bath. The gel solution is obtained by boiling in water for at least 1.3 minutes. In the kitchen, oil can be easily chilled by placing it in a bowl over ice water. The oil should be very cold, but not congealed. The dropping must be quick enough to eliminate setting of the fluid in the syringe. The droplets form spherical beads because oil and the water-based/hydrophilic gel do not mix. Thus, solid gel beads are produced. The set/gelled beads are then removed from the oil with a perforated spoon. To remove the oily residue from the surface of the gel beads, they should be rinsed with cold high-proof alcohol (Myhrvold et al. 2011). Agarose beads can be successfully manufactured in a single stirred vessel by emulsification of hot agarose solution and subsequent freezing/gelling of the aqueous drops into soft solid particles. The particle size and size distribution can be controlled by selecting an appropriate energy dissipation rate and/or surfactant concentration (Mu et al. 2005). Despite the fact that flow in a stirred vessel is transitional, the relationship between drop/particle mean size and energy dissipation rate is similar to that for turbulent flow (Baldyga et al. 2000). A different technique to make small frozen spherical beads is to dispense the gel base drop by drop into liquid nitrogen. This only works well for gels that remain stable after thawing (Myhrvold et al. 2011). Cold-set gel beads that have the flattened shape of lentils or split

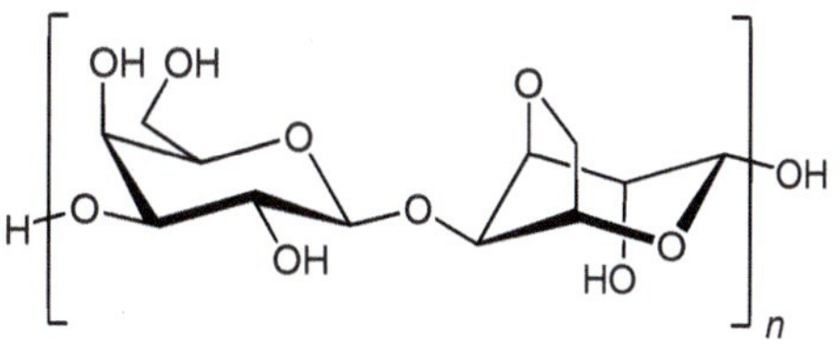

Figure 1.11 The structure of the repeating unit of an agarose polymer. *Source:* Yikrazuul/Wikimedia Commons/Public domain.

peas can be prepared by dispensing gel droplets onto the chilled surface of a pan. Once they set, water is poured over the solidified gel droplets and they are easily dislodged using a silicone spatula; the water is drained, leaving these cold-set semisolid moieties (Myhrvold et al. 2011).

A widespread and simple method for producing beads is to crosslink an alginate solution. It is not surprising that this is the model system of choice for numerous encapsulations and for bead manufacturers. The manufacture of different alginates, and their chemical and physical properties have been reviewed in many textbooks and manuscripts (among them Davidson 1980; Nussinovitch 1997; Phillips and Williams 2000), and the reader is referred to these for further details. The vibrating nozzle technique can be used to produce a monodispersion of beads that are larger than 200 μm in diameter, depending on the application (Serp et al. 2000). Furthermore, their manufacture can be scaled up. In particular, reasonably high production rates, between 1 and 15 mL/min depending upon the desired bead size and polymer solution rheology, can be achieved, producing an alginate bead-size distribution of less than 5% (Brandenberger et al. 1997; Seifert and Phillips 1997).

1.6.2 Parameters to Be Considered Upon Formation of Beads Through Capillary Jet Breakage

The breakup of a liquid jet by applying vibration was proposed by Savart in 1833 and interpreted by Lord Rayleigh in 1878 as a resonance effect (Savart 1833; Lord Rayleigh 1878). A liquid jet preferentially breaks into droplets at a defined frequency which is a function of the jet diameter and the linear jet velocity. If all vibrations are eliminated except for the optimum frequency, quasi-monodispersed droplets are produced. Droplets as small as a few hundred micrometers and yields of liters per hour may be achieved. Lord Rayleigh (Figure 1.12) showed that the breakage takes place when the length of the jet is longer than its circumference (Lord Rayleigh 1878). For Newtonian and low-viscosity liquids, the optimum breakup wavelength, λ_{opt}, to obtain monodispersed droplets is given by:

$$\lambda_{opt} = \pi\sqrt{2}\, d_j \tag{1.22}$$

where d_j is the jet diameter. The frequency, f_{opt} to be applied is equal to:

$$f_{opt} = \lambda_{opt} U_j \tag{1.23}$$

where U_j is the linear velocity of the jet corresponding to the flow rate:

$$F = \frac{\pi}{4} d_j^2 U_j \tag{1.24}$$

The volume of the droplets may be predicted by:

$$v = \frac{F}{f_{opt}} = \frac{\pi}{6} d^3 \tag{1.25}$$

Figure 1.12 John William Strutt, 3rd Baron of Rayleigh, British physicist. *Source:* Unknown author/Wikimedia Commons/Public Domain.

and by combining these equations, one obtains:

$$d = \sqrt[3]{v\frac{6}{\pi}} = 1.89 d_j \tag{1.26}$$

Equation (1.26) predicts that the droplet size must be 1.89 times the jet diameter and independent of the frequency. Weber extended Eq. (1.22) to viscous liquids (Weber 1931):

$$\lambda_{opt} = \pi\sqrt{2}\, d_{j2}\sqrt{1 + \frac{3\eta}{\sqrt{\rho \sigma d_j}}} \tag{1.27}$$

where η is the fluid viscosity, ρ is the liquid density, and σ is the surface tension. However, for usual encapsulation solutions (i.e. alginate solution, melts), the theoretical droplet size is affected by less than 1%. On the other hand, the jet diameter, d_j, is generally assumed equal to the internal nozzle diameter, d_n (Schneider and Hendricks 1964). However, at the exit of the nozzle, the jet is subjected to pressure relaxation, leading to jet expansion that can be expressed by (Brandenberger et al. 1997):

$$\frac{d_j}{d_n} = 4.33 W_e^{-0.337} \tag{1.28}$$

with

$$W_{e_n} = \frac{\rho d_n u^2}{\sigma} \tag{1.29}$$

where W_e is the Weber number and W_{e_n} is the Weber number related to flow inside the nozzle. Recent experiments recorded droplet formation and jet breakup using a fast camera. The experimental data well fit Lord Rayleigh's equation if the jet expansion at the exit of the nozzle was considered. In addition, it appeared that good transmission of the vibration (i.e. by a metallic membrane) is needed to guarantee low dispersion (Chevallier et al. 2017).

1.6.3 Bead Shape and Its Improvement

The shape of an object in a space is the part of that space occupied by the object, as determined by its external boundary. Another definition of shape is the geometrical information from an object that remains when location, scale and rotational effects are filtered out (Kaye 1993). The sphere is one of the most common shapes in nature (Musser and Burger 1997), probably due to its ability to enclose the greatest volume for a given surface area. Most beads are produced in a sphere-like shape. Archimedes (287–212 BCE) found that the surface area (volume) of a sphere is two-thirds the surface area (volume) of the smallest cylinder containing that sphere. Figure 1.13 shows a sphere of radius r contained within a cylinder with radius base r and height $2r$. The surface area of the cylinder is $2\pi r^2 + 2\pi r(2r) = 6\pi r^2$. Based upon Archimedes' observation, the formula for the surface area of a sphere of radius r is $\frac{2}{3}(6\pi r^2) = 4\pi r^2$. The volume of the sphere is $\frac{2}{3}(2\pi r^3) = \frac{4}{3}\pi r^3$. Of course, the r in $4\pi r^2$ is squared, a unit of area, whereas the r in $\frac{4}{3}\pi r^3$ is cubed, a unit of volume (Musser and Burger 1997). The shape of a bead can be described by borrowing terms from different fields. These include: round – approaching spheroid; oblong – vertical diameter greater than the horizontal diameter; ovate – egg-shaped; obvate – inverted ovate;

Figure 1.13 A sphere of radius r contained within a cylinder with radius base r and height $2r$.
Source: André Karwath aka/Wikimedia Commons/CC BY-SA 2.5.

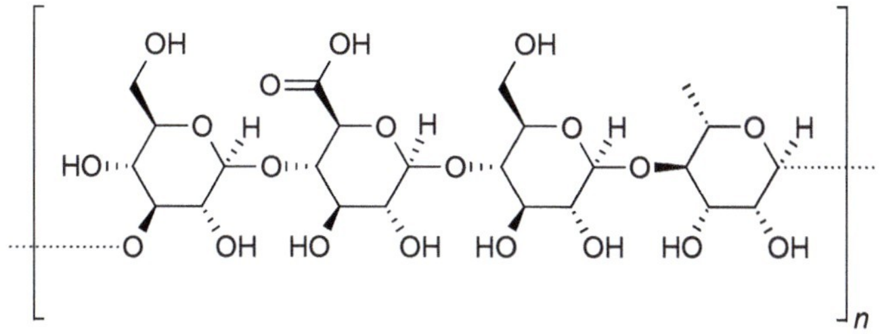

Figure 1.14 Chemical structure of gellan gum. *Source:* Edgar181/Wikimedia Commons/ Public domain.

elliptical – approaching ellipsoid; truncate – having both ends squared or flattened; ribbed – in cross section, sides are more or less angular; regular – horizontal cross section approaches a circle; irregular – horizontal cross section departs considerably from a circle (Mohsenin 1980).

Gelling agents [i.e. agar–agar, alginate, carrageenan, gellan, and low-methoxyl pectin (LMP)] can be used to create various gel shapes. For instance, κ-carrageenan's property of being easily induced into a gel form by contact with metal ions, amines, amino acid derivatives, and water-miscible organic solvents was used to tailor the gel into various shapes, such as beads, cubes and membranes (Tosa et al. 1979). Numerous bead manufacturers want to generate beads in a spherical or sphere-like shape. For instance, gellan beads can be formed in a spherical shape, enabling mathematical modeling studies of their spherical geometry. To manufacture such shapes, a dispersion of gellan gum (Figure 1.14) was extruded into a solution containing a mixture of calcium and zinc ions (counterions). By changing the experimental variables, for instance pH, of the counterion solution, process variables could be optimized to achieve control over the bead size and morphology (Agnihotri et al. 2004). Different bead shapes sometimes result from the ingredients used in the manufacturing process, as well as the entrapped moieties. Dropping a 3.5% κ-carrageenan-containing papain solution into 0.5 M potassium chloride using the ionotropic gelation method, with a hardening time of 20 minutes, it was possible to obtain beads characterized by a spherical disc shape with a collapsed center, and an absence of aggregates (Sankalia et al. 2006a). Dissimilar combinations of water-soluble polymers and production techniques can lead to control of bead size, shape and other parameters (Agrawal et al. 2004). For instance, beads containing chitosan, fine-particle ethylcellulose, HPMC and caffeine were manufactured. Spherical beads with good mechanical properties could be created without microcrystalline cellulose. Furthermore, beads with high sphericity could be obtained with high chitosan content, and low HPMC content, water content, spheronizer speed and extruder speed (Agrawal et al. 2004).

A study on shape improvement of prepared alginate beads described injection of an aqueous solution of 2% alginate into a novel solvent consisting of layers of hexane and n-butanol, n-butanol with 1% $CaCl_2$, or n-butanol with 2% $CaCl_2$ in water. Beads of up to 3.5 mm in diameter obtained with this method had a roundness which was similar or up to 5% better than comparable beads prepared by just dropping alginate solution into a $CaCl_2$ hardening bath (Buthe et al. 2004). In addition, the novel solvent-based method allowed for highly reproducible preparation of alginate beads with precisely predictable sizes.

The biggest beads obtained with this method were 9 mm in diameter. Thus, with the solvent-based preparation of alginate beads, it was possible to obtain beads of exactly the type needed for specific analytical purposes (Buthe et al. 2004). Another study described the production of alginate hydrogel microbeads of various shapes with narrow size distributions (94–150 µm) in non-silanized/silanized poly(dimethylsiloxane) devices (Tan and Takeuchi 2007). The method combined an internal gelation method with T-junction droplet formation in a microfluidic device (Okushima et al. 2004). The use of calcium-carbonate nanoparticles enabled an internal gelation method for microscale production. The method allowed easy control over bead size by varying flow parameters, and better monodispersity and control of the shape of the hydrogel beads compared to conventional external gelation methods performed in microfluidic devices (Tan and Takeuchi 2007). Modifications to the design (e.g. shortening the channel length) or switching to devices that can withstand higher pressures or silicon glass-based devices may enable the production of smaller hydrogel beads (Takeuchi et al. 2005). Another report claimed that the optimal conditions for calcium-alginate bead sphericity were a concentration of 2.24% sodium alginate, a flow rate of 0.059 mL/s for the sodium-alginate solution, and a 459 rpm rotation for the calcium-chloride solution. The predicted and experimental bead sphericities under optimal conditions were 94.5 and 96.7%, respectively, showing close agreement (Woo et al. 2007). Immersion in hot water slightly decreased bead size and rupture strength. Sodium-chloride treatment increased bead size and decreased rupture strength. While the pH of the calcium-chloride solution had little effect on bead sphericity, bead size and gel strength decreased with longer times in each solution with different pH. Beads coated with pectin and glucomannan showed no significant changes in sphericity, but their sizes decreased with time. The coated beads showed higher rupture strengths than the non-coated ones (Woo et al. 2007).

Elongation ratio (ER) – the quotient of the bead's length to width (Wong and Nurjaya 2008) – can be calculated to characterize bead shape. An ER of unity represents a perfect sphere, while higher values represent greater elongation or more divergence from a spherical shape (Das and Ng 2010). Almost all of the tested calcium-pectinate beads (except the dried beads prepared with 3% pectin), which included the following parameters: crosslinking with calcium chloride at a concentration of 2.5–20.0%, crosslinking pH of 1.5–5.5, crosslinking time of up to 24 hours, pectin concentration of 3–6%, and a pectin-to-drug ratio of 1:1 to 4:1, were spherical in shape (i.e. ER < 1.25) (Das and Ng 2010). In the case of κ-carrageenan, the sphericity of the formed beads was calculated as the ratio of the surface area of an equivalent sphere to the surface area of the bead, which in the case of oblate spheroids can be measured (Keppeler et al. 2009). When the conventional bead manufacturing method of dripping was used, highly spherical beads with a homogeneous surface were observed within a carrageenan concentration range of 1–6% (w/w). Higher carrageenan concentrations did not produce spherical beads, probably due to the high viscosity of the dropping solution. By adjusting the dripping rate (10 drops/min), the carrageenan solution temperature (50±4 °C) and the distance between the syringe tip and the hardening solution (optimal at 2 cm), a constant bead size could be obtained for the entire range of carrageenan concentrations, with a diameter of 3.1±0.1 mm and excellent sphericity. An increase in carrageenan concentration led to a significant increase in dropping solution viscosity and to larger beads (Sankalia et al. 2006b). Nevertheless, the bead size was not significantly different among concentrations used, in agreement with a previous report (Sipahigil and Dortunc 2001). The explanation for this phenomenon is probably the constant surface tension of the dripping solution, which leads to homogeneous bead drops independent of carrageenan concentration in the solution (Keppeler et al. 2009).

1.6.4 Shape and Size of Hydrocolloid Beads and Their Estimation

Precise assessments of shape and size are vital in the study of hydrocolloid beads. Most of them are produced in a spherical shape. However, deviations from sphericity occur, with spheroid and ellipsoid beads being quite common (Nussinovitch 2010). Size is a measure of how big or small a bead is. It is easier to specify the size of a spherical or ellipsoid-shaped bead; for irregularly shaped beads, the term size must be arbitrarily defined (Nussinovitch 2010). Beads can be formed in the millimeter, micrometer or nanometer size ranges. Overall, particle size can be determined by sieve analysis, passage through an electrically charged opening or settling rate (Sahin and Sumnu 2006). Numerous techniques have been developed for bead manufacture. These techniques include dripping (Walsh et al. 1996; Romo and Perez-Martinez 1997), emulsification and coacervation (Poncelet et al. 1993, 1994; Green et al. 1996), rotating-disc atomization (Bégin et al. 1991; Ogbonna et al. 1991), air jet (Klein et al. 1983; Levee et al. 1994), atomization (Siemann et al. 1990), electrostatic dripping (Bugarski et al. 1994; Halle et al. 1994; Poncelet et al. 1994), mechanical cutting (Prüsse et al. 1998), and the vibrating nozzle technique (Ghosal et al. 1993; Brandenberger et al. 1997; Seifert and Phillips 1997). Another common technique for manufacturing spherical polymer beads is suspension polymerization. In this process, minute droplets of monomer stabilized by small amounts of surface-active polymers are suspended in a continuous non-solvent medium, most commonly water, and polymerization is carried out by the addition of monomer-soluble initiators (O'Connor and Gehrke 1997). Suspension-crosslinking technique was used to yield thermally responsive gels of HPMC in a spherical form. The spherical beads had diameters ranging from 500 to 3000 μm. Bead size commonly decreased with the use of a larger impeller, swinging at high stirring speeds, or a lower phase ratio. As bead size decreased, the size distribution also narrowed (O'Connor and Gehrke 1997). Some of these methods suffer from limitations, particularly with respect to difficulties in achieving the simultaneous production of beads at a high production rate with a satisfactory level of material utilization under mild and non-toxic conditions, or under completely sterile conditions, in the ability to scale up the process, and in particular, in obtaining beads with a narrow size distribution (Nussinovitch 2010).

1.7 Assorted Specially Shaped and Sized Hydrocolloid Foods

1.7.1 Ham Consommé with Alginate Melon Beads

Melon juice beads are an example of a crosslinked gel product that can be included in a consommé. This recipe was created by Ferran Adrià i Acosta, the head chef of El Bulli restaurant and considered one of the best chefs in the world (https://en.wikipedia.org/wiki/Ferran_Adri%C3%A0). To produce melon juice beads, 500 g of melon juice (extracted from

800 g of fresh melon) were mixed with 0.4% sodium alginate, until the hydrocolloid dissolved. The blend was vacuum-sealed to remove trapped air and transferred into a syringe. For the crosslinking solution, calcium-chloride salt (2.5 g) was dissolved in 500 g water. The melon–alginate fluid was dropped into the calcium-chloride bath. The resulting beads were removed from the bath after three minutes. Then, the crosslinked beads were rinsed in a clean water bath and 10 g was added to champagne flutes that included 50 g each of cold ham consommé. The beads remained suspended in the product (Myhrvold et al. 2011, p. 48).

1.7.2 Extruded Gel Noodles

Ferran Adrià's revolutionary cooking has made El Bulli the most famous restaurant in the world (McInerney 2010). One of his achievements is extruded gel noodles. In general food-grade tubing is filled with liquid gel and after setting by placing the tubing in a cold bath, air (or gas) is used to push out the spaghetto (i.e. the singular form of spaghetti). For this, a bicycle pump, whipping siphon charged with nitrous oxide and the like can be used (Myhrvold et al. 2011). The reader is also referred to Section 1.9.

1.7.3 Cold Gels

Cold gels can have different shapes. Hydrocolloids for manufacturing such gels could be vegetative-originated gums, such as agar, carrageenan, gellan and LMP, and gelatin, which originate from animals. To prepare a cold gel, selection of texture and firmness is important, ranging from elastic to brittle or rubbery, and from very soft to very firm. Then the gelling agent needs to be chosen. Many of these formulas use a mixture of two or more gelling agents. The quantities of the gelling agents are measured relative to the weight of the liquid, followed by dry blending of the gelling agents, and dispersing them into the cold flavorful liquid. Hydration is fully achieved by blending while heating to ensure that the gelling agents are evenly distributed during hydration. The gel is then cast in a mold (Myhrvold et al. 2011).

1.7.4 Knot Foie

The Ashley Book of Knots, an encyclopedia of knots written and illustrated by the American sailor and artist Clifford W. Ashley, was first published in 1944. This book was written over 11 years, and contains thousands of numbered entries and about 7000 illustrations. It remains one of the leading wide-ranging books on knots (Ashley 1944). Sixty years later, Wylie Dufresne discovered that knots (Figure 1.15) can also be used in foods. Wylie Dufresne is a leading American proponent of molecular gastronomy, and the creator of the magnificent dish knot foie (https://en.wikipedia.org/wiki/Wylie_Dufresne). In 2007, Dufresne stated that the addition of hydrocolloid gums to numerous products can create products such as fried mayonnaise and foie gras that can be tied into a knot (Chang 2007). He also mentioned that his reading material included "Water-Soluble Polymer Applications in Foods" and "Hydrocolloid Applications: Gum Technology in the Food and Other Industries," both written by A. Nussinovitch, one of the authors of this book (https://www.starchefs.com/about_us/press_kit/images/NewYorkTimes110607.pdf; https://blog.ideasinfood.com/

1.7 Assorted Specially Shaped and Sized Hydrocolloid Foods

Figure 1.15 Some knots. *Source:* Nordisk Familjebok/Wikimedia Commons/Public Domain.

ideas_in_food/2007/07/not-our-idea.html). In this recipe, the hydrocolloids konjac mannan, agar and xanthan are used. The hydrocolloids are dry blended with the warm foie gras which was first heated to 82 °C to melt the torchon fat and split the terrine. Then water heated to 80 °C and a beaten egg yolk are added to the foie gras, and the mixture is blended until fully incorporated. The mixture is poured into a non-stick mold and refrigerated for about three hours to set. The set product is cut into strips which are tied carefully into a knot, paying attention not to break them. The full recipe and instructions for garnishing can be found in Myhrvold et al. (2011, p. 144).

1.7.5 Shapes of Gummy Worms

Fishermen use mold systems to produce artificial lures from liquid plastic. These molds come in numerous shapes and sizes, and are suitable for casting gels. If metal molds are used, they require a covering (sprayed or wiped) of a very thin film of cooking oil (Myhrvold et al. 2011).

1.7.6 Gel Films

Gel films can be created by casting the gel solution on an acetate sheet, petri dish or silicone mat. Acetate sheets are a flexible, transparent, ultra-versatile material with an extensive variety of purposes that are also suitable for chocolate and confectionery work. Very thin gel layers can be used to mimic other foods, e.g. a cloudy gel that includes coconut milk can mimic a slice of cheese; and by using skill, creativity and technique, it is possible to create films that look like lacy veils, or fruit-based gels that resemble stained glass (Myhrvold et al. 2011).

1.8 Foods for the Elderly

Hydrocolloids are used to control and adjust food texture. It is essential to identify how these polysaccharides affect swallowing behavior when planning foods for the elderly. Since the texture of foods is influenced by aroma and taste, and vice versa, a familiarity with the relationship between texture and flavor release is critical in manipulating foods to diminish the risk of aspiration (Nishinari et al. 2016). The difficulty in tailor-making foods for the elderly is the fact that knowledge in various research fields, such as food rheology, oral physiology, processing and psychology, is required to make any significant contribution to this endeavor (Nishinari 2006; Chen 2009). The number of elderly suffering from difficulties in chewing and deglutition (i.e. the action or process of swallowing) is growing. As a result, pneumonia due to aspiration is currently the third largest cause of death in Japan (Nishikubo et al. 2015). Dysphagia (i.e. difficulty or discomfort in swallowing, as a symptom of disease) also results in undernourishment, and conversely, malnutrition causes dysphagia (Via and Mechanick 2013; Carrión et al. 2015).

Enzyme technology was used to develop more or less soft foods, easily crushed between the tongue and palate (Nakatsu et al. 2014; Takei et al. 2015). Appearance, flavor and texture were chief attributes (Delwiche 2012; Spence et al. 2016) for the palatability of these food products, as their shape, size and color needed to be attractive to increase the elderly's desire for food, thereby improving their undernourishment and quality of life. Variety is also essential for good taste (Nishinari et al. 2016).

Milled foods are commonly used for the elderly and/or individuals with trouble masticating. Heated and pressurized gels were found to be appropriate for this purpose since they are deformable and adaptable for dysphagic patients (Tokifuji et al. 2012). For such meat gels, the transit time in the vallecular area was short. The epiglottic vallecula is a depression just behind the root of the tongue in the medial and lateral glossoepiglottic folds (i.e. a mucous membrane fold that extends from the margin of the epiglottis to the pharyngeal wall and base of the tongue on each side, forming the lateral boundary of the epiglottic valleculae) in the throat. Very little food residue was found in the oropharynx (i.e. the part of the throat at the back of the mouth behind the oral cavity) (Tokifuji et al. 2012, 2013). Not much mastication effort is required for minced foods; nevertheless, food fragments disintegrate, and the establishment of a coherent bolus becomes difficult, leading to the possibility of aspiration (Nishinari et al. 2016). It was remarkably demonstrated that for thinly sliced cucumber, cutting into smaller parts does not automatically cause a decrease in chewing time (Kohyama et al. 2003). For agar gel cubes, the size distribution was more important than the actual dimensions of the gel pieces for bolus formation (Kitade et al. 2013). Furthermore, the gel bolus was easier to assemble with gel pieces of various sizes than with those of similar sizes. It was also demonstrated that the mastication effort was not necessarily less for smaller gels (Kitade et al. 2014), and that the elapsed time from the end of mastication to swallowing as a function of chewing time indicated that structural breakage by mastication was effective for the gel cubes that were up to 20 times larger than gels with smaller sizes in the initial stage of mastication (Kitade et al. 2014).

1.8.1 Effects of Hydrocolloid Addition on the Mastication of Minced Foods

The effects of xanthan gum, locust bean gum and guar gum on the mastication of boiled minced carrots or mackerel fish were studied (Yoshimura et al. 2003, 2004). Addition of these thickening agents reduced both the time and number of chews of the minced foods. The age of the panel influenced the results. The young panel favored minced food with 0.6% xanthan gum, while those with 2.0% xanthan were not highly rated due to the marked residual that had to be swallowed. The minced foods fortified with the thickening agents were judged by the elderly panel as viscous, easy to swallow and tasty. Thus, the viscoelasticity of the thickening agents enhanced the palatability of the minced foods, particularly for the elderly panel (Yoshimura et al. 2003, 2004). The effect of xanthan gum with different intrinsic viscosities on the mastication properties of minced (0.5 cm) and cubed (1.0 cm) steamed carrots was studied (Yoshimura et al. 2005). With young panelists, the number of chews and the duration of chewing for these carrots were not considerably different; nevertheless, the inclusive evaluation of the carrots mixed with xanthan at the highest intrinsic viscosity was lowest (Yoshimura et al. 2005). Another report (Ichimura et al. 2000) described that patients who suffer from aspiration (i.e. when something you swallow "goes down the wrong way" and enters your airway or lungs) favored thickened fluid foods, whereas individuals without this problem preferred less viscous liquid foods (Ichimura et al. 2000). Xanthan gum with the lowest intrinsic viscosity was also used with cooked carrots served to an elderly panel (Yoshimura et al. 2005). They found that the addition of xanthan gum improved mouthfeel and made the carrots easy to swallow. If members of this panel had their natural teeth or partial dentures, they suggested that minced carrots (0.5 cm) with the addition of xanthan gum were easier to chew. Those members with full dentures were suggested that cut carrots (1.0 cm) were best for chewing, the minced carrots being too small. It was concluded that the condition of the teeth was important in the mastication of minced foods for the elderly panel members (Yoshimura et al. 2005).

The effects of mixed polysaccharides – xanthan/high acyl gellan – as viscosity-forming agents to control the consistency of steamed minced carrots was also studied. The young panel favored the addition of xanthan gum on its own. The elderly panel members preferred mixed hydrocolloids because they felt that they decreased the number of chews and duration of chewing; they felt that use of these mixed thickening agents results in a more coherent bolus. In addition, the number of chews increased with deteriorating dental condition, meaning that they had to invest extra effort (Yoshimura et al. 2008). In brief, addition of viscosity formers to minced foods can lessen the number of chews and chewing duration, convert the minced foods into a more coherent mass and make the mastication more effective. For more efficient mastication, the size of each fragment should not be too small (Yoshimura et al. 2008).

1.8.2 Hydrocolloids for the Design of Food for the Elderly

Starch is largely used as a viscosity-forming agent in numerous commercial items for consumption (Nishinari et al. 2016). Since dispersed starch can exhibit thixotropic behavior (i.e. the apparent viscosity decreases with time of shearing but the change is reversible) and its viscosity changes throughout storage, it is difficult to use it for handling the viscosity of products for treatment of dysphagia. Solutions of guar gum and xanthan do not retrograde (like the starch dispersion) and therefore could serve for such thickening purposes (Nishinari et al. 2016).

A xanthan-based solution (4.1% w/w) demonstrated shear-thinning behavior over a wide range of experimental shear rates, i.e. between 0.03 and 1000 per second. The xanthan solution showed the highest viscosity at the lowest shear rate and the lowest viscosity at the highest shear rate. Similarly, solutions of guar gum (2.0 and 2.6% w/w) exhibited shear thinning (Nishinari et al. 2016). These solutions also demonstrated identical steady shear viscosity at 50 per second but dissimilar shear-thinning performance. A shear rate 50 per second was chosen because it was strongly believed to be the shear rate in the oral cavity, despite much debate on this value (Nishinari et al. 2016). Shear-thinning occurrence depended on the polydispersity Mw/Mn. Mn is the number averaged MW (molecular weight), and Mw is the weight averaged MW of the hydrocolloid (Nishinari et al. 2016). Those solutions served in a videofluorographic (VF) observation of 32 patients swallowing different liquids (Nishinari et al. 2016). The objectives of VF swallowing studies (VFSSs) are both diagnostic and therapeutic (Palmer et al. 1993). A comprehensive narrative of the VFSS method includes the position of the patient, the foods offered, the views achieved, amendments of ordinary feeding and swallowing behavior, a standardized set of interpretations, and reporting the results (Palmer et al. 1993). The VFSS does not automatically conclude when a patient aspirates. In fact, the purpose of the test is an overall assessment of aspiration, and the effects of maneuvers designed to reduce it. Modifications of feeding and swallowing are tested empirically throughout the VFSS. The amendments consist of therapeutic and compensatory techniques that may advance the patient's well-being and swallowing efficacy (Palmer et al. 1993). The results of the VFSS (Nishinari et al. 2016) demonstrated that for 32 performed observations, the ratio of aspirations was 4/32 for 4.0% xanthan, 5/32 for 2.0% guar gum and 7/32 for 2.6% guar gum solutions. The viscosity at the lower shear rate appeared to be more important rather than that at 50 per second or higher shear rate. The likelihood of aspiration seemed to decline with growing viscosity at lower shear rate. The authors also concluded that the number of observations was not sufficiently high, and that this hypothesis should be confirmed by a larger number of data in the future (Nishinari et al. 2016).

1.9 Demonstrating the Use of Hydrocolloids in Controlling Food Size and Shape

The use of hydrocolloids in controlling the shape and/or size of various products is demonstrated by a few examples and photographs. Pasta is a food that comes in various different shapes and sizes (roughly 350 different types of pasta worldwide). Chefs use these different shapes and sizes for diverse purposes, as some pasta shapes hold some types of sauce better than others. Spaghetti is America's favorite pasta shape. It can be used with nearly any variety of sauce in pots or stir-fried dishes. Spaghetti is a long, thin, solid, cylindrical pasta (https://www.dictionary.com/browse/spaghetti). It is a main food in traditional

Italian cuisine. Similar to other pasta, spaghetti is prepared from milled wheat and water, sometimes supplemented with vitamins and minerals. Italian spaghetti is characteristically prepared from durum wheat (Chihak 2020). Due to its popularity, it is not surprising that chefs have tried to use gelling agents to mimic the shape and size of spaghetti (http://lilrizzos.blogspot.com/2015/01/why-does-pasta-come-in-so-many.html).

Figure 1.16 demonstrates the preparation of agar spaghetti. The gelling cooks at the restaurant ElBulli, who create appetizing gels in noodle form, designed this procedure. Briefly, food-grade tubing is filled with a fluid gel composition and then left to set. Once ready to serve, air is forced into one end of the tube, extruding a long strand of spaghetti. A peristaltic pump can accelerate the process and increase the yield (Myhrvold et al. 2011). Agar was chosen as a gelling agent for this recipe due to its high gelling hysteresis, that is, the difference between gelling (~38 °C) and melting (~85 °C) temperatures (Nussinovitch and Hirashima 2014). This difference allows mixing the ingredients into the gelling solution before it sets. In addition, there is a marked dependence of gel melting point and sol setting point on the concentration of different agars, raising the importance of choosing a suitable agar type and concentration for a well-defined task (Davidson 1980; Kojima et al. 1960).

Being one of the most common shapes in nature (https://www.mathnasium.com.hk/2015/01/math-in-nature-perfect-spheres), ball-shaped foods are very popular and many

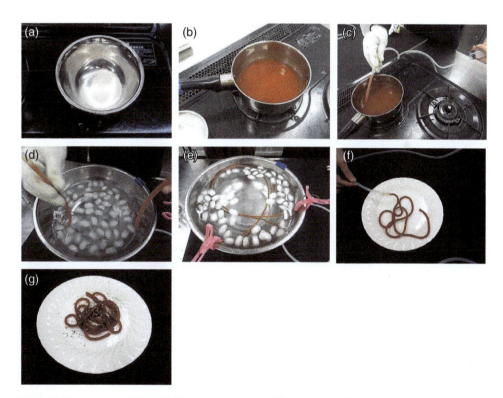

Figure 1.16 Agar spaghetti. (a) Agar–agar powder. (b) Tomato soup. (c) Drawing tomato soup by syringe into attached silicone tube, (d) cooling the silicone tube filled with the tomato soup, (e) cooling the silicone tube using clothespins, (f) extruding the jelly. (g) Serve.

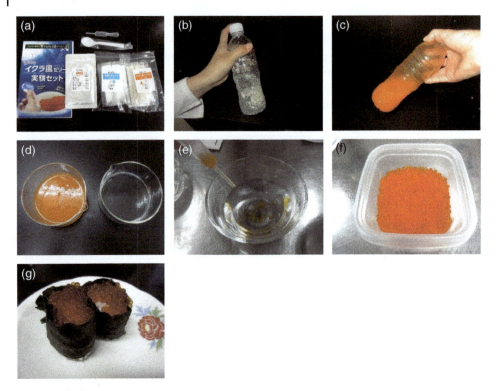

Figure 1.17 Comercial experimental set for producing artificial salmon roe. (a) Ingredients in commercial set. (b) Shaking the sodium alginate aqueous dispersion, (c) adding orange food coloring. (d) The two solutions. (e) Dropping the sodium-alginate solution into the calcium-lactate solution. (f) Artificial salmon roe. (g) Artificial salmon roe sushi.

foods are spherical. A few examples of spherical foods are eggnog popcorn balls, feta-stuffed falafel, boudin balls, cake pops, cinnamon-oat truffle treats, the turducken of cheese balls, tomato–basil arancini, chevré truffles, and brigadeiros, to name a few (https://www.chowhound.com/food-news/179944/have-a-ball-spherical-food-we-love/). Therefore, it is not surprising that chefs have tried to mimic the shape, size, hue and texture of the spherical salmon roe (i.e. salmon eggs). Figure 1.17 demonstrates the preparation of artificial salmon roe using an experimental set that is marketed in Japan by the Kimica Corporation. Alginate was chosen for this preparation because it forms a gel with a number of divalent cations (McDowell 1960). Calcium is particularly suited to food purposes because it is non-toxic. Spontaneous crosslinking takes place via carboxyl groups by primary valences and via hydroxyl groups by secondary valences (Nussinovitch 1997). It is important to note that when the colored alginate drops fall into the setting bath (which usually includes calcium chloride or calcium lactate), their surface starts to gel (i.e. becomes crosslinked). Rapid removal of the formed spheroids from the calcium setting bath leaves us with droplets of liquid surrounded by a membrane. If the formed spheres are left for a longer time in the setting bath, the whole bead solidifies. Obtaining "real" liquid-core beads requires another technique (Peschardt 1946), eventually dubbed reverse spherification (Myhrvold et al. 2011).

Figure 1.18 Real salmon roe (left) and commercial artificial salmon roe (right).

Figure 1.18 presents the similarities between real and artificial salmon roe. The shape, size and appearance of the products are the same, and the artificial roe are quite appealing, as well as costing much less.

1.9.1 Agar Spaghetti

(Serves 5)

3 g agar–agar powder (Figure 1.16a)
250 g tomato soup (Figure 1.16b, *Hint 1*)
Syringe (50 mL), silicone tube (4 mm diameter, 1–2 m length)
Cheese powder, herbs (optional)
pH = 3.98

For preparation, see Figure 1.16

1) Add agar–agar powder to tomato soup, and heat at 90–100 °C until completely dissolved.
2) Attach silicone tube to the syringe, and fill it with the agar tomato soup mixture (Figure 1.16c).
3) Remove the silicone tube from the syringe, and put it in a bowl of ice water for 10 minutes (Figure 1.16d and e, *Hint 2*).
4) Fill the syringe with air, attach the cold tube with the set agar jelly and extrude the jelly (Figure 1.16f).
5) Garnish with cheese powder and herbs (Figure 1.16g).

Preparation hints:

1) If there are solid lumps in the soup, strain them out. You can make tomato soup using soup stock, salt, pepper, sugar, etc.
2) Use clothespins to prevent the tube ends from sinking (Figure 1.16e).

1.9.2 Commercial Experimental Set to Produce Artificial Salmon Roe

(1 experimental set) (Figure 1.17a, *Hint 1*)

2 g sodium alginate (Figure 1.17a, right, orange package A)
2 g calcium lactate (Figure 1.17a, second from right, blue package B)
Orange food coloring (Figure 1.17a, third from right)
Dropper, spoon
Tap water

For preparation, see Figure 1.17

1) Put 200 mL tap water in a plastic bottle, add sodium alginate, close the lid, and shake the bottle (Figure 1.17b) until the lumps disappear.
2) Add orange food coloring to the sodium-alginate solution (1), shake the bottle (Figure 1.17c), and pour the solution into a bowl (Figure 1.17d).
3) Pour 200 mL tap water into another plastic bottle, add calcium lactate, close the lid, and shake it.
4) Pour the calcium-lactate solution (3) into another bowl (Figure 1.17d).
5) Suck the sodium-alginate solution (2) with a dropper, and drop one by one into the calcium-lactate solution (4) (Figure 1.17e).
6) Scoop out the artificial salmon roe (Figure 1.17f, *Hint 2*).

Preparation hints:

1) Made by Kimica Corporation (https://kimica-algin.com/, https://www.kimica.jp/ikurajelly/).
2) Eating suggestion: marinate the artificial salmon roe in soy sauce and sugar, and use for sushi (Figure 1.17g).

References

Agnihotri, S.A., Jawalkar, S.S., and Aminabhavi, T.M. (2004). Controlled release of cephalexin through gellan gum beads: effect of formulation parameters on entrapment efficiency, size, and drug release. *European Journal of Pharmaceutics and Biopharmaceutics* 63: 249–261.

Agrawal, A.M., Howard, M.A., and Neau, S.H. (2004). Extruded and spheronized beads containing no microcrystalline cellulose: influence of formulation and process variables. *Pharmaceutical Development and Technology* 9: 197–217.

Alton, M.W. and Browne, E.C. (2020). Atmospheric chemistry of volatile methyl siloxanes: kinetics and products of oxidation by OH radicals and Cl atoms. *Environmental Science and Technology* 54: 5992–5999.

Ashley, C.W. (1944). *The Ashley Book of Knots*. New York: Doubleday.

Baldyga, J., Bourne, J.R., Pacek, A.W. et al. (2000). Effects of agitation and scale-up on drop size in turbulent dispersions: allowance for intermittency. *Chemical Engineering Science* 56: 3377–3387.

Bégin, F., Castaigne, F., and Goulet, J. (1991). Production of alginate beads by a rotative atomizer. *Biotechnology Techniques* 5: 459–464.

Berk, Z. (2009). Size reduction. In: *Food Process Engineering and Technology*, 153–170. New York: Elsevier.

Bond, F.C. (1952). The third theory of comminution. *Transactions of the Metallurgical Society of the American Institute of Mining* 193: 484–494.

Brandenberger, H., Nüssli, D., Piëch, V., and Widmer, F. (1997). Monodisperse particle production: a new method to prevent drop coalescence using electrostatic forces. *Journal of Electrostatics* 45: 227–238.

Brennan, J.G., Butters, J.R., Cowell, N.D., and Lilley, A.E.V. (1990). *Food Engineering Operations*, 3e. London and New York: Elsevier Applied Science.

Bugarski, B., Li, Q.L., Goosen, M.F.A. et al. (1994). Electrostatic droplet generation: mechanism of polymer droplet formation. *AIChE Journal* 40: 1026–1031.

Buthe, A., Hartmeier, W., and Ansorge-Schumacher, A.B. (2004). Novel solvent-based method for preparation of alginate beads with improved roundness and predictable size. *Journal of Microencapsulation* 21: 865–876.

Carrión, S., Cabré, M., Monteis, R. et al. (2015). Oropharyngeal dysphagia is a prevalent risk factor for malnutrition in a cohort of older patients admitted with an acute disease to a general hospital. *Clinical Nutrition* 34: 436–442.

Chang, K. (2007) Food 2.0: chefs as chemists. *The New York Times* (6 November). https://www.nytimes.com/2007/11/06/science/06food.html (accessed 12 June, 2021).

Chen, J. (2009). Food oral processing – a review. *Food Hydrocolloids* 23: 1–25.

Chevallier, S., Mazzocato, M.C., Favaro-Trindade, C.S., and Poncelet, D. (2017). Monitoring the capillary jet breakage by vibration using a fast video camera. *25th International Conference on Bioencapsulation*, La Chapelle sur Erdre, France (3–6 July 2017), pp. 54–55.

Chihak, S. (2020). How to Make Spaghetti That's Cooked to Al Dente Perfection. BHG.com. https://www.bhg.com/recipes/ethnic-food/italian/how-to-make-spaghetti/ (accessed 6 November 2021).

Correa, M.J., Añón, M.C., Pérez, G.T., and Ferrero, C. (2010). Effect of modified celluloses on dough rheology and microstructure. *Food Research International* 43: 780–787.

Curray, J.K. (1951). Analysis of sphericity and roundness of quartz grains. MSc thesis. The Pennsylvania State University, University Park, PA.

Das, S. and Ng, K.-Y. (2010). Resveratrol-loaded calcium-pectinate beads: effects of formulation parameters on drug release and bead characteristics. *Journal of Pharmaceutical Sciences* 99: 840–860.

Davidson, R.L. (1980). *Handbook of Water-Soluble Gums and Resins*. New York: McGraw-Hill.

De Villiers, M. (1996). An extended classification of quadrilaterals. In: *Some Adventures in Euclidean Geometry*. South Africa: University of Durban-Westville.

Delwiche, J.F. (2012). You eat with your eyes first. *Physiology & Behavior* 107: 502–504.

Duggal, G. (2021). Lovely triangle. *Hindustan Times*. https://www.hindustantimes.com/india/lovely-triangles/story-aIL2GLawECV084GXmGqshI.html (accessed 12 November, 2021).

Earle, R.L. (1983). *Unit Operations in Food Processing*, 2e. Oxford, UK: Pergamon Press.

Fellows, P.J. (2000). *Food Processing Technology Principles and Practice*, 2e. Cambridge, UK: Woodhead Publishing Limited.

Field, M. and Field, F. (1970). *A Quintet of Cuisines, Foods of the World*. New York: Time-Life Books.

Fromme, H., Witte, M., Fembacher, L. et al. (2019). Siloxane in baking molds, emission to indoor air and migration to food during baking with an electric oven. *Environment International* 126: 145–152.

Ghosal, S.K., Talukdar, P., and Pal, T.K. (1993). Standardization of a newly designed vibrating capillary apparatus for the preparation of microcapsulses. *Chemical Engineering & Technology* 16: 395–398.

Gómez, M., Moraleja, A., Oliete, B. et al. (2010). Effect of fibre size on the quality of fibre-enriched layer cakes. *LWT – Food Science and Technology* 43: 33–38.

Green, K.D., Gill, I.S., Khan, J.A., and Vulfson, E.N. (1996). Microencapsulation of yeast cells and their use as a biocatalyst in organic solvents. *Biotechnology & Bioengineering* 49: 535–543.

Griffiths, J.C. (1958). Petrography and porosity of the cow run sand at St. Marys, West Virginia. *Journal of Sedimentary Petrology* 28: 15–30.

Gross, J.H. (2015). Analysis of silicones released from household items and baby articles by direct analysis in real time-mass spectrometry. *Journal of the American Society for Mass Spectrometry* 26: 511–521.

Halle, J.P., Leblond, F.A., Pariseau, J.F. et al. (1994). Studies on small (less than 300 μm) microcapsules. II. Parameters governing the production of alginate beads by high-voltage electrostatic pulses. *Cell Transplantation* 3: 365–372.

Hayashi, T. (1989). Structure, texture and taste. In: *Food Technology International Europe* (ed. A. Turner), 53–56. London: Sterling Publications International.

Helstosky, C. (2008). *Pizza: A Global History*. London: Reaktion Books.

Houston, R.K. (1957). New criterion of size for agricultural products. *Agricultural Engineering* 39: 856–858.

Ichimura, K., Kawanami, K., Yasuoka, T. et al. (2000). The relationship between the viscosity and the palatableness in easy-to-swallow foods (liquid/sol foods). *Medicinal Journal Ibaraki Prefecture Hospital* 18: 9–17.

Kattan, A.A., Ogle, W.G., and Kramer, A. (1956). Effect of process variables on quality of canned tomato juice. *Proceedings of the American Society for Horticultural Science* 68: 470–481.

Kaye, B. (1993). *Chaos and Complexity: Discovering the Surprising Patterns of Science and Technology*. Weinheim, NY: VCH.

Keppeler, S., Ellis, A., and Jacquier, J.C. (2009). Cross-linked carrageenan beads for controlled release delivery systems. *Carbohydrate Polymers* 78: 973–977.

Kick, F. (1883). The law of proportional resistance and its application to sand and explosions. *Dingler's Polytechnisches Journal* 250: 141–145.

Kitade, M., Kobayashi, N., and Moritaka, H. (2013). Relationship between number of chewing cycles and fragment size of agar gels. *Nippon Shokuhin Kagaku Kogaku Kaishi* 60: 554–562. (in Japanese with English summary and figure captions).

Kitade, M., Sagawa, A., Fuwa, M., and Moritaka, H. (2014). Properties of the masseter and digastric muscles during swallowing of agar gels of different sizes. *Nippon Shokuhin Kagaku Kogaku Kaishi* 61: 293–301. (in Japanese with English summary and figure captions).

Klein, J., Stock, J., and Vorlop, K.D. (1983). Pore size and properties of spherical Ca-alginate biocatalysts. *European Journal of Applied Microbiology and Biotechnology* 18: 86–91.

Kohyama, K., Nakayama, Y., Fukuda, S. et al. (2003). Thinly sliced cucumber requires more mastication. *Nippon Shokuhin Kagaku Kogaku Kaishi* 50: 339–343. (in Japanese with English summary and figure captions).

Kojima, Y., Tagawa, S., and Yamada, Y. (1960). Studies on the new method of preparation of agar-agar from *Ahnfeltia plicata*. On the properties of Itani agar. *Journal Shimonoseki University Fisheries* 10: 43–56.

Lassen, C., Hansen, C.L., Mikkelsen, S.H. and Maag, J. (2010) Siloxanes - Consumption, Toxicity and Alternatives. Environmental Project no. 1031 2005. Miljøprojekt. Report Prepared for the Danish Ministry of the Environmental Protection Agency. https://www2.mst.dk/udgiv/publications/2005/87-7614-756-8/pdf/87-7614-757-6.pdf (accessed 15 July, 2021).

Levee, M.G., Lee, G.M., Paek, S.H., and Palsson, B.O. (1994). Microencapsulated human bone-marrow cultures: a potential culture system for the clonal outgrowth of hematopoietic progenitor cells. *Biotechnology & Bioengineering* 43: 734–739.

Lynch, A.J. and Rowland, C.A. (2005). *The History of Grinding*, 15–16. Society for Mining, Metallurgy, and Exploration (SME). Littleton, CO.

McDowell, R.H. (1960). Applications of alginates. *Reviews of Pure and Applied Chemistry* 10: 1–5.

McInerney, J. (2010). It was delicious while it lasted. *Vanity Fair Magazine* (October). https://www.vanityfair.com/culture/2010/10/el-bulli-201010 (accessed 16 September 2021).

*Testing and Characterization of Powders and Fine Particles*Medalia, A.I. (1980). Three-dimensional shape parameters. In: (ed. J.K. Beddow and T. Meloy), 66. London: Heyden & Son Ltd.

Mohsenin, N.N. (1970). *Physical Properties of Plant and Animal Materials*, 51–54. New York: Gordon and Breach.

Mohsenin, N.N. (1980). *Physical Properties of Food and Agricultural Materials*. New York: Gordon and Breach.

Mu, Y., Lyddiatt, A., and Pacek, A.W. (2005). Manufacture by water/oil emulsification of porous agarose beads: effect of processing conditions on mean particle size, size distribution and mechanical properties. *Chemical Engineering and Processing* 44: 1157–1166.

Mudireddy, R.R. (2019). The attractive shape of the foods. https://www.linkedin.com/pulse/attractive-shape-foods-rakesh-reddy-mudireddy/?articleId=6527880881862897664 (accessed 12 November 2021).

Musser, G.L. and Burger, W.F. (1997). *Mathematics for Elementary Teachers, a Contemporary Approach*, 4e, 507–641. Upper Saddle River, NJ: Prentice Hall.

Myhrvold, N., Young, C., and Bilet, M. (2011). *Modernist Cuisine, the Art and Science of Cooking*, vol. 4. Bellevue, WA: The Cooking Lab, LLC.

Nakatsu, S., Kohyama, K., Watanabe, Y. et al. (2014). A trial of human electromyography to evaluate texture of softened foodstuffs prepared with freeze-thaw impregnation of macerating enzymes. *Innovative Food Science and Emerging Technologies* 21: 188–194.

Nishikubo, K., Mise, K., Ameya, M. et al. (2015). Quantitative evaluation of age-related alteration of swallowing function: videofluoroscopic and manometric studies. *Auris Nasus Larynx* 42: 134–138.

Nishinari, K. (2006). Polysaccharide rheology and in-mouth perception. In: *Food Polysaccharides and their Applications*, 2e (ed. A.M. Stephen, G.O. Phillips and P.A. Williams), 541–588. New York: Taylor & Francis.

Nishinari, K., Takemasa, M., Brenner, T. et al. (2016). The food colloid principle in the design of elderly food. *Journal of Texture Studies* 47: 284–312.

Nussinovitch, A. (1997). *Hydrocolloid Applications: Gum Technology in the Food and Other Industries*. London and Weinheim, UK: Blackie Academic and Professional.

Nussinovitch, A. (2010). Physical properties of beads and their estimation. In: *Polymer Macro-and Micro-Gel Beads: Fundamentals and Applications*, 1–22. New York: Springer.

Nussinovitch, A. and Gershon, Z. (1996). A rapid method for determining sphericity of hydrocolloid beads. *Food Hydrocolloids* 10: 263–266.

Nussinovitch, A. and Hirashima, M. (2014). *Cooking Innovations, Using Hydrocolloids for Thickening, Gelling and Emulsification*. Boca Raton, FL: CRC Press, Taylor & Francis Group.

O'Connor, S.M. and Gehrke, S.H. (1997). Synthesis and characterization of thermally-responsive hydroxypropyl methylcellulose gel beads. *Journal of Applied Polymer Science* 66: 1279–1290.

Ogbonna, J.C., Matsumura, M., and Kataoka, H. (1991). Effective oxygenation of immobilized cells through reduction in bead diameter: a review. *Process Biochemistry* 26: 109–121.

Okushima, S., Nisisako, T., Torii, T., and Higuchi, T. (2004). Controlled production of monodisperse double emulsions by two-step droplet breakup in microfluidic devices. *Langmuir* 20: 9905.

Onyango, C., Unbehend, G., and Lindhauer, M.G. (2009). Effect of cellulose-derivatives and emulsifierson creep-recovery and crumb properties of gluten-free bread prepared from sorghum and gelatinised cassava starch. *Food Research International* 42: 949–955.

Palmer, J.B., Kuhlemeier, K.V., Tippett, D.C., and Lynch, C. (1993). A protocol for the videofluorographic swallowing study. *Dysphagia* 8: 209–214.

Peschardt, W.J.S. (1946). Manufacture of artificial edible cherries, soft sheets and the like. United States Patent 2,403,547.

Phillips, G.O. and Williams, P.A. (2000). *Handbook of Hydrocolloids*. Cambridge, UK: Woodhead Publishing Limited.

Poncelet, D., Leung, R., Centomo, L., and Neufeld, R.J. (1993). Microencapsulation of silicone oils within polyamide polyethylenimine membranes as oxygen carriers for bioreactor oxygenation. *Journal of Chemical Technology & Biotechnology* 57: 253–263.

Poncelet, D., Bugarski, B., Amsden, B.G. et al. (1994). A parallel-plate electrostatic droplet generator: parameters affecting microbead size. *Applied Microbiology and Biotechnology* 42: 251–255.

Prüsse, U., Fox, B., Kirchhof, M. et al. (1998). New process (jet cutting method) for the production of spherical beads from highly viscous polymer solutions. *Chemical Engineering & Technology* 21: 29–33.

Quenoville, M.H. (1952). *Associated Measurements*. London: Butterworth-Sprinter.

Rayleigh, L. (John William Strutt)(1878). On the capillary phenomena of jets. *Proceedings of the London Mathematical Society* 10: 71–79.

Reza, S. (2015). Food's holy triangle. *DAWN* (18 January). https://www.dawn.com/news/1157291 (accessed 15 August, 2021). Archived from the original on 28 October 2018.

Rheon (2016). *Encrusting Machine*. Rheon Automatic Machinery Co. Ltd www.rheon.com/en/products/list.php. (accessed September 2021).

Richards, R.H. and Locke, C.E. (1940). *Textbook of Ore Dressing*, 3e. New York: McGraw-Hill.

Richardson, J.F., Harker, J.H., and Backhurst, J.R. (2002). *Chemical Engineering Series*, Particle Technology and Separation Processes, 5e, vol. 2, 95–145. Oxford, UK: Butterworth Heinemann.

Romo, S. and Perez-Martinez, C. (1997). The use of immobilization in alginate beads for long-term storage of *Pseudanabaena galeata* (Cyanobacteria) in the laboratory. *Journal of Phycology* 33: 1073–1076.

Rücker, C. and Kümmerer, K. (2015). Environmental chemistry of organosiloxanes. *Chemical Reviews* 115: 466–524.

Saha, D. and Bhattacharya, S. (2010). Hydrocolloids as thickening and gelling agents in food: a critical review. *Journal of Food Science and Technology* 47: 587–597.

Sahin, S.S. and Sumnu, S.G. (2006). *Physical Properties of Foods*. New York: Springer.

Sankalia, M.G., Mashru, R.C., Sankalia, J.M., and Sutariya, V.B. (2006a). Physicochemical characterization of papain entrapped in ionotropically cross-linked kappa-carrageenan gel reads for stability improvement using Doehlert shell design. *Journal of Pharmaceutical Sciences* 95: 1994–2013.

Sankalia, M.G., Mashru, R.C., Sankalia, J.M., and Sutariya, V.B. (2006b). Stability improvement of alpha-amylase entrapped in kappa-carrageenan beads: physicochemical characterization and optimization using composite index. *International Journal of Pharmaceutics* 312: 1–14.

Savart, F. (1833). Mémoire sur la constitution des veines liquides lancées par des orifices circulaires en mince paroi. *Annals of Chemistry* 53: 337–386.

Schneider, J.M. and Hendricks, C.D. (1964). Source of uniform-sized liquid droplets. *Review of Scientific Instruments* 35: 1349–1350.

Seifert, D.B. and Phillips, J.A. (1997). Production of small, monodispersed alginate beads for cell immobilization. *Biotechnology Progress* 13: 562–568.

Serp, D., Cantana, E., Heinzen, C. et al. (2000). Characterization of an encapsulation device for the production of monodisperse alginate beads for cell immobilization. *Biotechnology Bioengineering* 70: 41–53.

Siemann, M., Müller-Hurtig, R., and Wagner, F. (1990). Characterization of the rotating nozzle-ring technique for the production of small spherical biocatalysts. Physiology of immobilized cells. In: *Proceedings of the International Symposium of Physiology of Immobilized Cells* (ed. J.A.M. de Bont, J. Visser, B. Matiasson and J. Tramper), 275–282. Wageningen: Elsevier Science.

Sipahigil, O. and Dortunc, B. (2001). Preparation and in vitro evaluation of verapamil HCl and ibuprofen containing carrageenan beads. *International Journal of Pharmaceutics* 228: 119–128.

Soukoulis, C., Chandrinos, I., and Tzia, C. (2008). Study of the functionality of selected hydrocolloids and their blends with κ-carrageenan on storage quality of vanilla ice cream. *LWT – Food Science and Technology* 41: 1816–1827.

Spence, C., Okajima, K., Cheok, A.D. et al. (2016). Eating with our eyes: from visual hunger to digital satiation. *Brain Cognition* 110: 53–63.

Steiner, H.J. (1991). The significance of the Rittinger equation in present-day comminution technology. In: *Proceedings of the XVIIth International Mineral Processing Congress, Dresden*, vol. 1, 177–188. Bergakademie Freiberg/Sa, Freiberg: Polygraphischer Bereich.

Surak, J.G., Matthews, F.R., Wang, V. et al. (1979). Particle size distribution of commercial tomato juices. *Proceedings of the Florida State Horticultural Society* 92: 159–163.

Szczesniak, A.S. (1963). Classification of textural characteristics. *Journal of Food Science* 28: 385–389.

Szczesniak, A.S. (1983). Physical properties of foods: what they are and their relation to other food properties. In: *Physical Properties of Foods* (ed. M. Peleg and E.B. Bagely), 1–41. Westport, CT: Avi Publishing Company.

Takei, R., Hayashi, M., Umene, S. et al. (2015). Changes in physical properties of enzyme-treated beef before and after mastication. *Journal of Texture Studies* 46: 3–11.

Takeuchi, S., Garstecki, P., Weibel, D.B., and Whitesides, G.M. (2005). An axisymmetric flow-focusing microfluidic device. *Advanced Materials* 17: 1067.

Tan, W.H. and Takeuchi, S. (2007). Monodisperse alginate hydrogel microbeads for cell encapsulation. *Advanced Materials* 19: 2696.

Tashiro, Y. (1989). Method for shaping a spherical body. United States Patent 4,883,678A.

Tashiro, Y. (1993). Apparatus for cutting and shaping a spherical body. United States Patent 5,190,770.

Toft, K., Grasdalen, H., and Smidsrød, O. (1986). Synergistic gelation of alginates and pectins. *ACS Symposium Series* 310: 117.

Tokifuji, A., Matsushima, Y., Hachisuka, K., and Yoshioka, K. (2012). Physical properties of pressurized and heat treated meat gels and their suitability as dysphagia diet based on swallowing dynamics. *Japanese Journal of Comprehensive Rehabilitation Science* 3: 18–25.

Tokifuji, A., Matsushima, Y., Hachisuka, K., and Yoshioka, K. (2013). Texture, sensory and swallowing characteristics of high-pressure-heat-treated pork meat gel as a dysphagia diet. *Meat Science* 93: 843–848.

Tosa, T., Sato, T., Mori, T. et al. (1979). Immobilization of enzymes and microbial-cells using carrageenan as matrix. *Biotechnology & Bioengineering* 21: 1697–1709.

Via, M.A. and Mechanick, J.I. (2013). Malnutrition, dehydration, and ancillary feeding options in dysphagia patients. *Otolaryngologic Clinics of North America* 46: 1059–1071.

Walsh, P.K., Isdell, F.V., Noone, S.M. et al. (1996). Growth patterns of *Saccharomyces cerevisiae* microcolonies in alginate and carrageenan gel particles: effect of physical and chemical properties of gels. *Enzyme and Microbial Technology* 18: 366–372.

Weber, C. (1931). Zum zerfall eines flussikeitstahles. *Zeitschrift fur angewandte Mathematik and Mechanik* 11: 136–155.

Weisstein, E.W. (2003). *CRC Concise Encyclopedia of Mathematics*, 2e. Boca Raton, FL: CRC Press.

Wong, T.W. and Nurjaya, S. (2008). Drug release property of chitosan-pectinate beads and its changes under the influence of microwave. *European Journal of Pharmaceutics and Biopharmaceutics* 69: 176–188.

Woo, J.W., Roh, H.J., Park, H.D. et al. (2007). Sphericity optimization of calcium alginate gel beads and the effects of processing conditions on their physical properties. *Food Science and Biotechnology* 16: 715–721.

Yoshimura, M., Kuwano, T., Tanaka, M., and Nishinari, K. (2003). Rheological properties and sensory evaluation of the minced foods with thickening agents. *Nihon Soshaku Gakkai Zasshi [Journal of Japanese Society for Mastication and Health Promotion]* 13: 22–29. (in Japanese with English summary and figure captions).

Yoshimura, M., Kuwano, T., Morisaki, R., and Nishinari, K. (2004). Rheological properties and sensory evaluation of the minced foods with xanthan gum. *Nihon Soshaku Gakkai Zasshi [Journal of Japanese Society for Mastication and Health Promotion]* 14: 50–61. (in Japanese with English summary and figure captions).

Yoshimura, M., Kuwano, T., Morisaki, R., and Nishinari, K. (2005). Rheological properties of xanthan gum of different intrinsic viscosity and mastication of minced carrots. *Nihon Soshaku Gakkai Zasshi [Journal of Japanese Society for Mastication and Health Promotion]* 15: 48–57. (in Japanese with English summary and figure captions).

Yoshimura, M., Kuwano, T., Morisaki, R., and Nishinari, K. (2008). The properties of mixture solutions of xanthan gum/gellan gum and mastication of minced foods for aged person. *Nihon Soshaku Gakkai Zasshi [Journal of Japanese Society for Mastication Science and Health Promotion]* 18: 49–59. (in Japanese with English summary and figure captions).

2

Use of Hydrocolloids to Modulate Food Color and Gloss

2.1 Introduction

Many foods are difficult to identify when their texture or color is concealed, leaving flavor as the only recognizable attribute. Both color and gloss can be formulated and applied to compound or processed foods. Control of food gloss and color through coating with sauces that include hydrocolloids as stabilizers can be easily achieved using the suitable hydrocolloid(s) in combination with other food ingredients. Colors of baked and fried foods can also be changed through the inclusion of hydrocolloids in their composition. The controlled introduction of changes in foods will be explained and exemplified for special food coatings containing hydrocolloid films, for example, cooking fish with the addition of pullulan to produce a neutral mirror glaze (*nappage neutre*) with pectin.

2.2 Appearance of Objects

Appearance is the principal aspect for demonstrating the value of an object. Any manufacturing business is greatly concerned about the appearance of their merchandise. Appearance is a characteristic of the optical experience by which things are acknowledged (Hunter 1975). The chromatic presence of entities stems from the manner in which they reflect and transmit light. The color of items is determined by those parts of the incident white light that are reflected or transmitted, rather than absorbed. Additional appearance characteristics are defined in terms of properties such as clear, dull, glossy, matte, shiny, and turbid. The appearance of reflective objects is determined by the way the surface reflects the incident light (Choudhury 2014). The reflective properties of the surface can be characterized by microlevel observation of the topography of that surface. Structures on the surface typically have dimensions of between 10 and 0.1 mm (the detection limit of the human eye is ~0.07 mm). Smaller structures and topographies of the surface cannot be directly observed by the naked eye, but their effect becomes obvious in objects or images reflected in their surface (Choudhury 2014). Structures at and below 0.1 mm decrease the clarity of an image (DOI), structures in the range of 0.01 mm induce haze, and even smaller structures have an impact on surface gloss (Choudhury 2014).

Use of Hydrocolloids to Control Food Appearance, Flavor, Texture, and Nutrition, First Edition.
Amos Nussinovitch and Madoka Hirashima.
© 2023 John Wiley & Sons Ltd. Published 2023 by John Wiley & Sons Ltd.

2.3 Optical Properties

From a qualitative standpoint, the most important optical properties are the visible color and gloss of the product. These are due to reflected light, even if some spectrophotometers quantify light in both reflectance and transmission modes. Transmitted light may be used to distinguish inner defects, for instance blood spots in eggs (i.e. the remnants of a blood vessel that ruptured during the egg's formation) and watercore in apples (Francis 1983; Szczesniak 1983; Harker et al. 1999; Wada et al. 2021), a severe physiological disorder that occurs on the tree, characterized by water-soaked areas of the cortex, which cause the tissue to turn translucent (Figure 2.1). (https://www.asiaresearchnews.com/content/hidden-mechanisms-apple-watercore-formation).

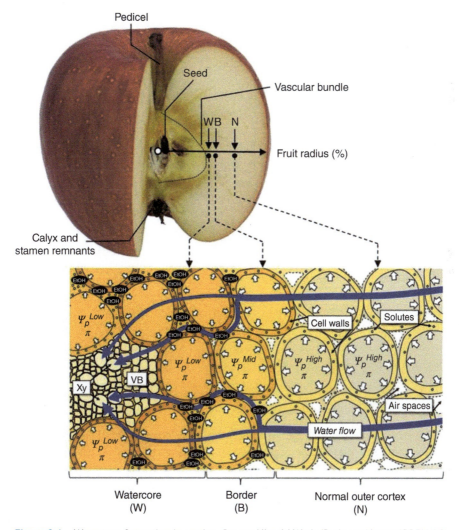

Figure 2.1 Watercore formation in apples. *Source:* Hiroshi Wada/Springer Nature/CC BY 4.0.

2.4 Color

2.4.1 Color of Food Commodities

Color is one of the most significant quality characteristics. In addition, it has key psychological associations that affect humans' mood and emotional state. Several research manuscripts have dealt with the effect of color on consumer acceptance of foods. For example, a mass consumer survey was performed to assess the response to the color of canned cling peaches. Fruit samples of three peach varieties representing pale-yellow, medium-orange, and deep-orange fruit were viewed by almost 3000 individuals. The medium-orange colored fruit was liked significantly better than deep-orange, and highly significantly better than pale-yellow fruit. The pale-yellow fruits were disliked by 51–56% of the people who participated in the tests (Czerkaskyi 1971). The influence of potato chip freshness and color on sensory preferences was also investigated (Maga 1973), as well as the color of canned peas as a factor in sales patterns (von Elbe and Johnson 1971). The effect of color on perceived flavor was reviewed by Kostyla and Clydesdale (1978). Although numerous researchers have examined the qualitative effect of color on flavor, quantification of these effects is limited. Kostyla and Clydesdale (1978) summarized available psychophysical techniques for such studies. They found that in general, there is disagreement regarding the need for colorants in foods. If color affects flavor quantitatively, it will affect consumption and, consequently the nutritional status of the public will be quantifiable. Therefore, such arguments should be determined by facts, not intuition. Several tests were conducted to evaluate the influence of food color on flavor identification of noncarbonated beverages and to weigh the interactive effects of food color and flavor intensities on the apparent flavor intensity and hedonic quality of beverages and cake (Du Bose et al. 1980). Color masking drastically reduced flavor identification of fruit-flavored beverages, whereas non-characteristic colors prompted incorrect flavor responses that were typically associated with the uncommon color (Du Bose et al. 1980). Furthermore, the color intensity of the beverages had a significant effect on their overall acceptability, satisfaction with color and flavor, and flavor intensity. Similar results were found with cake samples, except that a significant interactive effect of color and flavor level was observed on overall acceptability. The inclusive acceptability of both the beverage and cake products was more closely related to the ratings of flavor acceptability than to satisfaction ratings of color. Furthermore, a test of the effect of colorant safety information showed that such information did not change any aspect of a product's acceptability (Du Bose et al. 1980).

2.4.2 Expressing Color Numerically

The strategy used to rapidly and precisely describe a color is "color by the numbers." In 1905, an artist and professor of art at the Massachusetts Normal Art School, Albert H. Munsell (Figure 2.2), initiated a color-ordering system – or color scale, which remains in use today. Munsell wanted to generate a "sensible method to define color" that would use decimal notation instead of color names. He first started work on the system in 1898 and published it in its full form in *A Color Notation* (Munsell 1905). The Munsell system allocates numerical values to the three properties of color: hue, chroma, and value. Adjacent color samples represent

Figure 2.2 Prof. Albert H. Munsell. *Source:* Unknown author/Wikipedia Commons/Public Domain.

equal intervals of visual perception. Munsell's description of his color system, from a lecture to the American Psychological Association, was published in 1912 (Munsell 1912) and the deficiencies of the theoretical system as a physical representation were improved significantly in the 1929 *Munsell Book of Color* and through an extensive series of experiments carried out by the Optical Society of America in the 1940s, resulting in the notations for the modern *Munsell Book of Color* (i.e. Munsell, Book of Color Glossy Collection from X-rite, first available in 2012). The Munsell color system consists of a series of color charts which are intended to be used for visual comparison with the specimen. Colors are defined in terms of the Munsell hue (H), Munsell value (V; indicates lightness), and Munsell chroma (C; indicates saturation) and is written as HVC (Konica Minolta 2007).

Additional ways and means for communicating color numerically were developed by an international organization, the CIE (Commission Internationale de l'Eclairage – The International Commission on Illumination), the principal international organization concerned with light, color, and color measurements. The two most widely known of these methods are the *Yxy* color space, devised in 1931 and based on the tristimulus values *XYZ* defined by the CIE, and the *L*a*b** color space, devised in 1976 to provide more uniform color differences in relation to visual differences. After various improvements, color spaces such as these are now used worldwide to communicate colors (Konica Minolta 2007). Colors are frequently defined by three components, not only in the CIELAB (the *L*a*b** color space) but also in the RGB color model. The RGB color model is an additive color model in which the primary colors of light – red (R), green (G), and blue (B) – are added together in various ways to reproduce a broad array of colors (Figure 2.3). The RGB color model is *additive* in the sense that the three light beams are added together, and their light spectra are summed, wavelength for wavelength, to make up the final color's spectrum. The RGB color model is based on the Young–Helmholtz theory of trichromatic color vision, developed by Thomas Young and Hermann von Helmholtz in the early to mid-19th

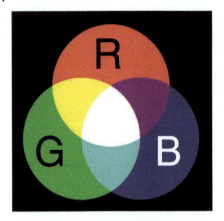

Figure 2.3 Additive color mixing: adding red to green yields yellow; adding green to blue yields cyan; adding blue to red yields magenta; adding all three primary colors together yields white. *Source:* Mike Horvath/Wikimedia Commons/Public domain.

Figure 2.4 Engraving of James Clerk Maxwell by the author GJ. Stodart from a photograph by Fergus of Greenock. *Source:* George J. Stodart/Wikipedia Commons/Public Domain.

century, and on James Clerk Maxwell's color triangle which elaborated that theory (c. 1860). The first experiments with RGB in early color photography were performed in 1861 by James Clerk Maxwell (Figure 2.4), the Scottish scientist responsible for the first theory to describe electricity, magnetism, and light as different manifestations of the same phenomenon. It involved the process of combining three separate color-filtered intakes. To reproduce the color photograph, three matching projections over a screen in a darkroom were necessary (Figure 2.5) (https://en.wikipedia.org/wiki/RGB_color_model).

The CIE color systems utilize three coordinates to locate a color in a color space. These color spaces include CIE XYZ; CIE $L^*a^*b^*$ and CIE $L^*C^*h^0$ (X-rite Inc. 2016). To obtain these values, our eyes need three things to see color: a light source, an object, and an observer/processor. No single parameter is correct for color measurement instruments. They perceive by collecting and then filtering the wavelengths of light reflected from an object. The instrument perceives the reflected light wavelengths as numerical values. These values are recorded as points across the visible spectrum and are termed spectral data, represented as a spectral curve. This curve is the color's fingerprint (X-rite Inc. 2016). As soon as color spectra, or reflectance curves, are obtained, mathematics can be applied to map the

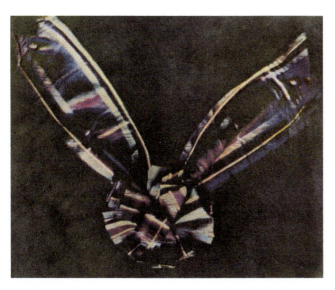

Figure 2.5 Tartan ribbon. The first permanent color photograph, taken by J.C. Maxwell in 1861 using three filters: red, green, and violet-blue. *Source:* James Clerk Maxwell/Wikipedia Commons/ Public Domain.

color onto a color space. The reflectance curve data are multiplied by a CIE standard illuminant, which is a graphical representation of the light source under which the samples are viewed. Each light source has a power distribution that affects how we see color (X-rite Inc. 2016). The result of this calculation is multiplied by the CIE standard observer. CIE-commissioned work in 1931 and 1964 derived the concept of a standard observer, which is based on the average human response to wavelengths of light. Briefly, the standard observer characterizes the way in which an average person sees color across the visible spectrum (X-rite Inc. 2016). The color response of the eye changes in relation to the angle of view (object size). The CIE first defined the standard observer in 1931 using a 2° field of view, termed 2° Standard Observer. In 1964, the CIE defined an accompanying standard observer, this time based upon a 10° field of view, and referred to as the 10° Supplementary Standard Observer (Konica Minolta 2007). Once these values are calculated, the data can be converted into the tristimulus values of XYZ and these values can now identify a color numerically (X-rite Inc. 2016). Because the color response of the eye changes in relation to the angle of view (object size), color-matching functions are the tristimulus values of the equal-energy spectrum as a function of wavelength, intended to correspond to the response of the human eye. Separate sets of three color-matching functions are specified for the 2° Standard Observer and 10° Supplementary Standard Observer (Konica Minolta 2007).

The $L^*a^*b^*$ color space is one of the uniform color spaces that was well-defined by the CIE in 1976 (Figure 2.6). The values of L^*, a^*, and b^* are calculated as follows:

Lightness variable L^*:

$$L^* = 116\left(\frac{Y}{Y_n}\right)^{1/3} - 16 \qquad (2.1)$$

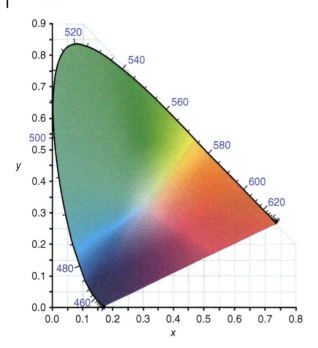

Figure 2.6 CIE 1931 xy color space diagram. *Source:* BenRG/Wikimedia Commons/Public domain.

Chromaticity coordinates:

$$a^* = 500\left[\left(\frac{X}{X_n}\right)^{1/3} - \left(\frac{Y}{Y_n}\right)^{1/3}\right] \quad (2.2)$$

$$b^* = 200\left[\left(\frac{Y}{Y_n}\right)^{1/3} - \left(\frac{Z}{Z_n}\right)^{1/3}\right] \quad (2.3)$$

where X, Y, Z are the tristimulus values for the 2° Standard Observer or X10Y10Z10 for the 10° Supplementary Standard Observer of the specimen (Konica Minolta 2007).

The L^*C^*h color space uses the same diagram as the $L^*a^*b^*$ color space, but the coordinates are cylindrical. Lightness L^* is the same as L^* in the $L^*a^*b^*$ color space; metric chroma C^* and metric hue angle h are described by the formulas:

Metric chroma:

$$C^* = \left[\left(a^*\right)^2 + \left(b^*\right)^2\right]^{1/2} \quad (2.4)$$

Metric hue angle:

$$h = \tan^{-1}\left(\frac{b^*}{a^*}\right)[\text{degrees}] \quad (2.5)$$

where a^*, b^* are the chromaticity coordinates in the $L^*a^*b^*$ color space.

For difference measurements, the metric hue angle difference is not calculated; instead, the metric hue difference Δ*H** is calculated according to:

$$\Delta H^* = \sqrt{\left[\left(\Delta E^*_{ab}\right)^2 - \left(\Delta L^*\right)^2 - \left(\Delta C^*\right)^2\right]} = \sqrt{\left[\left(\Delta a^*\right)^2 + \left(\Delta b^*\right)^2 - \left(\Delta C^*\right)^2\right]} \quad (2.6)$$

The metric hue difference is positive if the metric hue angle *h* of the specimen is greater than that of the target, and negative if *h* is less than that of the target (Konica Minolta 2007). The newest CIE DE2000 color difference formula is intended to correct the differences between the measurement results and visual evaluations, which was the weak point in the *L*a*b** color space. The calculation is based on the lightness difference, saturation difference, and hue difference with correction using weighted coefficients and constants called parametric coefficients. For further information and exact formulas, the reader is referred to https://www.konicaminolta.com/.

2.4.3 The Kubelka–Munk Concept

The Kubelka–Munk theory, formulated by Paul Kubelka and Franz Munk, is a fundamental methodology for modeling the appearance of paint films (Kubelka and Munk 1931; Kubelka 1948). Even though Kulbeka entered this field through an interest in coatings, his work has influenced researchers in other areas as well. In the original manuscript, the special case of interest to many fields was "the albedo of an infinitely thick coating" (https://en.wikipedia.org/wiki/Kubelka-Munk_theory).

The Kubelka–Munk concept is defined by:

$$\frac{K}{S} = \frac{(1 - R_\infty)^2}{2R_\infty} \quad (2.7)$$

where *K* is absorbed light, *S* is scattered light, and R_∞ is the reflectance of an infinitely thick sample. This function demonstrates that the reflectance of an opaque layer is determined by the ratio of *K* to *S* and not by their individual values (Kubelka and Munk 1931; Kubelka 1948). The Kubelka–Munk concept is useful in descriptions of the optical properties of foods.

Knowing the *K* and *S* values enables predicting whether the color of a certain sample should be measured by reflectance or transmittance. This methodology also demonstrates the potential for measuring the internal color and quality characteristics of foods without destroying the sample (Birth 1978, 1979). In general, the associations among instrument readings defined by reflectance at a single wavelength, *XYZ* or *L*a*b** data and visual responses are determined by transforming the instrument readings into *K/S* units (Francis 1983). This approach has been used for several food products, for example squash and carrot puree (Huang et al. 1970), orange juice (Gullett et al. 1972), an applesauce–berry mixture (Little 1964), powdered milk (Mackinney et al. 1966), and meats (Hunt 1980). In the textile and paint industries, the formulation of recipes for dye and pigment mixtures to produce specific colors is a very important activity. Most of the procedures that have been worked out in those industries use the Kubelka–Munk analysis as a base, but numerous elaborations have been established to give enhanced accuracy of prediction in certain positions (Berns and Mohammadi 2007). Very beneficial discussions of these topics are given in some manuscripts (McDonald 1997).

2.5 Gloss

2.5.1 General Approach

The surface is the outside part or uppermost layer of a physical object or space. Its look, whether it is glossy or dull, is a vital physical characteristic of food quality perceived by human vision. In general, shiny appearances are preferred, although this differs based on the item for consumption and seems to be a learned response (Szczesniak 1983). Shiny surfaces are valued for apples, cherries, cucumbers, fruit pie fillings, red bell pepper (Figure 2.7), and rice (Figure 2.8). Dull surfaces are expected for green beans and mushrooms. The appearance of a surface cannot be defined by any organized coordinate measurement. We identify light leaving an object from the illuminated side, i.e. reflected light; and light passing from one side of an object and leaving from the other side, i.e. transmitted light. These may be further divided into diffused and non-diffused light (Szczesniak 1983).

The four categories of light distribution from objects have the following general relationship to surface appearance: (i) diffuse reflection – shiny; (ii) specular reflection – glossy, mirror-like; (iii) diffuse transmission – cloudy, opaque; (iv) specular transmission – translucent (Hunter 1975; Szczesniak 1983).

2.5.2 What Is Gloss and Why Is It Measured?

Gloss is a visually perceived characteristic of an object. It is as important as color when considering the psychological impact of products on a consumer. In other words, "gloss sells." Gloss has been defined as the attribute of a surfaces that causes it to have a shiny or lustrous, metallic appearance. The gloss of a surface can be critically influenced by several factors, e.g. smoothness

Figure 2.7 A red bell pepper. One of the glossiest fruit. *Source:* Daniel Sone.

Figure 2.8 A bowl of cooked white rice.

accomplished through polishing, an applied coating, or the quality of the substrate. Industrialists plan their merchandise for maximal attractiveness. A few examples are highly reflective car body panels, glossy magazine covers, or satin black designer furniture. If, without warning, products suddenly look different, customers might consider this a defect, or poor quality. Using a glossmeter and having good-quality control practices eliminates this problematic variability. It is essential that gloss intensities be consistent for every product or across different batches of products. Gloss can also define the quality of a surface; for example, a drop in the gloss of a coated surface may specify problems such as poor adhesion or lack of protection for the coated surface (Imbotec Group 2006). Gloss effects are based on the interaction of light with the physical properties of the sample surface. More reflected direct light leads to a stronger impression of gloss. The human eye is a very good tool for assessing differences in gloss. Nevertheless, when evaluation conditions are not clearly defined, people can perceive and judge in different ways. Furthermore, the personal perception of appearance is dependent on the subjective experience, and vision and mood might play a decisive role in visual judgment. To provide consistent quality assurance, measurable objective criteria of appearance should be defined, as a precise classification of appearance can improve quality and production practices (Imbotec Group 2006).

2.5.3 Gloss Units and What Differences in Gloss Can Be Detected by Humans

The amount of light reflected from a surface of an object at the specular angle is quantified via gloss measurements and consequently, gloss quantifies the extent of shine of an object (Choudhury 2014). A glossmeter is an instrument used to measure the gloss of materials such as paint, paper, plastics, and foods. Glossmeters are configured such that the light from a light source is incident on the material surface at a certain angle relative to normal or perpendicular to the surface. The detector is placed at the same angle on the other side of the normal, so that the light reflected at the specular angle can be measured. A green filter corresponding to the CIE luminosity function is positioned in front of the detector to allow the instrument to closely simulate human vision. Many international technical standards outline the means of use and specifications for different kinds of glossmeters measuring different types of products (Choudhury 2014). Glossmeter measurement results are associated with the extent of light reflected from a black glass standard with a well-defined refractive index. The value (upper point calibration) for this standard is equal to 100 gloss units (GU). Materials with a higher refractive index (e.g. films) can have values above 100 GU (Imbotec Group 2006). The lower endpoint for the established standard is 0 on a perfectly matte surface. This scaling is suitable for most non-metallic coatings and materials such as paints and plastics, which commonly fall within this range. For other materials that are excessively reflective in appearance (i.e. mirrors, plated/raw metal components), higher values can be reached, up to 2000 GU. For transparent materials, these values can also increase due to multiple reflections within the materials. For such applications, it is recommended to use percent reflection of incident light rather than GU (Choudhury 2014). Where two different coatings are applied to a sample object, the important questions are, what number of gloss units will be detectable by the human eye, and how many units will be perceived as significantly different? If gloss is measured at 60°, then detectable differences depend on the gloss level of the sample. For example, a 3 GU difference measured on a very matte surface (perhaps 5 GU) would be seen by the human eye, but on a higher gloss coating (perhaps 60 GU), the difference would be very hard to notice. Therefore, under these circumstances, the only way in which product appearance can be determined is to prepare experimental printed samples at different gloss levels that can be presented to end users of these coatings or to internal "experts." Another option is to switch from a 60° GU instrument to a 20/60/85° instrument. As further elaborated in Section 2.5.4, the 85° glossmeter is more sensitive to differences in gloss below 10 GU (at 60°) and the 20° instrument has higher resolution on high gloss coatings (above 70 GU at 60°). The benefit of using the three angles is that the gloss differences are more equal, in most cases 5 GU, and when measured with the correct geometry are just visible to a trained observer (Imbotec Group 2006).

2.5.4 How Gloss Is Measured and Glossmeter Types

As already described, gloss is measured by directing a light at a specific angle onto the test surface and simultaneously measuring the amount of reflection. The type of examined surface governs what glossmeter angle is used and consequently, the glossmeter model. In the case of coatings and plastics (i.e. non-metals), the extent of reflected light rises with increasing illumination angle. The remaining illuminated light penetrates the material and is absorbed or diffusely scattered, dependent on the color (Imbotec Group 2006). To achieve clear differences across the complete measurement range from high gloss to matte, three different geometries, i.e. three different ranges, were defined using a 60° glossmeter (Table 2.1). For better-quality resolution of low gloss (less than 10 GU at 60°), an angle of 85° is used to measure the surface. This angle also has a larger measurement spot which will balance out variances in the gloss of textured or somewhat irregular surfaces. Thus, although all gloss levels can be measured using the standard measurement angle of 60°, sometimes a different angle is preferred. For extreme gloss, one gets better measurements with the angles of 85° or 20°, often used for low and high gloss levels, respectively. Surfaces that measure 70 GU and above at the standard angle of 60° are frequently measured with 20° geometry. The 20° angle is more sensitive to haze effects that affect the appearance of a surface, and better shows the different sample glosses (Imbotec Group 2006).

For regular gloss measurements, a flat surface of approximately 50×10 mm is needed so that the glossmeter can be suitably positioned on it. If the surface area is smaller (e.g. 2×2 mm), a unique instrument with custom-designed optics for accurate measurement of

Table 2.1 Different angles and recommendations for measuring gloss.

Gloss range with 60° glossmeter	Measure with
If semi-glossy — 10–70 GU	60°
If high gloss > 70 GU	20°
If low gloss < 10 GU	85°

the gloss of curved surfaces and small areas is required. Such an instrument can also be used to measure flat surfaces. Most commodities fall into the semi-glossy range. Some are matte and envisioned to have very low gloss and others, such as metals or car finishes, have very high gloss.

2.6 On the Psychological Impact of Food Color and Gloss

Color is but one aspect of the vision's influence on taste and flavor perception (Spence 2015). Scientists have found that people have a tendency to judge the freshness of fish, to some extent, based on the glossiness of its eyes (Murakoshi et al. 2013). The distribution of shine also seems to be a significant signal for the perceived freshness of strawberries and carrots (Péneau et al. 2007). Researchers have also studied the influence of visual food texture on people's sensory perception and consumer behavior (Lawless and Klein 1991; Prinz and de Wijk 2004; Okajima and Spence 2011). The earliest reports that altering the color of a food might change its perceived taste/flavor were published about 80 years ago (Moir 1936; Duncker 1939). Most of the more recent studies tend to emphasize color's effect on taste/flavor proof of identity (Spence 2010). Food color can be considered the single most important product-intrinsic sensory indicator, governing the consumer's sensory and hedonic anticipation of the food and drinks that they are seeking, buying and eating (Schaefer and Schmidt 2013). It is hypothesized that humans' experience of taste and flavor is determined to a large extent by their expectations (frequently spontaneous) prior to tasting (Piqueras-Fiszman and Spence 2015). Food coloring plays a significant part in driving consumer attraction to, and acceptability of food and beverage varieties. Although more color variety in foods can lead to enhanced consumption (Piqueras-Fiszman and Spence 2014), what we see can just as easily suppress our appetite due to an association with off-colors or our own perceptions. As a final point, given the practical difficulties associated with delivering flavors while a participant lies in a brain scanner (Spence and Piqueras-Fiszman 2014), it is understandable that there have not been many neuroimaging studies of the influence of color on flavor perception (Österbauer et al. 2005; Skrandies and Reuther 2008). Thus, further research is needed to understand the various psychological effects of color on our sensitivity and actions toward foods (Kappes et al. 2006; Spence 2016).

2.7 Where and When Are Hydrocolloids Utilized to Modulate Food Color and Gloss?

2.7.1 Color of Fruit Leathers and Bars

Fruit leather is a delicious, chewy, dried fruit item for consumption. Fruit leathers are prepared by pouring puréed fruit onto a flat surface for drying. Once dried, the fruit is pulled from the surface and rolled. Its name comes from the dried fruit's shine and texture, resembling leather (Harrison and Andress 2021). Food dehydrators allow for fruit leathers to be produced year round at home. Fruit leather is regarded as a confectionery product manufactured by the dehydration of fruit pulp into leathery sheets (Raab and Oehler 2000). Fruit

Figure 2.9 *Manilkara zapota* (sapodilla, chikoo, chico, chicle) fruit. Location: Paia, Maui, Hawaii. *Source:* Forest Starr & Kim Starr/CC BY 4.0.

Figure 2.10 *Artocarpus heterophyllus* (jackfruit) fruit. Location: Pali o Waipio, Maui, Hawaii. *Source:* Forest Starr & Kim Starr/CC BY 4.0.

such as apple, mango, sapodilla (Figure 2.9), jackfruit (Figure 2.10) and others are used in the preparation of fruit leather (Owen et al. 1991; Summers 1994; Che Man and Taufik 1995). The composition of the fruit leather depends mainly on the processed fruit, being rich in pectic and cellulitic substances, which in great measure determines the textural properties of these products (Torley et al. 2008; Valenzuela and Aguilera 2013).

Mango is a very important fruit crop in India and mango leather is a traditional product. It is prepared by spreading ripe pulp with added sugar (and sometimes other ingredients) on bamboo mats for sun drying. However, the sun-dried product is discolored, and the process is unsanitary and lengthy (Rameshwar 1979; Srinivas et al. 1997). Cabinet drying for mango leather production leads to a product with better color and flavor (Heikal et al. 1972; Mir and Nath 1995). Some trademarks of mango leather available in the Indian market (i.e. Sunrays, Concept Foods Pvt. Ltd., Hyderabad) use the hydrocolloid pectin as one of their ingredients. It is assumed that pectin thickens the fruit pulp and improves the texture of the dried product (Gujral and Brar 2003). However, in general, information on the influence of hydrocolloid addition to the fluid-like mixture of fruit leather is limited; studies are needed to determine the effects of different added hydrocolloids, such as carboxymethylcellulose (CMC), pectin, guar gum, gum acacia, and sodium alginate, on dehydration rate, color, texture, and equilibrium relative humidity properties of mango leather (Gujral and Brar 2003). One such study looked at the addition of several hydrocolloids to the mango pulp at concentrations of 1, 2, or 3% (w/w). Drying was carried out using hot air at a temperature of $60 \pm 1°C$ and 15% relative humidity in a cabinet dryer. The color of the mango leather samples was measured with a HunterLab colorimeter using a yellow reference tile as a standard. Mango leather covered by a glass plate was placed beneath the optical sensor and L^*, a^* and b^* values were recorded (Gujral and Brar 2003).

The color of mango leather can be described as yellowish orange and is a significant factor in consumer choice. The L^*, a^* and b^* values of control mango leather were 35.0, 12.3, and 20.7, respectively. Increasing hydrocolloid concentration decreased the a^* value, representing a decrease in sample redness. Nevertheless, 1–3% guar gum and 1–2% CMC did not significantly change the redness of the sample (Gujral and Brar 2003). The b^* value defines sample yellowness and it decreased with increasing hydrocolloid concentration. However, 1–3% guar gum, 1–2% CMC, 1–3% gum acacia and 1% pectin did not change the b^* value significantly. The reduction in yellowness was most pronounced (19.7%) with the inclusion of sodium alginate. Increasing hydrocolloid concentrations did not considerably affect the L^* values, i.e. the lightness of samples (Gujral and Brar 2003). Another manuscript reached partially similar results for a different leather. They concluded that the addition of 1% pectin and 1% CMC provides the best product properties in terms of color, flavor, texture, and general acceptability of date–mango leather. As a result of using hydrocolloids, the product had a more gummy and harder texture (Patil et al. 2017). These studies provide us with concentrations of hydrocolloids that might thicken the fluidic mass of the leather before drying without adversely affecting the color of the dried product.

A considerable proportion of fruit and vegetable crops is lost due to lack of storage facilities, among other reasons. To exploit the surplus production and avoid postharvest losses, crops may be given the suitable shape of a fine product which could potentially serve as a decent source of carbohydrate, fiber, minerals, vitamin C, and β-carotene, will have an extended shelf life, and will allow for using low-grade fruit (Ahmad et al. 2005). The physicochemical and microbiological changes during storage of papaya fruit (*Carica papaya* L.) bars (termed tandra) were studied for nine months (Reddy et al. 1999). The bar was stored at different temperatures and organoleptic changes were also evaluated. Sensory evaluation of the fruit bar revealed higher deterioration in color, appearance, and texture at six and nine months of storage at higher temperatures. Mango cheese/leather is a tropical confectionary prepared from fresh or frozen mango pulp (Cariri, The Caribbean Industries Research Institute 1993). Another type of bar, jackfruit tandra, was developed in accordance with Farmer Producer Organization (FPO) specifications (Manimegalai et al. 2001).

The FPO is a legal entity formed by a group of farmers or primary producers that can be a producer company, a cooperative society or any other legal form that provides for sharing of profits/benefits among its members (Marbaniang and Kharumnuid 2019). Two varieties of jackfruit were used, and the product was packed in either butter paper (BP), polypropylene pouches (PP), or metallized polyester low-density polyethylene laminated pouches (MPP) and stored at room temperature (30–36 °C) for six months. The bar samples store in MPP had a higher percentage of nutrient retention and the lowest microbial count compared to the samples in BP and PP at the end of 180 days (Manimegalai et al. 2001). The quality attributes of fruit bars made from papaya and tomato by incorporating hydrocolloids was studied (Ahmad et al. 2005). The included hydrocolloids were pectin, starch, and ethyl cellulose. A study of sensory characteristics revealed that all fruit bar samples were acceptable in terms of taste, color, and aroma, but they differed significantly in their texture. The samples were packed in low-density polyethylene bags (100 μ thickness) and stored at 35–45 °C for four months. Significant ($p < 0.05$) changes in physicochemical properties, such as acidity and vitamin C content, were observed throughout storage (Ahmad et al. 2005). However, no significant ($p < 0.05$) effect on either browning index (defined as logarithmic ratio of intensity of incident light to that of emergent light) or color, taste or aroma of these samples was observed. During storage, changes in color and texture were not uniform for all treatments. Addition of 0.5 and 1% of each starch plus ethyl cellulose was effective at preserving the color; addition of 0.5, 1, and 1.5% of each pectin plus starch was effective at improving the texture (Ahmad et al. 2005).

2.7.2 Gloss and Transparency of Edible Films

There is growing interest in developing edible films from hydrocolloids (Krochta and Mulder-Johnston 1997). Numerous studies have demonstrated their ability to extend the shelf life of fresh and processed fruit and vegetables (Wong et al. 1994; Baldwin et al. 1995a; Nussinovitch and Lurie 1995; Debeaufort et al. 1998; Park 1999; Sason and Nussinovitch 2020). Hydroxypropylcellulose (HPC) and methylcellulose (MC) have outstanding film-forming properties (Nisperos-Carriedo 1994; Debeaufort and Voilley 1997), as well as very efficient oxygen, carbon dioxide, and lipid barriers, but reduced resistance to water vapor transport. However, the water vapor barrier properties can be improved by the addition of hydrophobic ingredients, for example lipids, into the film-forming solution (Kamper and Fennema 1984; Kester and Fennema 1986). Many studies have dealt with the permeability properties of cellulose derivative-based films (Hagenmaier and Shaw 1990; Donhowe and Fennema 1993; Park and Chinnan 1995; Sebti et al. 2002; Coma et al. 2003) and their mechanical properties (Donhowe and Fennema 1993; Park et al. 1993; Debeaufort and Voilley 1995, 1997; Sebti et al. 2002), but only a few have studied their optical properties, such as gloss and transparency (Villalobos et al. 2005). For example, the optical (gloss and transparency) and microstructural properties of nine edible films prepared with mixtures of hydroxypropyl methylcellulose (HPMC) and surfactant (Span 60 and sucrose ester P-1570) with different hydrophilic–lipophilic balances (HLB) were evaluated (Villalobos et al. 2005). The gloss was measured at 20, 60, and 85° using a glossmeter and transparency was evaluated through Kubelka–Munk coefficients obtained from the films' reflection spectra (400–700 nm). Film microstructure was also analyzed using light microscopy, atomic force microscopy and scanning electron microscopy. Film transparency and gloss increased with increasing hydrocolloid–surfactant ratio. The greatest HLB values gave rise to the lowest film transparency, but increased gloss. A great impact of the film's internal and surface arrangement of the different phases was detected on its optical properties (Villalobos et al. 2005).

2.7.3 High-Gloss Edible Coating

Edible coatings contribute to the gloss and improve the appearance of apples and citrus. Shellac, which is included in almost all high-gloss fruit coatings and confectioneries (Groves 1977; Dreier 1991), has some disadvantages. Shellac is a natural insect exudate that is sometimes in limited supply (Martin 1982). Coatings that include shellac are prone to whiten upon contact with moisture. In addition, shellac has low gas permeability, and therefore coated citrus fruit are vulnerable to pitting (Petracek et al. 1998) and may show increased content of off flavors (Baldwin et al. 1995b). Aside from shellac, which is considered generally recognized as safe (GRAS) for coatings, wood rosin ester has also been permitted, but only for citrus coatings (FDA 1995). Therefore, there is a strong need for additional constituents to prepare high-gloss edible coatings. Polyvinyl acetate (PVA) was investigated as an ingredient in fungicide-carrying coatings for cheese (Fente-Sampayo et al. 1995), in sausage casings (Stiem 1990), and as an egg coating (Lin et al. 1984). PVA can also be used in pharmaceutical coatings designed for controlled-release applications (Baichwal and McCall 1998). Edible coating preparations were manufactured by dissolution of food-grade PVA in alcohol–water mixtures. PVA coatings had high gloss and relatively high permeance to oxygen and water vapor. They formed glossy surfaces on citrus fruit and apples (Hagenmaier and Grohmann 1999). Addition of propylene glycol facilitated maintenance of the coating's gloss when alcohol content of the solvent was reduced to less than 70%. Fresh apples or citrus fruit with PVA coatings had higher internal oxygen levels than fruit with shellac and resin coatings and therefore, less inclination to ferment and produce alcohol during storage. Fruit with PVA coatings did not whiten or "blush" following contact with water (Hagenmaier and Grohmann 1999).

2.7.4 Gloss and Transparency of HPMC Films Containing Surfactants as Affected by Their Microstructure

As we have already established, gloss and transparency have a great impact on the appearance of coated products. As soon as light strikes a food surface, it is reflected, absorbed, or transmitted, determining its color, gloss, and transparency (Hutchings 1999; Villalobos et al. 2005). Trezza and Krochta (2000) found that the gloss of edible coatings is affected by surfactants, lipids, relative humidity, and storage time. They observed that for lipid/surfactant-dispersion coatings, the surfactant level greatly decreased gloss relative to the pure hydrocolloid film, depending on the surfactant type and the particle size in the dispersed phase (Trezza and Krochta 2000). Gloss and transparency of edible films prepared from HPMC and surfactant with different HLB were evaluated (Villalobos et al. 2005). The gloss was measured at 20°, 60°, and 85° using a glossmeter and transparency was evaluated through Kubelka–Munk coefficients (see Section 2.4.3). It was concluded that film transparency and gloss increase with increasing hydrocolloid-to-surfactant ratio. The greatest HLB values gave rise to the lowest film transparency, but increased gloss. A great impact of the film microstructure (internal and surface arrangement of the different phases) on film optical properties was also observed (Villalobos et al. 2005).

2.7.5 Hydrocolloids in Forming Properties of Cocoa Syrups

Chocolate syrup is a sweet, chocolate-flavored condiment. It is frequently used as a topping or a sauce for ice cream; mixed with milk to make chocolate milk; or blended with milk and ice cream to make a chocolate milkshake. Chocolate syrup is manufactured and sold in a variety of consistencies, ranging from a thin liquid to a thick sauce to be spooned onto the dessert item (https://en.wikipedia.org/wiki/Chocolate_syrup). Hydrocolloids are used extensively for cocoa syrup production, but this has been less explored in the literature. The applicability of selected hydrocolloids for cocoa syrup production was assessed, and some commercial syrups' properties were compared with those of syrups obtained under laboratory conditions (Sikora et al. 2003). Increasing concentrations of agar, carrageenan, CMC, and xanthan gum were used in cocoa syrup as stabilizers and thickeners, and to improve consistency (Sikora et al. 2003). The highest score was given to syrups obtained in the laboratory with the addition of 0.3% xanthan gum and 0.3% agar, proving xanthan gum's ability to stabilize emulsions and suspensions (Hennock et al. 1984; Yilmazer et al. 1991), and agar's ability to improve consistency by increasing viscosity and adhesiveness and to improve gloss (Dziezak 1991). The syrups produced under laboratory conditions were compared to the commercial ones. Sensory evaluation of commercial and laboratory syrups was similar, with the latter acquiring a slightly higher assessment (Sikora et al. 2003).

2.7.6 Color of Deep-Fat-Fried Products

Deep-fat frying is a widely used method for preparing foods. It causes crust formation, starch gelatinization, protein denaturation and water vaporization (Rimac-Brnčić et al. 2004). The high temperature of frying causes oil movement into the product along with water transfer from the product into the oil (Khalil 1999; Moyano et al. 2002). The soft and moist interior and porous crispy crust enhance the fried foods' palatability (Mallikarjunan et al. 1997; Mellema 2003; Akdeniz et al. 2006). Hydrocolloids are multifunctional food additives of special interest to fried products because they retain a good barrier against oxygen, carbon dioxide, and lipids, which could reduce oil absorption during deep-fat frying (Mallikarjunan et al. 1997; Williams and Mittal 1999; Albert and Mittal 2002; Rimac-Brnčić et al. 2004). The influence of the hydrocolloids alginate, CMC and pectin on oil absorption by banana chips was studied (Singthong and Thongkaew 2009). The bananas were first washed, hand-peeled, cut into strips and washed with water. The strips were blanched in an aqueous solution of calcium chloride at different concentrations at 85 °C for 30 seconds. They were then immediately immersed in the hydrocolloid solution at 37 °C for two minutes. Then, all pieces were drained and dried in a hot-air oven at 135 °C for three minutes to reduce the surface moisture. The frying process was carried out in a thermostatically controlled fryer at 150 °C for five minutes in palm oil with a constant product weight-to-oil volume ratio. All fried samples were drained and cooled to

Table 2.2 Effect of coating with the hydrocolloids alginate, CMC and pectin on color attributions of banana chips.

Hydrocolloids	Color		
	L^*	a^*	b^*
Control	39.7 ± 1.7^b	14.6 ± 0.4^a	8.4 ± 1.9^b
Alginate	52.6 ± 1.8^a	13.8 ± 3.9^a	18.2 ± 2.0^a
CMC	37.4 ± 1.8^b	15.3 ± 2.4^a	5.7 ± 3.1^b
Pectin	36.8 ± 1.7^b	12.9 ± 1.4^a	6.8 ± 2.4^b

Source: Adapted with changes from Singthong and Thongkaew (2009).
[a,b] Values with different letters in the same column are significantly different ($p \leq 0.05$). Each value in the table is the mean $\pm$ SD of three replications.

room temperature before analysis (Singthong and Thongkaew 2009). The color of the banana chips was determined using a chroma meter. The Hunter color-scale parameters – lightness (L^*), redness (a^*), and yellowness (b^*) – were used to estimate color changes during frying. The results (Table 2.2) showed that redness among all samples was not significantly different ($p \geq 0.05$). Similarly, lightness and yellowness of all samples were not different significantly ($p \geq 0.05$), except with alginate, where the banana chips exhibited higher ($p \leq 0.05$) lightness (L) and yellowness (b) than with the other coatings (Singthong and Thongkaew 2009).

Another study dealt with the effect of pre-drying and hydrocolloid type on color and textural properties of coated fried yam chips, which is a popular street food in Sub-Saharan Africa countries (Alimi et al. 2013). It was previously reported that addition of hydrocolloids to a coating reduces coating loss during frying (Maskat et al. 2005); on the other hand, it increases water retention, which could result in reduced crispiness of the fried food (Primo-Martín et al. 2010). White yam tubers were peeled, cut, and washed using distilled water to remove surface starch. Yam slices were then blanched in water at 75°C for five minutes. A paper towel was used to remove excess water and loose materials adhering to the surface. The blotted blanched yam slices were dried at 75°C for 15 minutes, then yam chips were dipped into the coating formulations, drained to remove excess coating material and deep fried at 180°C for five minutes. Excess oil was allowed to drain off the chips after removal from the fryer for about 50 seconds (Alimi et al. 2013). The coating solutions included xanthan gum, CMC or gum tragacanth mixed with egg white in a commercial blender at a proportion of 0.05 g/kg until the mixture was uniform and free of lumps (Alimi et al. 2013). The color parameters studied were lightness index (L^*), hue angle (h) and browning index (BI). Hydrocolloid type did not have any significant effect on the lightness index of the egg white-coated fried yam chips. However, pre-drying and interaction of hydrocolloid type and pre-drying significantly influenced the lightness index of the coated chips ($p < 0.05$). Pre-drying of yam slices led to paler coated fried chips as evidenced by the lower L^* values (Alimi et al. 2013). The hue angle (h) values of all chips was above 70°, demonstrating a very clear transition from red to yellow. This indicates the development of golden brown color (Maskan 2001), which is highly favored by consumers. Pre-dried samples had significantly ($p<0.05$) higher h values compared to samples that were not pre-dried. This could also be due to the reduction in moisture interference with browning reactions (Alimi et al. 2013). BI, which characterizes the transparency of the brown color and has been reported as a significant parameter when enzymatic and non-enzymatic browning take place (Maskan 2001), was significantly higher than in the non-pre-dried counterparts. This could be a consequence of the lower moisture content of the pre-dried samples (Alimi et al., 2013).

2.7.7 Spray-Dried Products

Spray drying is a well-established and extensively used method for manufacturing powders from fluids (Goula and Adamopoulos 2005; Loksuwan 2007). The conversion of fluid into powder involves atomization of a liquid feed which undergoes heat treatment to decrease its moisture content to the required level (Sivarajalingam 2009). Its advantages are a short processing time and precise operational conditions. Those properties make spray drying a unique operational means for the manufacture of products with extraordinary value properties, such as color, flavor, and nutrients (Cano-Chauca et al. 2005; Goula and Adamopoulos 2005; Rodríguez-Hernández et al. 2005). Furthermore, spray-dried food powders have some instant properties, which also makes them much more attractive to consumers (Barbosa-Cánovas and Juliano 2005). Numerous manuscripts have reported on the conditions for use and qualitative properties of spray-dried juice/extract powders (Cano-Chauca et al. 2005; Goula and Adamopoulos 2005, 2008; Rodríguez-Hernández et al. 2005; Chegini and Ghobadian 2007; Quek et al. 2007; Georgetti et al. 2008). Many studies have been performed on a number of additives' process parameters and the properties of spray-dried products (Goula and Adamopoulos 2005, 2008; Namaldi et al. 2006; Vehring et al. 2007). The use of additives via carriers in spray-drying processes, and its effects on physicochemical properties such as hygroscopicity, flavor retention, and color indexing are of the utmost significance (Lee et al. 2018). The physical appearance of spray-dried powder is highly valued, as the color of the powder strongly affects consumer perceptions. The inclusion of additives can preserve or harm the physical appearance of the powders, especially the color index, depending upon the nature of the feed material and concentration of the additives used (Grabowski et al. 2006). Carrier materials such as maltodextrins, modified starches, and gum arabic are generally used in those spray-drying applications to provide good product recovery and stability. It was found that an increasing concentration of maltodextrin dilutes the color of the feed solution in spray drying and the dried product significantly, as maltodextrin is customarily bright white in its natural form (Du et al. 2014). However, another group suggested that adding maltodextrin at up to 15% does not affect the appearance of spray dried pineapple juice powder (Abadio et al. 2004). Other reports have stated that atomization of the feed solution into tiny droplets increases the surface area exposed to rapid pigment oxidation, leading to lower $a*/b*$ value and higher h (Desobry et al. 1997).

There have not been enough studies on the effect of additives on powder color index; most studies are dedicated to yields, moisture content, and other parameters (Lee et al. 2018). Spray drying of mountain tea (*Sideritis stricta*) water extract using different hydrocolloid carriers is a good example, especially because instant (soluble) herbal tea

Figure 2.11 Hydroxymethylfurfural ($C_6H_6O_3$). The molecule consists of a furan ring with both aldehyde and alcohol functional groups. *Source:* Emeldir/Wikimedia Commons/Public domain.

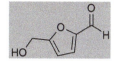

powder, with typical quality and good stability, is an attractive product to many consumers (Nadeem et al. 2011). Spray drying is also an appropriate process for yoghurt drying because of the very short duration of heat contact, in addition to the high rate of evaporation, resulting in quality products (Koc et al. 2011; Sarala et al. 2011; Yousefi et al. 2011). Dry powdered yoghurt is very popular due to its extended shelf life, low transportation cost and smaller storage capacity. It is utilized as an ingredient in various food products, and is directly consumed for its high content of yoghurt starter bacteria. The optimal spray-drying conditions for yoghurt powder production are: inlet air temperature of 171 °C, outlet air temperature of 60.5 °C and feed temperature of 15 °C, determined by targeting the maximal survival rate of lactic acid bacteria, maximum overall sensory attributes, minimum color change, and acceptable moisture content (Koc et al. 2010).

Pekmez (grape molasses) may be spray-dried to produce pekmez powder. Storage of pekmez in dry form is a useful way to prevent or delay excessive Maillard browning, consequently decreasing hydroxymethylfurfural (Figure 2.11) development during storage (Whistler and Daniel 1985; Yılmaz et al. 2009). Diluted pekmez powder (DPP; i.e. 70 g pekmez + 30 mL distilled water) had total and soluble dry matter content, ash content, and pH value similar to those reported for mulberry and white grape pekmez (Sengül et al. 2005; Yogurtçu and Kamisli 2006). The L^*, a^*, and b^* values of DPP were 50.7, +4.4 and +10.2, respectively. A high redness (a^*) value was not expected as this would reflect too much caramelization of sugars. Low redness (a^*) and high brightness (L^*) values indicate good quality parameters for pekmez (Aksu and Nas 1996). DPP, which was regarded to be equivalent to pekmez, demonstrated high L^* values and low a^* values, revealing a good-quality DPP product (Yılmaz et al. 2009).

The color values of DPP and mixed solutions: control – diluted pekmez powder and wheat starch (DPP–WS), and DPP–WS with locust bean gum, guar gum or gum tragacanth is presented in Table 2.3. The lightness (L^*) value of the DPP solution was higher ($p < 0.01$) than those of the DPP–WS control and the solutions with the various gums, whereas redness (a^*) and yellowness (b^*) values demonstrated the opposite trend, with lowest redness and yellowness values for the DPP solution. This was expected due to the heating process at 95 °C for the various solutions, whereas the DPP solution was prepared without heating. The color of the DPP–WS and DPP–WS–gum solutions darkened; consequently L^* values decreased and a^* and b^* values increased in parallel with probable development of the Maillard reaction throughout heating. Increasing temperatures increase this reaction rate and development of hydroxymethylfurfural in thermally processed fruit juice (Whistler and Daniel 1985).

2.7.8 Interaction of Anthocyanins with Food Hydrocolloids

Among the water-soluble natural pigments, anthocyanins are extensively manifested as striking hues and bright colors in numerous flowers, fruit and grains. Nevertheless, their

Table 2.3 Color properties of pekmez–hydrocolloid solutions.

Solutions	L^*	a^*	b^*
DPP solution	50.68 ± 1.07^a	4.44 ± 0.71^c	10.19 ± 0.40^d
Mixed solutions			
Control model solution	39.75 ± 0.95^b	8.23 ± 0.58^b	14.44 ± 0.75^b
DPP-WS mix			
Main model solutions	36.92 ± 1.29^b	8.28 ± 0.54^b	19.31 ± 0.74^a
DPP-WS-LBG mix			
DPP-WS-GT mix	29.73 ± 1.19^c	9.36 ± 0.45^a	11.05 ± 0.52^{cd}
DPP-WS-GG mix	30.96 ± 1.36^c	9.21 ± 0.45^a	12.14 ± 1.45^d

Source: Adapted with changes from Yılmaz et al. (2009).
a–d Values in a column with a different letter are significantly different ($p < 0.01$) as determined by Duncan multiple range test. Means ± SD for six determinations.
DPP, diluted pekmez powder; GG, guar gum; GT, gum tragacanth; LBG, locust bean gum; WS, wheat starch.

restricted stability upon processing and storage poses a major challenge for manufacturers (Kammerer 2016). Bayer et al. (1966) described a pigment from blue cornflower. It was composed of a metalloanthocyanin based on cyanidin glucoside, which interacted with a pectin-like substance, by means of carboxylic functions of the hydrocolloid acting as a ligand for the anthocyanin–metal ion chelate. In further studies, it was hypothesized that anthocyanins may only be adsorbed onto polymeric plant constituents, avoiding chemical binding (Asen et al. 1970). Additional studies of the interaction of polysaccharides with anthocyanins demonstrated a stabilizing effect of sodium alginate, pectin, and cornstarch when added to certain anthocyanin solutions (Hubbermann et al. 2006). The stabilizing potential of pectins when added to black currant anthocyanins in model solutions revealed the most notable effects with amidated pectins. Citrus pectins performed better than apple pectins with regard to anthocyanin retention upon storage (Buchweitz et al. 2013). A specific pectin portion isolated from sugar beet pectin brought about a bathochromic shift of up to 50 nm, thus yielding appealing intense gentian-blue colors due to the formation of metalloanthocyanins, which were stabilized by interaction with the pectic compounds, thus preventing the precipitation of anthocyanin complexes (Buchweitz et al. 2012a, b). Studies of anthocyanin interactions in processed foods with polymeric hydrocolloids and metal ions have uncovered means to yield stable anthocyanin-based blue food colorants (Buchweitz et al. 2012a, b, 2013). Consequently, this type of interaction and stabilization may be exploited for the application of anthocyanin-based blue food coloring (Kammerer 2016).

2.8 Demonstrating the Use of Hydrocolloids to Prepare Colored and Glossy Products/Recipes

The use of hydrocolloids in the preparation of glossy products is demonstrated in this section with three examples. In the first glossy preparation, pullulan was used (Figure 2.12). Pullulan is a non-ionic exopolysaccharide obtained from the fermentation medium of the

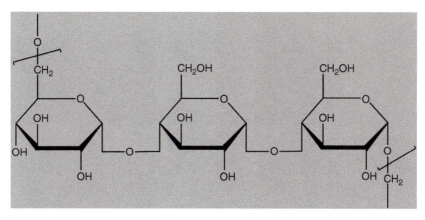

Figure 2.12 Pullulan (food additive E1204). *Source:* Klever/Wikimedia Commons/Public domain.

fungus-like yeast *Aureobasidium pullulans*; it is a biocompatible, biodegradable, non-toxic, non-mutagenic, non-carcinogenic, non-hygroscopic, and edible water-soluble biopolymer (Singh et al. 2017). Pullulan can be produced from agroindustrial waste (Zarei et al. 2020). The chief commercial use of pullulan is in the manufacture of edible films. Pullulan and HPMC blends can be used as a vegetarian substitute for drug capsules (instead of gelatin). Pullulan is a food additive, with the E number E1204 (https://en.wikipedia.org/wiki/Pullulan). Compared to other polysaccharides, pullulan offers supplementary benefits, such as high gloss and transparency, heat sealability, and high mixing capacity with other biopolymers, in addition to its derivatives being tasteless and colorless (Biliaderis et al. 1999; Farris et al. 2014; Silva et al. 2018). Pullulan is also used as a glazing agent, and its edible films can serve as matrices to hold flavors (Ullah et al. 2021). The use of pullulan in a home preparation of almond cookies, and in cooking *teriyaki* chicken have been previously demonstrated (Nussinovitch and Hirashima 2019). Pullulan was included in the *teriyaki* fish recipe (Section 2.8.1) not only to improve the appearance of the grilled food (Figure 2.13), but also to serve as an adhesive for the sauce to the fish. The *teriyaki* sauce contains pullulan, white sugar (Figure 2.13a), *mirin* and *sake*.

Mirin (味醂 or みりん) is a type of rice wine and a common ingredient in Japanese cooking. It is a kind of *sake*, but with lower alcohol and higher sugar contents (Shimbo 2000). Mirin includes a complex carbohydrate that forms naturally during the fermentation process; no sugars are added. The alcohol content is further lowered when the liquid is heated (https://en.wikipedia.org/wiki/Mirin). *Hon mirin* is a typical Japanese liquor seasoning (https://www.honmirin.org/knowledge/). It is prepared from steamed glutinous rice, rice jiuqu, shochu or alcohol, and is saccharified and aged for 40–60 days. During this time, the enzymes in the rice jiuqu work to decompose the starch and protein of glutinous rice to produce various sugars (e.g. glucose, isomaltose, and oligosaccharides), amino acids (glutamic acid, leucine, aspartic acid, etc.), organic acids (lactic acid, citric acid, pyroglutamic acid, etc.) and aroma components (ethyl ferulate, ethyl phenylacetate, etc.), forming this *mirin*'s unique flavor (https://www.honmirin.org/knowledge/). There are "*mirin*-style" seasonings and fermented seasonings which are similar to *hon mirin*, but less costly; however, the industrial method and ingredients used to make them are completely different. For instance: *mirin*-style seasonings comprise virtually no alcohol, so you cannot expect the cooking effect of alcohol; fermented

Figure 2.13 *Teriyaki* fish with pullulan. (a) Pullulan and sugar. (b) Heating pullulan, white sugar, mirin and sake, (c) adding soy sauce and ginger juice, (d) marinating, (e) marinating fish in a plastic bag, (f) heating and turning over, (g) covering with the remaining sauce. (h, i) Serve.

seasonings contain salt, so the saltiness of the dish needs to be adjusted. There is a marked difference in the cooking effects of *hon mirin* and similar seasonings (https://www.honmirin.org/knowledge/). It is important to note that when the recipe (Figure 2.13) is prepared, the *mirin*, which contains multiple sugars, is effective in giving further luster (gloss and patina) to the surface of the treated fish. In addition, the amino acids and peptides produced from the glutinous rice are intricately entwined with sugars and other ingredients to create a deep richness and umami note (http://honmirin.sakura.ne.jp/). Further expanded uses of mirin are addition of shine to prepared foods so that the ingredients will display more attractive colors. *Mirin* is also a very important traditional seasoning in all kinds of Japanese-style foods, especially roasted meats and stews – the right flavor simply will not be brought out without *mirin*. *Teriyaki*, steamed, and roasted dishes such as *den-gakuyaki*, *miso-yaki*, require *mirin* to render flavor, color, and lightness to the foods (https://web.archive.org/web/20081221170622/http://www.taiwannews.com.tw/static/admilk/news/961001/index_en.html).

There are many kinds of canned *teriyaki* fish in Japan. Figure 2.14 shows two types of *teriyaki* sardine called *kabayaki*, sold by different manufacturers. Canned *teriyaki* sardine usually contains guar gum. The addition of guar gum to the product may be related to its viscosity-forming abilities, its ability to reach maximum viscosity rapidly at higher temperatures, its ability to be stable in solution over a wide pH range, and the possibility of

2.8 Demonstrating the Use of Hydrocolloids to Prepare Colored and Glossy Products/Recipes

Figure 2.14 (a, b) Canned *teriyaki* sardine with guar gum, (c) served with rice.

Figure 2.15 Neutral mirror glaze (nappage neutre). (a) Dissolving sugar, (b) low methoxy pectin powder and granulated sugar, (c) sterilizing neutral mirror glaze, (d) cooling to room temperature, (e) brushing glaze over a strawberry.

stabilizing the sauce (Glicksman 1969; Nussinovitch 1997). Figure 2.15 (and see preparation instructions, Section 2.8.2) includes a list of ingredients and preparation methodology for a neutral mirror glaze (nappage neutre). The involved hydrocolloid is pectin, which can be found in virtually all fruit. Pectin is a thickener and a gelling agent (https://chefiso.com/p/glaze-fruit-tart-recipe/) and in this recipe, the glaze has the characteristic of a mirror: it makes the coated fruit look shiny and glassy. The addition of sugar and corn syrup helps create that gorgeous silky and glossy look of coated strawberries or other foods, such as ganache (https://queen.com.au/10-ways-use-jar-queen-glucose-syrup/).

2.8.1 *Teriyaki* Fish with Pullulan (Figure 2.13)

(Serves 5)

Five slices (400–500 g) of yellowtail (*Hint 1*)
3 g (½ tsp) pullulan powder

18 g (2 Tbsp) white sugar
75 g (¼ cup) *mirin*
62.5 g (¼ cup) *sake*
75 g (¼ cup) soy sauce
5–10 g (1–2 tsp) ginger juice (optional)
Vegetable oil
pH of *teriyaki* sauce = 4.63

For preparation, see Figure 2.13

1) First make the *teriyaki* sauce: mix pullulan, white sugar (Figure 2.13a), *mirin* and *sake* in a pan, heat on a low flame until the powder dissolves (Figure 2.13b), add soy sauce and ginger juice (optional) (Figure 2.13c), heat an additional one minute, and cool to room temperature.
2) Marinate slices of yellowtail fish in *teriyaki* sauce for at least 20 minutes (Figure 2.13d, Hint 2).
3) Heat vegetable oil in a frying pan and fry the marinated fish (2) on high heat; turn the fish pieces over (Figure 2.13f), cover with the remaining sauce (Figure 2.13g), and fry until well done (Figure 2.13h, i).

Preparation hints:

1) Yellowtail fish is recommended, but you can also use other fish such as eel, sardine (Figure 2.14) or cod. Chicken is also good.
2) Marinating the slices in a plastic bag kept in the refrigerator for a few days, then cooking them right before eating, is optimal (Figure 2.13e).

2.8.2 Neutral Mirror Glaze (nappage neutre)

(Makes 200 g) (Figure 2.15)

180 mL (a little under ⅘ cup) water
155 g (a little over ⅔ cup) granulated sugar
10 g (1½ tsp) corn syrup
8 g (2½ tsp) low methoxy pectin powder
6 mL (a little over 1 tsp) lemon juice

For preparation, see Figure 2.15.

1) Put 60 g (4 Tbsp) water, 140 g (a little over ⅗ cup) granulated sugar, and corn syrup in a pan, and heat it to boiling to dissolve sugar (Figure 2.15a).
2) Mix low methoxy pectin powder and 15 g (1¼ Tbsp) granulated sugar in another pan (Figure 2.15b), pour 120 mL of water into it, and heat with stirring until the powder dissolves.
3) Add the sugar solution (1) to the pectin solution (2), and mix well with heating it until it boils.
4) Remove the pectin solution (3) from the heat, and add lemon juice.

5) Pour the neutral mirror glaze (4) into a glass bottle, boil it for 20 minutes to sterilize it (Figure 2.15c), and cool it at room temperature (Figure 2.15d).
6) Dilute the neutral mirror glaze (5) in an equal volume of water, warm in a microwave oven and brush over the fruit (Figure 2.15e).

References

Abadio, F.D.B., Domingues, A.M., Borges, S.V., and Oliveira, V.M. (2004). Physical properties of powdered pineapple (*Ananas comosus*) juice – effect of malt dextrin concentration and atomization speed. *Journal of Food Engineering* 64: 285–287.

Ahmad, S., Vashney, A.K., and Srivasta, P.K. (2005). Quality attributes of fruit bar made from papaya and tomato by incorporating hydrocolloids. *International Journal of Food Properties* 8: 89–99.

Akdeniz, N., Sahin, S., and Sumnu, G. (2006). Functionality of batters containing different gums for deep fat frying of carrot slices. *Journal of Food Engineering* 76: 522–526.

Aksu, I. and Nas, S. (1996). Dut Pekmezi Üretim Teknigi ve Çeşitli Fiziksel-Kimyasal Özellikleri. *Gida* 21: 83–88.

Albert, S. and Mittal, G.S. (2002). Comparative evaluation of edible coatings to reduce fat uptake in a deep-fried cereal product. *Food Research International* 35: 445–458.

Alimi, B.A., Shittu, T.A., Sanni, L.O., and Arowolo, T.A. (2013). Effect of pre-drying and hydrocolloid type on colour and textural properties of coated fried yam chips. *Nigerian Food Journal* 31: 97–102.

Asen, S., Stewart, R.N., Norris, K.H., and Massie, D.R. (1970). A stable blue non-metallic co-pigment complex of delphanin and C-glycosylflavones in Prof. Blaauw Iris. *Phytochemistry* 9: 619–627.

Baichwal, A. and McCall, T.W. (1998). Controlled release formulation (albuterol). United States Patent 5,455,046.

Baldwin, E.A., Nisperos-Carriedo, M.O., and Baker, R.A. (1995a). Edible coatings for lightly processed fruits and vegetables. *HortScience* 30: 35–37.

Baldwin, E.A., Nisperos, M., Shaw, P.E., and Burns, J.K. (1995b). Effect of coatings and prolonged storage conditions on fresh orange flavor volatiles, degrees brix, and ascorbic acid levels. *The Journal of Agricultural and Food Chemistry* 43: 1321–1331.

Barbosa-Cánovas, G.V. and Juliano, P. (2005). Physical and chemical properties of food powders. In: *Encapsulated and Powdered Foods* (ed. C. Onwulata), 40–66. Boca Raton, FL: CRC Press.

Bayer, E., Egeter, H., Fink, A. et al. (1966). Komplexbildung und Blütenfarben. *Angewandte Chemie* 78: 834–841.

Berns, R.S. and Mohammadi, M. (2007). Single-constant simplification of Kubelka-Monk turbid-media theory for paint systems – a review. *Color Research and Application* 32: 201–207.

Biliaderis, C.G., Lazaridou, A., and Arvanitoyannis, I. (1999). Glass transition and physical properties of polyol-plasticised pullulan–starch blends at low moisture. *Carbohydrate Polymers* 40: 29–47.

Birth, G.S. (1978). The light scattering properties of foods. *Journal of Food Science* 43: 916–925.

Birth, G.S. (1979). Radiometric measurement of food quality: a review. *Journal of Food Science* 44: 949–953.

Buchweitz, M., Carle, R., and Kammerer, D.R. (2012a). Bathochromic and stabilising effects of sugar beet pectin and an isolated pectic fraction on anthocyanins exhibiting pyrogallol and catechol moieties. *Food Chemistry* 135: 3010–3019.

Buchweitz, M., Nagel, A., Carle, R., and Kammerer, D.R. (2012b). Characterisation of sugar beet pectin fractions providing enhanced stability of anthocyanin-based natural blue food colorants. *Food Chemistry* 132: 1971–1979.

Buchweitz, M., Speth, M., Kammerer, D.R., and Carle, R. (2013). Impact of pectin type on the storage stability of black currant (*Ribes nigrum* L.) anthocyanins in pectic model solutions. *Food Chemistry* 139: 1168–1178.

Cano-Chauca, M., Stringheta, P.C., Ramos, A.M., and Cal-Vidal, J. (2005). Effect of carriers on the microstructure of mango powder obtained by spray drying and its functional characterization. *Innovative Food Science and Emerging Technologies* 6: 420–428.

Cariri, The Caribbean Industries Research Institute (1993). *Mango Cheese and Mango Leather*, 3–13. St. Augustine, Trinidad: University of West Indies.

Chegini, G.R. and Ghobadian, B. (2007). Spray dryer parameters for fruit juice drying. *World Journal of Agricultural Sciences* 3: 230–236.

Che Man, Y.B. and Taufik, B. (1995). Development and stability of Jackfruit leather. *Tropical Science* 35: 245–250.

Choudhury, A.K.R. (2014). *Principles of Colour Appearance and Measurement. Volume 1: Object Appearance, Colour Perception and Instrumental Measurement*, Chapter 3, 103–142. Cambridge, UK: The Textile Institute, Woodhead Publishing.

Coma, V., Sebti, I., Pardon, P. et al. (2003). Film properties from crosslinking of cellulosic derivatives with a polyfunctional carboxylic acid. *Carbohydrate Polymers* 51: 265–271.

Czerkaskyi, A. (1971). Consumer response to color in canned cling peaches. *Journal of Food Science* 36: 671–673.

Debeaufort, F., Quezada-Gallo, J.A., and Voilley, A. (1998). Edible films and coatings: tomorrow's packagings: a review. *Critical Reviews in Food Science and Nutrition* 38: 299–314.

Debeaufort, F. and Voilley, A. (1995). Effect of surfactants and drying rate on barrier properties of emulsified edible films. *International Journal of Food and Science Technology* 30: 183–190.

Debeaufort, F. and Voilley, A. (1997). Methylcellulose-based edible films and coatings: 2. Mechanical and thermal properties as a function of plasticizer content. *Journal of Agricultural and Food Chemistry* 45: 685–689.

Desobry, S.A., Netto, F.M., and Labuza, T.P. (1997). Comparison of spray-drying, drum-drying and freeze-drying for β-carotene encapsulation and preservation. *Journal of Food Science* 62: 1158–1162.

Donhowe, I.G. and Fennema, O. (1993). The effects of plasticizers on crystallinity, permeability, and mechanical properties of methylcellulose films. *Journal of Food Processing and Preservation* 17 (4): 247–257.

Dreier, W. (1991). The nuts and bolts of coating and enrobing. *Prepared Foods* 160: 47–48.

Du, J., Ge, Z.Z., Xu, Z. et al. (2014). Comparison of the efficiency of five different drying carriers on the spray drying of persimmon pulp powders. *Drying Technology* 32: 1157–1166.

Du Bose, C.N., Cardello, A.V., and Maller, O. (1980). Effects of colorants and flavorants on identification, perceived flavor intensity and hedonic quality of fruit-flavored beverages and cakes. *Journal of Food Science* 45: 1393–1399.

Duncker, K. (1939). The influence of past experience upon perceptual properties. *American Journal of Psychology* 52: 255–265.

Dziezak, J.D. (1991). A focus on gums. *Food Technology* 45: 128–129.

Farris, S., Unalan, I.U., Introzzi, L. et al. (2014). Pullulan-based films and coatings for food packaging: present applications, emerging opportunities, and future challenges. *Journal of Applied Polymer Science* 131: 1–12.

FDA (1995). *FDA Code of Federal Regulations (CFR)*. Washington, DC: Food & Drug Administration, U.S. Government Printing Office.

Fente-Sampayo, C.A., Vazquez-Belda, B., Franco- Abuin, C. et al. (1995). Distribution of fungal genera in cheese and dairies. Sensitivity to potassium sorbate and natamycin. *Archiv fuer Lebensmittelhygiene* 46: 62–65.

Francis, F.J. (1983). Colorimetery of foods. In: *Physical Properties of Foods* (ed. M. Peleg and E.B. Bagley), 105–123. Westport, CT: Avi Publishing Company.

Georgetti, S.R., Casagrande, R., Souza, C.R.F. et al. (2008). Spray drying of the soybean extract: effects on chemical properties and antioxidant activity. *Lebensmittel-Wissenschaft Und Technologie* 41: 1521–1527.

Glicksman, M. (1969). *Gum Technology in the Food Industry*, 139–152. New York: Academic Press.

Goula, A.M. and Adamopoulos, K.G. (2005). Spray drying of tomato pulp in dehumidified air. II. The effect on powder properties. *Journal of Food Engineering* 66: 35–42.

Goula, A.M. and Adamopoulos, K.G. (2008). Effect of maltodextrin addition during spray drying of tomato pulp in dehumidified air: II. Powder properties. *Drying Technology* 26: 726–737.

Grabowski, J.A., Truong, V.D., and Daubert, C.R. (2006). Spray-drying of amylase hydrolyzed sweet potato puree and physicochemical properties of powder. *Journal of Food Science* 71: E209–E215.

Groves, R.J. (1977). Pan coating – three basic varieties considered. *Candy and Stack Industry* 142 (33-34): 36–38.

Gujral, H.G. and Brar, S.S. (2003). Effect of hydrocolloids on the dehydration kinetics, color, and texture of mango leather. *International Journal of Food Properties* 6: 269–279.

Gullett, E.M., Francis, F.J., and Clydesdale, F.M. (1972). Colorimetry of foods. 4. Orange juice. *Journal of Food Science* 37: 389–393.

Hagenmaier, R.D. and Grohmann, K. (1999). Polyvinyl acetate as a high-gloss edible coating. *Journal of Food Science* 64: 1064–1067.

Hagenmaier, R.D. and Shaw, P.E. (1990). Moisture permeability of edible films made with fatty acid and (hydroxypropyl) methylcellulose. *Journal of Agricultural and Food Chemistry* 38: 1799–1803.

Harker, F.R., Watkins, C.B., Brookfield, P.L. et al. (1999). Maturity and regional Influences on watercore development and its postharvest disappearance in 'Fuji' apples. *Journal of the American Society for Horticultural Science* 124: 166–172.

Harrison, J.A. and Andress, E.L. (2021). *National Center for Home Food Preservation—Drying Fruits and Vegetables*. University of Georgia Cooperative Extension Service Food Preservation Publications https://www.fcs.uga.edu/extension/food-preservation (accessed 27 December, 2021).

Heikal, H.A., El-Sanafiri, N.Y., and Shooman, M.A. (1972). Some factors affecting quality of dried mango sheets. *Agricultural Research Review* 50: 185–194.

Hennock, R., Rahalkar, R., and Richmond, P. (1984). Effect of xanthan gum upon the rheology and stability of oil water emulsion. *Journal of Food Science* 5: 1271–1274.

Huang, I.L., Francis, F.J., and Clydesdale, F.M. (1970). Colorimetry of foods. 3. Pureed carrots. *Journal of Food Science* 35: 771–773.

Hubbermann, E.M., Heins, A., Stöckmann, H., and Schwarz, K. (2006). Influence of acids, salt, sugars and hydrocolloids on the colour stability of anthocyanin rich black currant and elderberry concentrates. *European Food Research and Technology* 223: 83–90.

Hunt, M.C. (1980). Meat color measurements. *Proceedings of the 33rd Annual Reciprocal Meat Conference*, Purdue University, West Lafayette, IN, June 22–25, 1980 pp. 41–46.

Hunter, R.S. (1975). *The Measurement of Appearance*. New York: Wiley.

Hutchings, J.B. (1999). *Food Color and Appearance*. Glasgow, UK: Blackie Academic and Professional.

Imbotec Group (2006) The newest technology for surface quality measurement. http://www.gloss-meters.com/. HORIBA, Ltd., Minami-ku, Kyoto, Japan.

Kammerer, D.R. (2016). Anthocyanins. In: *Handbook on Natural Pigments in Food and Beverages: Industrial Applications for Improving Food Color*, Series in Food Science, Technology and Nutrition Number, vol. 295 (ed. C. Reinhold and R.M. Schweiggert), Chapter 3, 61–80. Duxford, UK: Elsevier, Woodhead Publishing.

Kamper, S.L. and Fennema, O. (1984). Water vapour permeability of edible bilayer films. *Journal of Food Science* 49: 1478–1481, 1485.

Kappes, S.M., Schmidt, S.J., and Lee, S.-Y. (2006). Color halo/horns and halo-attribute dumping effects within descriptive analysis of carbonated beverages. *Journal of Food Science* 71: S590–S595.

Kester, J.J. and Fennema, O.R. (1986). Edible films and coatings: a review. *Food Technology* 40: 47–49.

Khalil, A.H. (1999). Quality of French fried potatoes as influenced by coating with hydrocolloids. *Food Chemistry* 66: 201–208.

Koc, M., Koc, B., Susyal, G. et al. (2011). Functional and physicochemical properties of whole egg powder: effect of spray drying conditions. *Journal of Food Science and Technology* 48: 141–149.

Koc, B., Sakin, M., Balkir, P., and Kaymak-Ertekin, F. (2010). Spray drying of yoghurt: optimization of process conditions for improving viability and other quality attributes. *Dry Technology* 28: 495–507.

Konica Minolta (2007). Precise color communication: color control from perception to instrumentation. https://www.konicaminolta.com/instruments/network/ (accessed 16 January, 2022).

Kostyla, A.S. and Clydesdale, F.M. (1978). The psychophysical relationships between color and flavor. *Critical Reviews in Food Science and Nutrition* 10: 303–321.

Krochta, J.M. and Mulder-Johnston, C. (1997). Edible and biodegradable polymer films: challenges and opportunities. *Food Technology* 51: 61–74.

Kubelka, P.J. (1948). New contributions to the optics of intensely light-scattering materials, Part I. *Journal of the Optical Society of America* 38: 448–457.

Kubelka, P. and Munk, F. (1931). Ein Bewitrag zur Optik der Farbenstriche. *Zeitschrift für Technische Physik* 12: 593–601.

Lawless, H.T. and Klein, B.P. (1991). *Sensory Science Theory and Applications in Foods*. New York: Marcel Dekker.

Lee, J.K.M., Taip, F.S., and Abdullah, Z. (2018). Effectiveness of additives in spray drying performance: a review. *Food Research* 2: 486–499.

Lin, C.K., Chu, B.C., and Wang, T.W. (1984). Effects of coating conditions on the quality of cooked salted egg during storage. *Journal of the Chinese Society of Animal Science* 13: 55–63.

Little, A.C. (1964). Color measurements of translucent food samples. *Journal of Food Science* 29: 782–789.

Loksuwan, J. (2007). Characteristics of microencapsulated betacarotene formed by spray drying with modified tapioca starch, native tapioca starch and maltodextrin. *Food Hydrocolloids* 21: 928–935.

Mackinney, G., Little, A.C., and Brinner, L. (1966). Visual appearance of foods. *Food Technology* 20: 1300–1308.

Maga, J.A. (1973). Influence of freshness and color on potato chip sensory preferences. *Journal of Food Science* 38: 1251–1252.

Mallikarjunan, P., Chinnan, M.S., Balasubramaniam, V.M., and Phillips, R.D. (1997). Edible coating for deep-fat frying of starchy products. *Lebensmittel-Wissenschaft Technologie* 30: 709–714.

Manimegalai, G., Kishnaveni, A., and Saravana Kumar, R. (2001). Processing and preservation of jack fruit. *Journal of Food Science and Technology* 38: 529–531.

Marbaniang, E. and Kharumnuid, P. (2019). Farmer Producer Organization (FPO): the need of the hour. *Agriculture & Food: e-Newsletter* 1 (12): 292–297.

Martin, J. (1982). Shellac. In: *Kirk-Othmer Encyclopedia of Chemical Technology*, vol. 20 (ed. J.I. Kroschwitz and M. Howe-Grant), 737–747. Hoboken, NJ: Wiley.

Maskan, M. (2001). Kinetics of colour change of kiwi fruits during hot air and microwave drying. *Journal of Food Engineering* 48: 169–175.

Maskat, M.Y., Yip, H.H., and Haryani, M.M. (2005). The performance of a methyl cellulose treated coating during the frying of a poultry product. *International Journal of Food Science and Technology* 40: 811–816.

McDonald, R. (ed.) (1997). *Colour Physics for Industry*, 2e, 209–291. Bradford, England: Society of Dyers and Colorists.

Mellema, M. (2003). Mechanism and reduction of fat uptake in deep-fat fried foods. *Trends in Food Science and Technology* 14: 364–373.

Mir, M.A. and Nath, N. (1995). Loss of moisture and sulphur dioxide during air cabinet drying of mango puree. *Journal of Food Technology* 32: 391–394.

Moir, H.C. (1936). Some observations on the appreciation of flavour in foodstuffs. *Journal of the Society of Chemical Industry: Chemistry and Industry Review* 14: 145–148.

Moyano, P.C., Ríoseco, V.K., and González, P.A. (2002). Kinetics of crust color changes during deep-fat frying of impregnated French fries. *Journal of Food Engineering* 54: 249–255.

Munsell, A.H. (1905). *A Color Notation*. Boston: G.H. Ellis Co.

Munsell, A.H. (1912). A pigment color system and notation. *The American Journal of Psychology* 23: 236–244.

Murakoshi, T., Masuda, T., Utsumi, K. et al. (2013). Glossiness and perishable food quality: visual freshness judgment of fish eyes based on luminance distribution. *PLoS One* 8 (3): e58994. https://doi.org/10.1371/journal.pone.0058994.

Nadeem, H.S., Torun, M., and Özdemir, F. (2011). Spray drying of the mountain tea (*Sideritis stricta*) water extract by using different hydrocolloid carriers. *LWT - Food Science and Technology* 44: 1626–1635.

Namaldi, A., Çalik, P., and Uludag, Y. (2006). Effects of spray drying temperature and additives on the stability of serine alkaline protease powders. *Drying Technology* 24: 1495–1500.

*Edible Coatings and Films to Improve Food Quality*Nisperos-Carriedo, M.O. (1994). Edible coatings and films based on polysaccharides. In: (ed. J.M. Krochta, E.A. Baldwin and M.N. Nisperos-Carriedo), Chapter 11, 305–335. Lancaster, PA: Technomic Publishing Co.

Nussinovitch, A. (1997). *Hydrocolloid Applications: Gum Technology in the Food and Other Industries*. London, UK: Blackie Academic & Professional.

Nussinovitch, A. and Hirashima, M. (2019). Pullulan. In: *More Cooking Innovations, Novel Hydrocolloids for Special Dishes*, Chapter 16, 189–197. Boca Raton, FL: CRC Press.

Nussinovitch, A. and Lurie, S. (1995). Edible coatings for fruits and vegetables. *Postharvest News and Information* 6: 53N–57N.

Okajima, K. and Spence, C. (2011). Effects of visual food texture on taste perception. *i-Perception* 3: 966.

Österbauer, R.A., Matthews, P.M., Jenkinson, M. et al. (2005). Color of scents: chromatic stimuli modulate odor responses in the human brain. *Journal of Neurophysiology* 93: 3434–3441.

Owen, S.R., Tung, M.A., and Durance, T.D. (1991). Cutting resistance of a restructured fruit bar as influenced by water activity. *Journal of Texture Studies* 2: 191–199.

Park, H.J. (1999). Development of advanced edible coatings for fruits. *Trends in Food Science and Technology* 10: 254–260.

Park, H.J. and Chinnan, M.S. (1995). Gas and water vapor barrier properties of edible films from protein and cellulosic materials. *Journal of Food Engineering* 25: 497–507.

Park, H.J., Weller, C.L., Vergano, P.J., and Testin, R.F. (1993). Permeability and mechanical properties of cellulose-based edible film. *Journal of Food Science* 58: 1361–1364, 1370.

Patil, S.H., Shere, P.D., Sawate, A.R., and Mete, B.S. (2017). Effect of hydrocolloids on textural and sensory quality of date-mango leather. *Journal of Pharmacognosy and Phytochemistry* 6: 399–402.

Péneau, S., Brockhoff, P.B., Escher, F., and Nuessli, J. (2007). A comprehensive approach to evaluate the freshness of strawberries and carrots. *Postharvest Biology and Technology* 45: 20–29.

Petracek, P.D., Dou, H.T., and Pao, S. (1998). The influence of applied waxes on postharvest physiological behavior and pitting of grapefruit. *Postharvest Biology and Technology* 14: 100–106.

Piqueras-Fiszman, B. and Spence, C. (2014). Colour, pleasantness, and consumption behavior within a meal. *Appetite* 75: 165–172.

Piqueras-Fiszman, B. and Spence, C. (2015). Sensory expectations based on product-extrinsic food cues: an interdisciplinary review of the empirical evidence and theoretical accounts. *Food Quality and Preference* 40: 165–179.

Primo-Martín, C., Sanz, T., Steringa, D.W. et al. (2010). Performance of cellulose derivatives in deep-fried battered snacks: oil barrier and crispy properties. *Food Hydrocolloid* 24: 702–708.

Prinz, J.F. and de Wijk R.A. (2004) Effects of flavor and visual texture on ingested volume. *Poster Presented at the 5th Meeting of the International Multisensory Research Forum*, Sitges, Spain (2–5 June 2004).

Quek, S.Y., Chok, N.K., and Swedlund, P. (2007). The physicochemical properties of spray-dried watermelon powders. *Chemical Engineering and Processing* 46: 386–392.

Raab, C. and Oehler, N. (2000). Making dried fruit leather. Reprint. Oregon State University Extension Services, Fact Sheet 1976, FS 232. https://catalog.extension.oregonstate.edu/user/login?destination=node/2240 (accessed 12 January, 2022).

Rameshwar, A. (1979). Tandra (mango leather) industry in Andhra Pradesh. *Indian Food Packer* 2: 11–12.

Reddy, V., Aruna, K., Vimla, V., and Dhana Lakshmi, K. (1999). Physico-chemical changes during storage of papaya fruit (*Carica papaya* L.) bar (*Thandra*). *Journal of Food Science* 36: 428–433.

Rimac-Brnčić, S., Lelas, V., Rade, D., and Šimundić, B. (2004). Decreasing of oil absorption in potato strips during deep fat frying. *Journal of Food Engineering* 64: 237–241.

Rodríguez-Hernández, G.R., González-García, R., Grajales-Lagunes, A. et al. (2005). Spray drying of cactus pear juice (*Opuntia streptacantha*): effect on the physicochemical properties of powder and reconstituted product. *Drying Technology* 23: 955–973.

Sarala, M., Velu, V., Anandharamakrishnan, C., and Singh, R.P. (2011). Spray drying of *Tinospora cordifolia* leaf and stem extract and evaluation of antioxidant activity. *Journal of Food Science and Technology* 49: 119–122.

Sason, G. and Nussinovitch, A. (2020). Selective protective coating for damaged pomegranate arils. *Food Hydrocolloids* 103 (5): 105647. https://doi.org/10.1016/j.foodhyd.2020.105647.

Schaefer, H.M. and Schmidt, V. (2013). Detect ability and content as opposing signal characteristics in fruits. *Proceedings of the Royal Society London B* 271 (Suppl): S370–S373.

Sebti, I., Ham-Pichavant, F., and Coma, V. (2002). Edible bioactive fatty acid-cellulosic derivative composites used in food-packaging application. *Journal of Agricultural and Food Chemistry* 50: 4290–4294.

Sengül, M., Ertugay, M.F., and Sengül, M. (2005). Rheological, physical and chemical characteristics of mulberry pekmez. *Food Control* 16: 73–76.

Shimbo, H. (2000). *The Japanese Kitchen: 250 Recipes in a Traditional Spirit*, 75. Boston: Harvard Common Press.

Sikora, M., Juszczak, L., and Sady, M. (2003). Hydrocolloids in forming properties of cocoa syrups. *International Journal of Food Properties* 6: 215–228.

Silva, N.H.C.S., Vilela, C., Almeida, A. et al. (2018). Pullulan-based nanocomposite films for functional food packaging: exploiting lysozyme nanofibers as antibacterial and antioxidant reinforcing additives. *Food Hydrocolloids* 77: 921–930.

Singh, R.S., Kaur, N., Rana, V., and Kennedy, J.F. (2017). Pullulan: a novel molecule for biomedical applications. *Carbohydrate Polymers* 171: 102–121.

Singthong, J. and Thongkaew, C. (2009). Using hydrocolloids to decrease oil absorption in banana chips. *LWT – Food Science and Technology* 42: 1199–1203.

Sivarajalingam, S. (2009) Modelling and control of a spray drying process. MSc Thesis. Technical University of Denmark, Kongens Lyngby, Denmark.

Skrandies, W. and Reuther, N. (2008). Match and mismatch of taste, odor, and color is reflected by electrical activity in the human brain. *Journal of Psychophysiology* 22: 175–184.

Spence, C. (2010). The color of wine – part 1. *The World of Fine Wine* 28: 122–129.

Spence, C. (2015). On the psychological impact of food color. *Flavor* 4: 21. https://doi.org/10.1186/s13411-015-0031-3.

Spence, C. (2016). The psychological effects of food colors. In: *Handbook on Natural Pigments in Food and Beverages: Industrial Applications for Improving Food Color*, Series in Food Science, Technology and Nutrition Number, vol. 295 (ed. C. Reinhold and R.M. Schweiggert), Chapter 2, 29–58. Duxford, UK: Elsevier, Woodhead Publishing.

Spence, C. and Piqueras-Fiszman, B. (2014). *The Perfect Meal: The Multisensory Science of Food and Dining*. Oxford: Wiley-Blackwell.

Srinivas, R.N., Venkatesh Reddy, T., Ravi, P.C. et al. (1997). Post harvest loss assessment of tota puri and alphonso mangoes. *Journal Food Technology* 34: 70–72.

Stiem, M. (1990). Tubular casing with improved peel ability. German Patent Application DE 3842969 A1.

Summers, S. (1994). Getting the most out of apple. *Cereal Food* 10: 746–749.

Szczesniak, A.S. (1983). Physical properties of foods: what they are and their relation to other food properties. In: *Physical Properties of Foods* (ed. M. Peleg and E.B. Bagley), 1–41. Westport, CT: Avi Publishing Company.

Torley, P.J., de Boer, J., Bhandari, B.R. et al. (2008). Application of the synthetic polymer approach to the glass transition of fruit leathers. *Journal of Food Engineering* 86: 243–250.

Trezza, T.A. and Krochta, J.M. (2000). The gloss of edible coatings as affected by surfactants, lipids, relative humidity and time. *Journal of Food Science* 65: 658–662.

Ullah, M.W., Ul-Islam, M., Khan, T., and Park, J.K. (2021). Recent developments in the synthesis, properties and applications of various microbial polysaccharides. In: *Handbook of Hydrocolloids*, 3e (ed. G.O. Phillips and P.A. Williams), 975–1008. Duxford, UK: Elsevier, Woodhead Publishing.

Valenzuela, C. and Aguilera, J.M. (2013). Aerated apple leathers: effect of microstructure on drying and mechanical properties. *Drying Technology* 31: 1951–1959.

Vehring, R., Foss, W.R., and Lechuga-Ballesteros, D. (2007). Particle formation in spray drying. *Journal of Aerosol Science* 38: 728–746.

Villalobos, R., Chanona, J., Hernández, P. et al. (2005). Gloss and transparency of hydroxypropyl methylcellulose films containing surfactants as affected by their microstructure. *Food Hydrocolloids* 19: 53–61.

Von Elbe, J.H. and Johnson, C.E. (1971). Sales pattern changing – color may be factor. *Canner/Packer* 140 (8): 10–11.

Wada, H., Nakata, K., Nonami, H. et al. (2021). Direct evidence for dynamics of cell heterogeneity in watercored apples: turgor-associated metabolic modifications and within-fruit water potential gradient unveiled by single-cell analyses. *Horticulture Research* 8: 187–202.

Whistler, R.L. and Daniel, J.R. (1985). Carbohydrates. In: *Food Chemistry*, 2e (ed. O.R. Fennema), 69–137. New York: Marcel Dekker.

Williams, R. and Mittal, G.S. (1999). Low fat fried foods with edible coatings: modeling and simulation. *Journal of Food Science* 64: 317–322.

Wong, D.W.S., Camirand, W.M., and Pavlath, A.E. (1994). Development of edible coatings for minimally processed fruits and vegetables. In: *Edible Coatings and Films to Improve Food*

Quality (ed. J.M. Krochta, E.A. Baldwin and M.N. Nisperos-Carriedo), Chapter 3, 65–88. Lancaster, PA: Technomic Publishing Co.

X-rite Inc. (2016). A Guide to Understanding Color. https://www.xrite.com/.

Yılmaz, M.T., Sert, D., and Karakaya, M. (2009). Rheological and sensory properties of spray dried pekmez mixtures with wheat starch-gum. *International Journal of Food Properties* 12: 691–704.

Yilmazer, G., Carillo, A., and Kokini, J. (1991). Effect of propylene glycol alginate and xanthan gum on stability of O/W emulsions. *Journal of Food Science* 2: 513–517.

Yogurtçu, H. and Kamisli, F. (2006). Determination of rheological properties of some pekmez samples in Turkey. *Journal of Food Engineering* 77: 1064–1068.

Yousefi, S., Emam-Djomeh, Z., and Mousavi, S.M. (2011). Effect of carrier type and spray drying on the physicochemical properties of powdered and reconstituted pomegranate juice (*Punica granatum* L.). *Journal of Food Science and Technology* 48: 677–684.

Zarei, S., Khodaiyan, F., Hosseini, S.S., and Kennedy, J.F. (2020). Pullulan production using molasses and corn steep liquor as agro-industrial wastes, physiochemical, thermal and rheological properties. *Applied Food Biotechnology* 7: 263–272.

3

Use of Hydrocolloids to Modify Food Taste and Odor

3.1 Introduction

Flavor, comprised of taste (perceived by the tongue) and odor (perceived by the olfactory center in the nose), results from the response of receptors in the oral and nasal cavities to chemical stimuli. These are called "the chemical senses." In this chapter, flavor, aroma, taste, and volatile compounds are discussed. Examples of foods whose flavor is governed, enhanced or changed by the addition of hydrocolloids, such as low-fat cheddar cheese, fish, and meat analog products, specialty breads, protein beverages, and spreads, are provided. A section is devoted to the detailed interactions of flavor and different food ingredients, including interactions between proteins and flavor compounds, hydrocolloids and flavor compounds, and a special discussion devoted to starch. Other distinct topics include the effect of hydrocolloids on sensory properties of selected model systems and beverages; the influence of hydrocolloids on the release of volatile flavor compounds; and the relationship between gels and flavor release. Most of the available literature emphasizes the effect of hydrocolloids on flavor release and perception. Nevertheless, from a practical perspective, it is just as important to understand the influence of flavor molecules on the behavior of hydrocolloids, and therefore, a section is devoted to this issue. Finally, the use of hydrocolloids to modify the taste/odor of foods is exemplified by two recipes. In the first, the batter for fried chicken includes methylcellulose (MC) to improve taste and flavor; in the second, hydroxypropyl methylcellulose (HPMC; also written as MHPC) is used for the creation of gluten-free bread, with good mouthfeel, satisfactory volume, and retained softness over time.

3.2 Flavor Perception: Aroma, Taste, and Volatile Compounds

The brain's perception of flavor is considered to combine two senses: smell and taste. Therefore, flavor can be broken down into two major components, the volatile compounds that are sensed by the olfactory epithelium (aroma) and the non-volatile compounds that are sensed by the taste buds on the tongue (taste). As the food is being consumed, various factors can influence the release of the volatile components as well as the tastants, including breakdown of the structure by mastication and saliva. In this chapter, the described studies address the role of these factors and how they are affected by hydrocolloid solutions and gels.

Flavor is the perceptual impression of food or other material as determined first and foremost by the chemical senses of the gustatory and olfactory system (Small and Green 2012; Wolfe et al. 2012). Of the chemical senses, smell is the major determinant of a food item's flavor (Figure 3.1). Five basic tastes – sweet, sour, bitter, salty, and *umami* (savory) – are generally documented, even though some cultures also take into account pungency (Oaklander 2015), i.e. a strong, sharp smell, or flavor. *Umami* (旨味), a term proposed by Kikunae Ikeda (Figure 3.2) in 1908 (Lindemann et al. 2002), is characteristic of broths and cooked meats (Torii et al. 2013). People sense umami through taste receptors that characteristically react to glutamates (Figure 3.3) and nucleotides, such as inosinic acid (Figure 3.4) or inosine monophosphate, which are pervasive in meat broths (Fleming 2013). Flavorings are products added to foods to affect their odor and/or taste (https://www.merriam-webster.com/dictionary/flavor). Several types of flavorings have been defined by EU legislation (European Commission website, Directorate General for Health and Food Safety, https://ec.europa.eu/food/safety/food_improvement_agents/flavourings_en). Three types of flavoring are used in foods: natural flavoring substances; nature-identical flavoring substances; and artificial flavoring substances. Numerous flavorings are composed of esters, which are often described as being "fruity" or "sweet." An ester is an organic compound in which the hydrogen in the compound's carboxyl group is replaced by a hydrocarbon group. Esters are derived from carboxylic acids and (frequently) alcohol (https://goldbook.iupac.org/terms/view/E02219). On the one hand, constituents that have an exclusively sweet, sour, or salty taste are not considered flavorings (EU legislation Article 2, EC Regulation No 1334/2008 of the European Parliament and of the Council of 16 December 2008 on flavorings and certain

Figure 3.1 The sense of smell. *Source:* Yale Center for British Art/Public Domain.

Figure 3.2 Photograph of Kikunae Ikeda. *Source:* Unknown author/Wikipedia Commons/Public Domain.

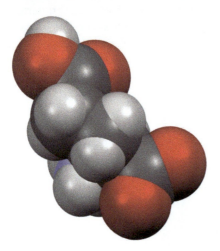

Figure 3.3 Space-filling model of the L-glutamic acid molecule, $C_5H_9NO_4$, in the zwitterion form found in the crystal structure determined by single crystal X-ray diffraction and reported in Ruggiero et al. (2016). Color code: carbon (C), gray; hydrogen (H), white; nitrogen (N), blue; oxygen (O), red. *Source:* Ben Mills/Wikimedia Commons/Public domain.

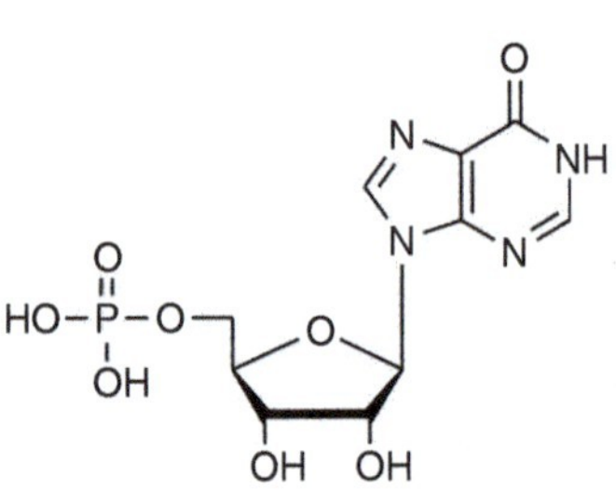

Figure 3.4 Structure of inosinic acid. *Source:* Su-no-G/Wikimedia Commons/Public Domain.

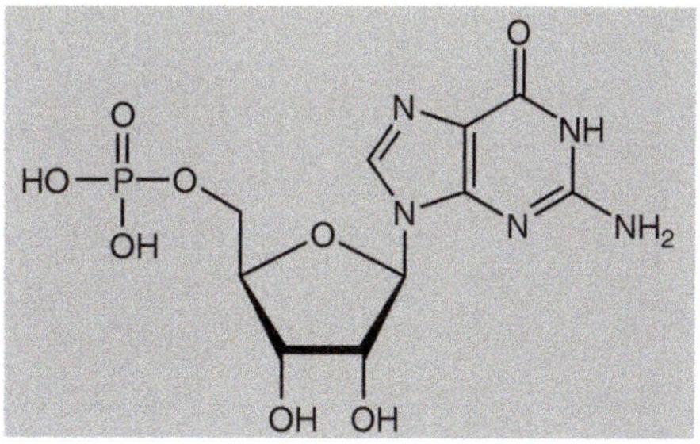

Figure 3.5 Structure of guanosine monophosphate (GMP). *Source:* NEUROtiker/Wikimedia Commons/Public Domain.

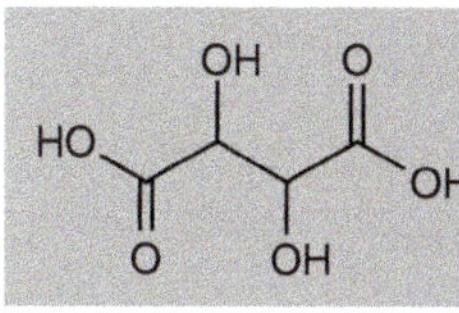

food ingredients with flavoring properties for use in and on foods: https://eur-lex.europa.eu/legal-content/EN/ALL/?uri=CELEX:32008R1334). On the other hand, *umami* or "savory" flavorants, which are ordinarily labeled as flavor or taste enhancers, are to a large extent based on amino acids and nucleotides. These are typically used as sodium or calcium salts. *Umami* flavorants that are documented and permitted by the EU include glutamic acid salts, glycine salts, guanylic acid salts (GMP) (Figure 3.5), inosinic acid salts, and 5′ ribonucleotide salts.

Figure 3.6 Tartaric acid. *Source:* JaGa/Wikimedia Commons/CC BY-SA 3.0.

Defined organic and inorganic acids (i.e. acetic acid, ascorbic acid, citric acid, fumaric acid, lactic acid, malic acid, phosphoric acid, and tartaric acid) (Figure 3.6) can be used to enhance sour taste, but similar to salt and sugar, they are not generally considered or regulated as flavorants under law. Each acid conveys a slightly different tart or sour taste that modifies the flavor of a food (https://en.wikipedia.org/wiki/Flavor). Food color can considerably influence the anticipation of a flavor (Shankar et al. 2010). For example, addition of supplementary red color to a drink enhances its perceived sweetness. Darker colored solutions were rated 2–10% better than lighter ones, even though they had less sucrose (1% concentration) (Johnson and Clydesdale 1982). Food manufacturers exploit this phenomenon; for instance, Froot Loops cereal ("froot" being a cacography of fruit) and some varieties of Gummy Bears frequently use the same flavorings (Locker 2014; Stevens 2018; https://en.wikipedia.org/wiki/Froot_Loops).

Food and beverage companies require flavorings for novel products, low-fat versions of prevailing products, or current foodstuffs with variations in formulation or processing. The global food flavoring market was evaluated at $12.71 billion in 2020 and is projected to reach $19.22 billion by 2030 (Allied Market Research 2021). A rise in demand for new flavors from the food & beverage sector and their continuous development are fueling the growth of the food flavor market. Moreover, a rise in demand from the fast-food industry is anticipated to offer growth opportunities in the food flavoring market, with the Asia-Pacific region expected to take the lead during the forecasted period (Allied Market Research 2021).

Aroma compounds are also acknowledged as odorant, aroma, fragrance, or flavor. These are chemical compounds with a smell or odor (https://en.wikipedia.org/wiki/Aroma_compound). To convey a smell or fragrance, a distinct chemical must be volatile, so that it can be transmitted through the air to the olfactory system in the upper part of the nose. Such volatile compounds have molecular weights of less than 310 Da (Rothe and Specht 1976). Aroma compounds can be found in various foods (El Hadi et al. 2013; Haugeneder et al. 2018), such as wines, which have

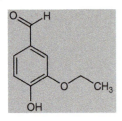

Figure 3.7 Structure of ethyl vanillin. *Source:* NEUROtiker/Wikimedia Commons/Public domain.

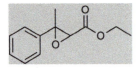

Figure 3.8 Chemical structure of the strawberry aldehyde ethyl methylphenylglycidate. *Source:* Edgar 181/Wikimedia Commons/Public domain.

more than 100 aromas formed as byproducts of fermentation (Ilc et al. 2016). Aroma compounds are classified by structure and include esters, linear terpenes, cyclic terpenes, aromatic compounds, and amines. Other aroma compounds include alcohols, aldehydes, esters, ketones, lactones, and thiols, to name a few (https://en.wikipedia.org/wiki/Aroma_compound). Most flavors embody a blend of aroma compounds. A single synthetic compound is rarely used in its pure form. The artificial vanilla flavors vanillin and ethyl vanillin (Figure 3.7) are a noteworthy exception, in addition to the artificial strawberry flavoring ethyl methylphenylglycidate (Figure 3.8). The abundant "green apple" aroma is based on hexylacetate (Luebke 2017). An odor is caused by one or more volatilized chemical compounds (https://en.wikipedia.org/wiki/Odor), and low concentrations of these compounds can be perceived by the sense of smell. Odor is also termed "smell" or "scent," and can refer to both pleasant and unpleasant odors (https://en.wikipedia.org/wiki/Odor). The perception of odor is mediated by the olfactory nerve, i.e. the first cranial nerve, which contains sensory nerve fibers relating to the sense of smell. The olfactory receptor cells are neurons in the olfactory epithelium that act as sensory signaling cells. Each neuron has cilia that come into direct contact with the air. Odorous molecules bind to receptor proteins extending from these cilia (Gardiner 2005) and act as chemical stimuli, initiating electrical signals that travel along the olfactory nerve's axons (i.e. a long, slender projection of a nerve cell) to the brain (de March et al. 2015). Once the electrical signal reaches a threshold, the signal travels along the axon to the olfactory bulb, a part of the limbic system (i.e. a set of brain structures located on both sides of the thalamus). There, the smell is deciphered, by linking it to previous experience and as related to the substance(s) inhaled (Axel 1995). Odor intensity is ranked as follows: 0 – no odor, 1 – very weak (odor threshold), 2 – weak, 3 – distinct, 4 – strong, 5 – very strong, 6 – intolerable. The intensity can be judged by laboratory-trained specialists (Jiang et al. 2006). Dissimilar categorizations of principal odors have been suggested, comprising the following seven primary odors: musky – perfumes; putrid – rotten eggs; pungent – vinegar; camphoraceous – mothballs; ethereal – dry cleaning fluid; floral – roses; and pepperminty – mint gum. Unpleasant odors play numerous roles in nature, frequently to warn of danger; nevertheless, this may not be known to the subject who smells it. Such unpleasant odors can be considered by some people or cultures as attractive (Trygg 1991).

3.3 Flavor of Hydrocolloid-Supplemented Value-Added Foods

3.3.1 Low-Fat Cheddar Cheese

Nearly one-third of the world's milk production is used for cheese manufacturing (Farkye 2004). Different types of cheeses are prepared from cow, camel, goat, sheep, and buffalo milk or their blends (Hussain et al. 2012). Fat contributes to the desirable flavor and texture of cheeses, and reduced-fat cheeses suffer from decreased acceptability and functionality (Murtaza et al. 2017). Addition of gums has been shown to improve the sensory acceptability of low-fat cheddar cheese (Figure 3.9), with the hydrocolloids serving as fat replacers (Koca and Metin 2004; Rogers et al. 2010). A low-fat cheddar cheese sample containing 0.45% guar gum was comparable to its full-fat counterpart in most of its sensorial and textural qualities. Hence, it was concluded that hydrocolloids – predominantly guar gum and xanthan gum (less effective) – could be used to increase the functionality and satisfactoriness of low-fat cheddar cheese (Murtaza et al. 2017). The hydrocolloids' success stems from a reduction in the hardness of low-fat cheeses, as well as binding and retention of water and perhaps flavor (Rahimi et al. 2007).

3.3.2 Wholegrain Sorghum Bread

Wholegrain products are rich in dietary fiber and other bioactive components, and are regarded as more desirable than refined products (Seal et al. 2006). Incorporation of more than 16% (w/w) whole sorghum flour in wheat bread led to a darker color, intense bitter taste, lower specific volume, and harder crumb. In short, this addition produced a negative consumer response (Mariera et al. 2017). The effect of dextran produced *in situ* by *Weissella confusa* A16 on flavor and texture perception of wholegrain sorghum bread containing 50% wheat flour was studied (Wang et al. 2020). Dextrans are natural hydrocolloids produced extracellularly by lactic acid bacteria using sucrose as the substrate (Monsan et al. 2001). High-molecular-weight linear dextrans are very important in bakery formulations (Rühmkorf et al. 2012), because they can increase loaf volume, moist mouthfeel, and crumb softness, and serve as staling inhibitors (Wang et al. 2018, 2019). Dextran-enriched sorghum sourdough bread showed a substantial decrease in perceived flavor intensity compared to the control sorghum sourdough bread, regardless of comparable levels of

Figure 3.9 Bravo wax-covered cheddar cheese from Bravo Farms, Traver, California. *Source:* Jon Sullivan/Wikipedia Commons/Public Domain.

acidification and polyphenols (Wang et al. 2020). The flavor suppression seemed to take place at above the critical overlap concentration (i.e. what sets apart dilute from semi dilute polymer solutions), but continued unaffected at lower concentrations. In summary, dextran produced by *W. confusa* A16 is a promising texture-enhancing and flavor-masking agent in wholegrain products, which may lead to innovations in this field (Wang et al. 2020).

3.3.3 Fish Fingers

Hydrocolloids have extensive uses in meat and fish products. Alginate, carboxymethylcellulose, and konjac are used in various ready-to-eat products such as frankfurters, patties, and sausages (Jiménez-Colmenero et al. 2010). They were included in a fish sausage formulation to improve its sensory characteristics (Santana et al. 2013). The effect of pectin, among other inclusions, on the quality of silver carp fish fingers was studied (Shaviklo and Fahim 2014). Incorporation of 1% chitosan in the batter formulation for the fish fingers led to improved biochemical properties, enhanced oil reduction, and better coating characteristics (Hauzoukim et al. 2019). The effect of using different hydrocolloids, i.e. sodium alginate, carrageenan, HPMC, xanthan gum or chitosan as part of the batter formulation, on the composition, and physical, chemical, microbiological, and sensory properties of carp fish fingers throughout storage at $-18\,°C$ for six months was studied (Talab and Abou-Taleb 2021). The batter containing the different hydrocolloids was prepared using a common method (Hauzoukim et al. 2019). The different added hydrocolloids did not significantly affect the flavor of the carp fish fingers. Chitosan and carrageenan significantly improved the taste, texture, and appearance of these fish fingers at the end of the storage period. No mechanistic explanations for this success were provided (Talab and Abou-Taleb 2021).

3.3.4 Meat Analogs

Meat analogs are cholesterol-free plant-based protein products that contain highly valuable essential amino acids and low saturated fat. Meat analogs are also termed meat substitutes, mock meat, faux meat, or soy meat (Omohimi et al. 2014). Extrusion is the best way to texturize plant proteins (Guo et al. 2020). A fibrous meat-like structure was produced from plant proteins via high-moisture extrusion technology (Palanisamy et al. 2018). Such extrusion gives the plant proteins – such as wheat gluten, soy or pea protein – a characteristic texture (Grabowska et al. 2014; Osen et al. 2014). The central factors affecting consumer acceptance of meat analog products are texture and flavor. These two parameters can be improved/modified by fine-tuning the process conditions or material appearance (Guo et al. 2020). Numerous flavor compounds, such as aldehydes, ketones, and esters, bind with proteins through hydrophobic interactions (Guichard and Langourieux 2000). In the case of high wheat gluten content, the amount of random coil structures in the meat analogs was high but the retention rate of total volatile flavor substances was low, possibly affected by the microstructure (Guichard 2002). As the moisture content of the raw material increased, the random coil structures first increased and then decreased. The wheat gluten and moisture contents of the raw materials potentially changed the flavor features of the meat analogs by affecting the microstructure, the binding capacity of internal proteins and water molecules, and the secondary structure of the proteins. The retention and release of volatile flavor substances can therefore be optimized by generating an appropriate microenvironment for the flavor compounds in the meat analogs (Guo et al. 2020).

3.3.5 Spreads

Spreads consist mostly of fat and sugar and can benefit from the substitution of oleogels for the fat (Demirkesen and Mert 2019); however, there are very few studies on the use of oleogels in spreads (Bascuas et al. 2021). The fat content of chocolate spreads can reach up to 60% (Espert et al. 2020). The substitution of solid fat with liquid oil can greatly affect the chocolate spread's performance and consequently, the quality of the product (Fayaz et al. 2017). The microstructure, rheological behavior, and oxidative stability of oleogels prepared using HPMC and xanthan gum were studied, and sunflower or olive oleogels were found to have potential food applications (Bascuas et al. 2021). The incorporation of wax- or shellac-based oleogels in chocolate spreads was also studied (Patel et al. 2014). Recently, the feasibility of using olive and sunflower oil oleogels prepared with biopolymer oleogelators to replace the saturated fat in chocolate spreads (Figure 3.10) was assessed by studying their structural properties (microstructure, rheology, and texture) and sensory attributes (Bascuas et al. 2021). The structural properties of the traditional spread were maintained. In fact, the formed structure was more homogeneous and when sunflower oleogel was used together with coconut fat, the spread had the same sensory attributes as the control spread (Bascuas et al. 2021).

3.3.6 Protein Beverages

Viscosity and interactions with oral surfaces play significant roles in the perception of taste, texture, and oral-tactile sensations of beverages. In contrast, how specific food properties affect the perception of taste and mouthfeel is still unknown (Wagoner et al. 2020). Previous studies have dealt with simple aqueous systems or viscosity levels that are outside

Figure 3.10 "Hashahar Haole" chocolate spread, Israel. *Source:* Hovev/Wikipedia Commons/Public Domain.

the range of many beverages (Christensen 1980). The role of macromolecules such as protein and fat in texture–taste interactions in beverages is still unclear. Therefore, the effects of structurally different hydrocolloids (λ-carrageenan and tapioca starch) on perceived taste and texture of three separate fluid systems (water, skim milk, and whole milk) at different viscosity levels (4–6 mPa·s, 25–30 mPa·s, and 50–60 mPa·s at 50 per second) were studied, along with an evaluation of how these systems alter perception over time of consumption (Wagoner et al. 2020). It was concluded that viscosity influences texture perception of protein beverages more than hydrocolloid type, and that milk beverages can be formulated with viscosities ranging from 3 to 74 mPa·s, covering a wide range of textures, without suppressing sweet or salty taste (Wagoner et al. 2020).

3.4 Interactions of Flavor Compounds with Different Food Ingredients

3.4.1 Interactions Between Proteins and Flavor Compounds

Flavor is considered to be one of the most important attributes in food acceptance by consumers (Guichard 2002). Food matrix components, for instance proteins, polysaccharides, and lipids, are known to interact with flavor compounds (Overbosch et al. 1991; Bakker 1995; Land 1996; Taylor 1999). Proteins are frequently used as food ingredients due to their emulsifying and stabilizing abilities in lipid-dispersed food systems. Proteins interact with flavor compounds by reversible or irreversible binding. Covalent irreversible binding of aldehydes with proteins, in addition to hydrophobic interactions, has been reported (Gremli 1974; O'Keefe et al. 1991). The processing of the food protein and its structure, and the flavor's chemical nature, in addition to other factors such as ionic conditions, temperature, and presence of ethanol, will define the magnitude of the protein–flavor compound interactions (O'Keefe et al. 1991; O'Neil 1996). The milk protein β-lactoglobulin interacts with numerous flavor compounds, for instance aldehydes and ketones (O'Neill and Kinsella 1987), ionones (Dufour and Haertlé 1990), and esters (Pelletier et al. 1998). Linear relationships between the logarithm of the binding constants and the hydrophobicity of flavor compounds were observed for homologous series of esters, aldehydes, ketones, and alcohols (Guichard and Langourieux 2000). The most likely binding site of these compounds is the hydrophobic pocket of the protein, which binds fatty acids as well (Wu et al. 1999). Binding of β-ionone, retinol, and tetradecanoic acid in the central cavity of β-lactoglobulin has been observed (Lubke et al. 1999). For other ligands (p-cresol, eugenol, and γ-decalactone), there were no detectable conformational changes in the protein and binding to either the protein's surface or central cavity was suggested (Lubke et al. 1999). The addition of sodium caseinate (0.1% in water) reduces the activity coefficients of flavor compounds in the following order: β-ionone, n-hexanol, ethyl hexanoate, and isoamyl acetate (Voilley et al. 1991). Lower volatilities of acetaldehyde and acetone were also detected in aqueous systems including egg albumin, casein, and gelatin (Maier 1970). Soy protein and bovine serum albumin exhibited the same binding properties (Beyeler and Solms 1974). On the whole, binding capacities of proteins increase from alcohols to ketones and aldehydes. As proved for β-lactoglobulin, the more hydrophobic compounds in a homologous series bind more strongly to the proteins (Damodaran and Kinsella 1980, 1981).

Chemical or physical modification of a protein's conformational state noticeably changes its flavor-binding features. Sulfur compounds interact with proteins by irreversible binding. Heating an aqueous solution comprised of thiols and disulfides in the presence of egg albumin resulted in reduction in a disulfide (2-furanmethyl disulfide) and formation of the corresponding thiol (Mottram et al. 1996). This effect was less evident when casein, which has a much lower proportion of sulfhydryl groups than egg albumin, was used. A change in the relative concentrations of disulfides and thiols resulted in considerable changes in sensory properties when these compounds were used to flavor foodstuffs enclosing proteins (Mottram and Nobrega 2000). The sulfur compound allyl isothiocyanate is responsible for mustard aroma. It interacts with cystine, and cystine-containing peptides and proteins to induce oxidized cleavage of all or part of their disulfide bonds (Kawakishi and Kaneko 1987). The presence of β-lactoglobulin or sodium caseinate at the oil–water interface increases resistance to the transferal of hydrophobic aroma compounds from oil to water and consequently, decreases flavor release and flavor perception (Harvey et al. 1995; Rogacheva et al. 1999). Interactions between proteins and flavor compounds may induce retention in emulsions of the compounds with higher binding constants (Rogacheva et al. 1999; Voilley et al. 2000).

Whey proteins demonstrated a greater degree of binding than casein and produced a notable reduction in the flavor intensities of vanillin, citral, benzaldehyde, and limonene in a 2.5% sucrose solution (Hansen and Heinis 1991, 1992). Adding 1% β-lactoglobulin to water substantially reduced the odor intensity of methyl ketones (Andriot et al. 2000) and eugenol (Reiners et al. 2000); no noticeable effect was observed for vanillin, the compound with the lowest affinity for the protein. Differences between these results and those acquired for vanillin may have been due to the different media used and differences in protein batches (Hansen and Booker 1996). A good correlation was also detected between physicochemical and sensory data for the retention of menthone by egg albumin (25 and 50% in water) (Ebeler et al. 1988).

3.4.2 Interactions Between Starch and Flavor Compounds

Inclusion of polysaccharides in a formulation might change flavor release and perception, in many cases due to modifications of viscosity (Godshall 1997). Starch contains linear amylose and branched amylopectin. In its native state, these are packed in well-organized starch granules (Escher et al. 2000). Amylose binds flavor compounds by establishing inclusion complexes (Rutschmann and Solms 1990a, 1990b). The increase in ligand binding is accompanied by an increase in diameter of the formed helices (Rutschmann and Solms 1990c). The amylose–ligand complexes give characteristic X-ray diffraction spectra (Buleon et al. 1990). Aroma retention which cannot be explained by complexation with amylose is hypothesized to be a result of interaction with amylopectin (Langourieux and Crouzet 1994; Godet et al. 1995). However, aroma–starch interactions mostly result from adsorption involving hydrogen bonds and not inclusion complexes (Boutboul et al. 2001). Granular starches, perhaps due to lower accessibility of the starch macromolecules, retain less aroma compounds. Retention increases when the surface area is enlarged (Hau et al. 1994). Increasing the water content of wheat starch samples increases the rate of diacetyl uptake (Hau et al. 1998). Glassy-state samples bound diacetyl more slowly than rubbery samples. A combination of sucrose with a starch extrudate has an insignificant effect on volatile binding performance (Hau et al. 1998). In general, retention by carbohydrates of flavor compounds in the same chemical class increases with molecular weight and decreases with growing polarity and volatility of the aroma compounds (Goubet et al. 1998). Interactions of flavors with dextran were exemplified in Section 3.3.2, and with xanthan and pectin in Section 3.3.3, and these are further discussed in Sections 3.5 and 3.7.1.

Phenolic compounds induce changes in flavor release and aroma characteristics (King and Solms 1982). In hydroalcoholic solutions, addition of catechin prompted a decrease in the activity coefficients of various wine aroma compounds (Dufour and Bayonove 1999), owing to hydrophobic interactions. In an alcohol medium, no binding was observed with limonene, which could be described by hindered approachability to the binding sites for this hydrophobic molecule (Langourieux and Crouzet 1997). Salts are regularly added to aqueous samples to intensify the concentrations of the aroma compounds in the vapor phase. This phenomenon is more distinct for alcohols than for aldehydes and less noteworthy for esters (Poll and Flink 1984). Sooner or later, these differences will alter the value of the perceived aroma. Lipids influence the flavor of foods through their effects on flavor perception (aroma, mouthfeel, and taste), generation, and stability. In systems containing lipids, flavor compounds are distributed between the lipid and aqueous phase, following the physical laws of partition. Further information on the influence of fat on flavor release; thermodynamic constants of flavor compounds in fat; the effect of fat content; the effect of the nature of the fat and flavor release from oil/water emulsions, as well as dynamic flavor release and flavor perception, can be located elsewhere (Guichard 2002).

3.4.3 Interactions Between Hydrocolloids and Flavor Compounds

Hydrocolloids are used in food processing due to their ability to serve as gelling agents and viscosity formers, in addition to numerous other virtues (Nussinovitch 1997). In most cases, a low concentration of hydrocolloid is sufficient to achieve a thickening effect. It is clear that understanding how hydrocolloids might influence the flavor of a product can assist in optimizing its value (Yven et al. 1998). The specific effects of hydrocolloids on flavor release involve a reduction in release related to increased viscosity. This phenomenon is associated with diffusion, which is inversely proportional to viscosity, in agreement with the Stokes–Einstein equation (Wilke and Chang 1995). In fact, this mechanism explains why the distance of diffusion of the molecule to be released in the mouth is smaller for more spreadable materials. The second mechanism includes molecular interactions of the flavor compounds with the macromolecule. A few studies have shown that an increase in solution viscosity causes a decrease in the rating of sweetness and additional flavor perceptions (Vaisey et al. 1969; Morris 1987). Other similar studies dealt with the influence of oat gum, guar gum, and carboxymethylcellulose on the perception of sweetness and flavor (Malkki et al. 1993); the effect of hydrocolloids and viscosity on flavor and odor intensities of aromatic flavor compounds (Pangborn and Szczesniak 1974); the effects of sucrose, guar gum, and carboxymethylcellulose on the release of volatile flavor compounds under dynamic conditions (Roberts et al. 1996); and the interaction of flavor compounds with microparticulated proteins (Schirle-Keller et al. 1992), to name a few.

The final means for hydrocolloid–taste interactions to occur is through the hydrocolloids' effect on mass transfer in the mouth. Even though hydrocolloids are tasteless, they are commonly utilized to add viscosity to a food, thereby reducing mass transfer from the food to the taste receptors (Reineccius 2005). A sample viscosity greater than 12–16 cP results in a significant reduction in sweetness (Pangborn et al. 1973). Hydrocolloids that readily undergo shear thinning (e.g. gellan gum) reduce sweetness less than hydrocolloids that do not (Vaisey et al. 1969). Proteins with the ability to increase viscosity or cause jellification might also influence aroma release from foods, due to their unique rheological properties (Reineccius 2005). If one wants to substitute one viscosity or gelling agent for another, it should be borne in mind that the interactions taking place are so complex that they cannot be modeled; modifications of the flavor compounds should not be attempted at all (Reineccius 2005). Thus, we know that substituting one protein or hydrocolloid for another will impact the flavor of a product, but we cannot quantify these effects and consequently, make changes in flavor formulations; we can only anticipate the effects and attempt to deal with them through empirical efforts (Reineccius 2005).

Interactions between hydrocolloids and flavor compounds have been assessed by sensory, headspace, and binding methodologies (Yven et al. 1998). As already noted, molecular interactions and viscosity can affect sensory perception. One sensory study indicated which flavor sensations are influenced by hydrocolloid inclusion using two methodologies: static headspace technique, to determine global flavor compound retention attributable to molecular interactions, and a chromatographic method to directly quantify the molecular interactions (Yven et al. 1998). A size-exclusion mechanism was used to estimate ligand–macromolecule interactions (Hummel and Dreyer 1962) by separating the ligand and macromolecule in a column (Sébille et al. 1987, 1990). Additional linking techniques allowed estimating the actual perceptual magnitude of the effects (Yven et al. 1998). Sensory analysis of flavored solutions [1-octen-3-ol (mushroom alcohol), diallyl disulfide/diallyl sulfide (garlic), and diacetyl (buttery)] was conducted. The flavor mixtures were prepared by mixing the flavor compound in ethanol with cold water and adding the hydrocolloid, followed by additional shear and storage at 20 °C until tasting (Yven et al. 1998). 1-Octen-3-ol is a secondary alcohol derived from 1-octene that exists as two enantiomers. Octenol is produced by several plants and fungi, including edible mushrooms and lemon balm. It is formed during the oxidative breakdown of linoleic acid and its odor has been described as green and moldy or meaty; it is used in certain perfumes (https://pdt.biogem.org/output.php?id=pdtdbl00176&db=ligand). Diallyl disulfide is an organosulfur compound derived from garlic and a few other *Allium* plants and diallyl sulfide is a sulfur-containing volatile compound present in garlic; diacetyl is an organic compound with an intense buttery flavor (https://en.wikipedia.org/wiki/Diallyl_disulfide; https://en.wikipedia.org/wiki/Diacetyl). The sensory evaluation demonstrated that the overall in-mouth and garlic flavors were greatest for water, intermediate for 0.1% xanthan, and lowest for 0.3% guar gum (Yven et al. 1998). This order corresponded to their viscosity under mouth shear rates. No significant differences were observed for mushroom and buttery attributes. Equilibrium headspace analysis established that addition of xanthan and guar gum decreased flavor release, with the largest reductions found for the diallyl sulfides (50%) (Yven et al. 1998). Molecular interactions between xanthan (0.01%) and 1-octen-3-ol were studied by exclusion chromatography. Weak reversible hydrogen-bonding interactions were observed with approximately one binding site per pentasaccharide repeating unit (Yven et al. 1998).

Sensory paired comparison tests were used to study differences in taste intensity in solutions of HPMC at concentrations above (1.0% w/w) and below (0.2% w/w) c^*, the coil-overlap concentration (the point at which viscosity changes abruptly with increasing thickener) (Cook et al. 2002). The sweetness intensities of aspartame (250 ppm), sucrose (5%), fructose (4.5%) and neohesperidin dihydrochalcone (39 ppm), and the saltiness of sodium chloride (0.35%) were all found to be significantly reduced in the more viscous HPMC solution (Cook et al. 2002). There was no significant effect of HPMC concentration on the acidity of citric acid (600 ppm) or the bitterness of quinine hydrochloride (26 ppm). The sweetness intensities of sucrose and aspartame were similarly investigated in two other hydrocolloid solutions, guar gum, and λ-carrageenan. Experiments were designed so that the ratios of the thickener concentrations (above and below c^*) to their measured c^* values remained constant (Cook et al. 2002). The sweetness of sucrose was found to be significantly reduced in the more viscous guar gum solution, and that of aspartame was reduced in the λ-carrageenan above c^*. The perceived sweetness of 6.5% sucrose in 1.0% HPMC did not differ significantly from that of 5% sucrose in 0.2% HPMC. The magnitude of the effect with aspartame was broadly analogous (Cook et al. 2002). The perceptions of sweetness and flavor were studied in viscous solutions containing 50 g/L sucrose, 100 ppm isoamyl acetate and varying concentrations of three hydrocolloid thickeners (guar gum, λ-carrageenan and HPMC). Zero-shear viscosity of the samples ranged from 1 to 5000 mPa·s. Perception of both sweetness and aroma was suppressed at thickener concentrations above c^* (Cook et al. 2003b). Sensory data for the three hydrocolloids were only loosely correlated with their concentration relative to c^* (c/c^* ratio), particularly above c^*. However, when perceptual data were plotted against Kokini oral shear stress (τ), calculated from rheological measurements, data for the three hydrocolloids aligned to form a master curve, enabling a prediction of flavor intensity in such systems. The fact that oral shear stress can be used to model sweetness and aroma perception supports the hypothesis that somatosensory tactile stimuli can interact with taste and aroma signals to modulate their perception (Cook et al. 2003b).

3.5 Effect of Hydrocolloids on Sensory Properties of Selected Model Systems and Beverages

Hydrocolloids are added to processed foods for many purposes: as thickening and bodying agents, and for stabilization, gelling and creating suspensions, to name a few (Nussinovitch 1997, 2003). In some distinct cases, hydrocolloids are used to mask unwelcome aftertastes (Glicksman and Farkas 1966). The effect of low concentrations of five selected hydrocolloids on the intensity of taste compounds in aqueous model systems was studied (Pangborn et al. 1973). In general, the sourness of citric acid and the bitterness of caffeine were suppressed, whereas the sweetness of saccharin was enhanced (Pangborn et al. 1973). Another study (Pangborn and Szczesniak 1974) dealt with the effect of hydrocolloids (hydroxypropyl cellulose, sodium alginate, xanthan, and sodium carboxymethylcellulose of low and medium viscosity(on a number of aromatic flavor compounds (Table 3.1 and Figure 3.11) found in natural and prepared foods (i.e. acetaldehyde, acetophenone, butyric acid, and dimethyl sulfide). On the whole, the addition of

Table 3.1 Characteristics of tested flavor compounds.

Compound	Polarity	Boiling point (°C)	Molecular mass (g/mol)	Flavor and odor
Acetaldehyde	Polar	21.0	44.05	Characteristic, pungent
Butyric acid	Polar	163.5	88.10	Unpleasant, rancid, sour
Dimethyl sulfide	Non-polar	37.5	62.13	Disagreeable, cabbage-like, beet-like
Acetophenone	Non-polar	202.3	120.14	Orange blossom-like, sweet

Source: Adapted with changes from Pangborn and Szczesniak (1974) and after Arctander (1969).

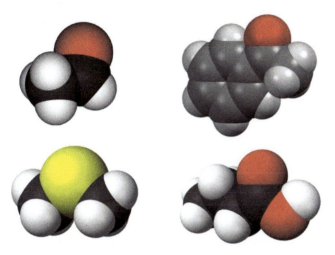

Figure 3.11 Acetaldehyde (top left), acetophenone (top right), dimethyl sulfide (bottom left) and butyric acid (bottom right).

hydrocolloids decreased both odor and flavor intensities, with dimethyl sulfide being affected the most and acetophenone the least. The flavor of butyric acid was more affected by the hydrocolloids than its odor. For acetaldehyde, the overall effect of the hydrocolloids was to decrease its odor but increase its flavor (Pangborn and Szczesniak 1974). In general, the effects were independent of viscosity and specific for the gum/odorant combination. A relationship could be postulated between the physicochemical nature of the odorant and the frequency/intensity of the observed effects (Pangborn and Szczesniak 1974).

The effect of hydrocolloids on the apparent viscosity and sensory properties of selected beverages was studied (Pangborn et al. 1978). The same hydrocolloids as in Pangborn and Szczesniak (1974) were used. Three commercial beverages – orange drink, instant coffee and tomato juice – were tested. The orange drink and tomato juice were tested at 0°C and 22°C, whereas the coffee was tested at 22°C and 60°C, by 11–14 highly trained tasters (Pangborn et al. 1978). As a result of precipitate formation, it was not possible to check sodium alginate in the orange drink nor hydroxypropyl cellulose in coffee at 60 °C. Aside from these, and without exception, increasing the hydrocolloid concentration considerably diminished the flavor and aroma intensities of all beverages at both test temperatures. Taste effects were specific for the gum/beverage blends. Overall, gums depressed the sourness and saltiness of tomato juice, the sourness of orange drink and the bitterness of coffee (Pangborn et al. 1978).

A more recent study dealt with the effect of some hydrocolloid blends on the viscosity and sensory properties of raspberry juice-milk (Abedi et al. 2014) – a mixed fruit-milk beverage with positive health effects (Pszczola 2005) – as well as to improve iron availability (Nebot et al. 2010). Hydrocolloids are frequently included in such acidic milk drinks as stabilizers, to prevent protein aggregation (Tuinier et al. 2002; Saha and Bhattacharya 2010; Sakhale et al. 2011; Sanyal et al. 2011). A major challenge in developing juice-milk beverages is their sensory grade upon testing (Lee and Coates 2003). As has been observed for different drinks, hydrocolloids influence sensory properties (Soukoulis et al. 2008; Azarikia and Abbasi 2009). The sensory properties evaluated for the raspberry juice-milk were taste, consistency, odor and overall acceptability (Abedi et al. 2014). It was observed that when viscosity increased, odor decreased. In the presence of pectin or κ-carrageenan, sedimentation of samples including carboxymethylcellulose decreased. In general, samples that included only pectin demonstrated higher viscosity, lower sedimentation and better sensorial evaluation of the beverage, and gum concentration affected the stability of the raspberry juice-milk drink (Abedi et al. 2014).

3.6 Influence of Hydrocolloids on the Release of Volatile Flavor Compounds

Hydrocolloids can modify flavor profile and/or perception (Cook et al. 2003a; van Ruth and King 2003; Juteau et al. 2004). Flavor perception can be determined by assessing the sensory responses to different aroma compounds and levels as a function of time (Taylor 1998). The aspects that govern flavor release from food products are mass transport and phase partitioning (De Roos 2000). The partition coefficient K_{ap} defines the maximum potential extent of flavor release:

$$K_{ap} = \frac{C_a}{C_p}$$

where C_a is the flavor concentration in the air phase above the product, and C_p is the flavor concentration in the product at equilibrium. The degree of interaction between flavor compounds and the matrix can be estimated from the analysis of the equilibrium headspace concentration above the product. Under non-equilibrium conditions, as occur during food consumption, mass transfer plays a part in flavor release, together with partitioning, and describes the rate at which aroma compounds are transferred from one environment to another (Bylaite et al. 2004). Consequently, familiarity with the aroma compounds' behavior within the food and of their partitioning rates is relevant in the flavoring of foods throughout product development, production, and consumption (Bylaite et al. 2004). The effects of a number of hydrocolloids on partitioning and release of aroma compounds have been studied (Rankin and Bodyfelt 1996; Roberts et al. 1996; Yven et al. 1998; Hansson et al. 2003), demonstrating their diverse impacts on flavor release – from no effect to large declines in aroma headspace concentrations (Rankin and Bodyfelt 1996; Bylaite et al. 2003). Nevertheless, the influence and magnitude of flavor retention by hydrocolloids differed depending on four parameters: hydrocolloid type; the involved aroma compounds; the composition of the model system; and the analytical methods used in the research (Bylaite et al. 2004). Perception studies showed that aroma and taste were less strongly perceived in carrageenan-thickened systems than in gelatin, gellan or starch systems with equivalent rheological properties (Guinard and Marty 1995; Costell et al. 2000). The effect of changes in λ-carrageenan and macroscopic viscosity (i.e. viscosity of a system in which water mobility remains nearly unchanged) on the release of aroma compounds under static and dynamic conditions was studied in viscous solutions (Bylaite et al. 2004). The differences in release behaviors of 43 aroma compounds (aldehydes, esters, ketones, and alcohols) with dissimilar carbon chain lengths, functional groups and chemical classes were estimated in water and viscous solutions thickened with λ-carrageenan at four different concentrations (0.1, 0.25, 0.4, and 0.5% w/w) and 10% (w/w) sucrose. Dynamic headspace analyses with only molecular diffusion were performed in combination with gas chromatography (Bylaite et al. 2004). The analytical methods included: (i) static headspace gas chromatography (i.e. liquid–vapor partition coefficients of the aroma compounds in water and thickened solutions) (Kolb and Ettre 1997); (ii) dynamic headspace gas chromatography (i.e. the release rates of the different aroma compounds) as described previously (Bylaite et al. 2003); and (iii) determination of release-rate constants (i.e. determining the dynamics of the flavor release throughout the collection of aroma compounds on traps) (Roberts and Acree 1995). The release-rate constants (k) were calculated from the rate of adsorption on the traps in accordance with Roberts and Acree (1995):

$$k = \frac{\left(\frac{d[A]}{dt}\right)_{trap}}{[A_0]}$$

where $[A]$ is the concentration of the volatile in the liquid phase (mg/L), $[A_0]$ is the initial concentration of the volatile in the liquid phase (mg/L), and t is time (minutes). The amounts of aroma compounds collected on the traps were negligible compared to the amounts remaining in the liquid phase, and it was therefore assumed that the concentration of aroma compounds in the liquid phase throughout the volatile collection was equal to the initial concentration. Thus, the slopes of the $\left(\frac{d[A]}{dt}\right)_{trap}$ graph from plotting micrograms of flavor compounds collected from the trap versus time were then determined (Roberts and Acree 1995). It was observed that k (37°C) increased as the carbon chain increased within each homologous series. Esters revealed the highest volatility, followed by aldehydes, ketones, and alcohols. At equilibrium, no inclusive effect of λ-carrageenan was observed, except with the most hydrophobic compounds (Bylaite et al. 2004). Analysis of flavor release under non-equilibrium conditions exposed the dominant effect of λ-carrageenan on the release rates of aroma compounds, and on the magnitude of the decrease in release rates depended on the physicochemical characteristics of the aroma compounds, with the strongest effect for the most volatile compounds (Bylaite et al. 2004). Nevertheless, none of these effects were comparable in extent to the achieved changes in macroscopic viscosity, and the suppressive effects were consequently attributable to the thickener and not to the physical properties of the increasingly viscous systems (Bylaite et al. 2004).

3.7 Gels and Flavor

3.7.1 Hydrocolloid Gels and Flavor Release

Numerous studies have been conducted to determine the effect of hydrocolloid gel texture on flavor release, with inconsistent results (Koliandris et al. 2008). Atmospheric pressure chemical ionization-mass spectrometry (APCI-MS) can be used to follow volatile release in the exhalation of an individual who is consuming food. Sensory perception is then related to the rate of change of the in-nose volatile concentration during the consumption of model gels (Baek et al. 1999; Weel et al. 2002). To perform these studies, a device that samples air from the nose is inserted into one nostril. The sampled volatile compounds are then ionized by APCI. The formed ions are then detected in a quadrupole mass spectrometer (Taylor et al. 2000). Use of this technique confirmed that soft gelatin gels release higher concentrations of volatile compounds than the harder gelatin gels (Linforth and Taylor 1999).

The influence of gelatin and pectin gel texture on strawberry flavor release and perception was studied (Boland et al. 2006). The type of hydrocolloid affected static and in-nose compound concentrations considerably. The pectin gels demonstrated lower air/gel partition coefficients than the gelatin gels, but increased flavor release (Boland et al. 2006). Increased gel rigidity gave rise to lower air/gel partition coefficients; higher maximum concentrations of volatiles and lower release rates throughout in-nose analysis; diminished perception of odor, strawberry flavor and sweetness; and higher intensity ratings for thickness in sensory analyses. Subsequently, both hydrocolloid type and rigidity of the sample significantly affected flavor release and perception (Boland et al. 2006). The release of 11 flavor compounds from gelatin, starch and pectin gels was investigated (Boland et al. 2004). The gels were characterized by Young's modulus of elasticity. Static headspace analysis was performed to define the partition coefficients of the compounds. Model mouth/proton transfer reaction-mass spectrometry analysis produced flavor-release profiles (Boland et al. 2004).

Gelatin gel, which was the most rigid moiety, had the lowest partition coefficient for six compounds and the lowest maximum concentrations released for all compounds. Even though the rigidity of the starch and pectin gels was not noticeably different, there was greater release of hydrophilic compounds from pectin gels compared to starch gels (Boland et al. 2004). The opposite was observed for hydrophobic compounds. These results specified the matrix–volatile interactions taking place in the starch and pectin gels. The gelatin gel showed large increases in flavor release in saliva, whereas the starch and pectin gels demonstrated a decrease in flavor release (Boland et al. 2004). In another study, it was also reported that firm gelatin and carrageenan gels release flavor with a lower I_{max} (i.e. maximum volatile intensity) than soft or medium gels, and that both texture and gelling agent affect flavor release (Guinard and Marty 1995). Apparent flavor was related to gel rheology (Morris 1994). A strong negative correlation between the magnitude of the strain and perception was detected. The proposed mechanism was accelerated release of tastants and volatiles as a result of the greater exposure of fresh surfaces by fracture of the gel during chewing for the more brittle gels (Morris 1994).

To provide further insight into the relationship between the structure of hydrocolloid gels and the perception of taste and flavor, gels were prepared from mixtures of high- and low-acetyl gellan as well as mixed κ-carrageenan/locust bean gum gels. They were flavored with ethyl butyrate. The gels were classified by rheological measurements into three categories: strong/brittle, intermediate and soft/elastic (Koliandris et al. 2008). Volatile release was measured by monitoring nose-space volatile concentrations during consumption by APCI-MS, as well as headspace concentrations. The latter did not exhibit significant differences between the gel systems but the release of ethyl butyrate in the nose was affected by the matrix, showing a higher intensity for the more brittle gels containing high levels of low-acetyl gellan (Koliandris et al. 2008). The more brittle gels containing high levels of low-acetyl gellan and a high amount of κ-carrageenan exhibited significantly higher release of Na^+. Strain at break was inversely correlated with salt release and more weakly inversely correlated with the maximum nose-space volatile concentration. It was concluded that the intensity of flavor perception in hydrocolloid gels is dominated by release of the tastant and by a low strain at break (Koliandris et al. 2008).

3.7.2 Phase-Separated Gels and Aroma Release

Microstructures and textures of gelled products stem from opposing processes – phase separation and gelation – which take place simultaneously (van de Velde et al. 2015). Textural properties and microstructure of phase-separated gelatin–starch gels have been described (Firoozmand and Rousseau 2013), but their impact on flavor release is less understood. Twenty years ago, Taylor et al. (2001) studied the link between phase-separated gels and aroma release. Release of compounds with different physicochemical properties were investigated *in vivo*. There was no noticeable difference in the maximum intensity of release and the time to maximum-intensity release among the gels with different microstructures (Taylor et al. 2001). In a more recent study, flavor distribution and release from gelatin–starch matrices were investigated (Su et al. 2021). It was concluded that modifying the phase volumes of a biphasic gelatin–starch gel to obtain different microstructures and textures has no notable influence on aroma release *in vivo*. Fortification of this mixture with sucrose affected the microstructure, as less aggregation of the polysaccharide was observed. On the whole, less hydrophobic and more volatile compounds were not significantly affected by changes in sucrose concentration compared to the more hydrophobic, less volatile compounds (Su et al. 2021). In other words, if a flavor mixture comprises more than a few key aroma compounds that are on the hydrophobic end of the scale, it is essential to consider how varying the structure of the matrix will affect the release of these compounds, and more importantly, perception of overall flavor of the altered product (Su et al. 2021).

3.8 The Influence of Flavor Molecules on the Behavior of Hydrocolloids

Most of the literature emphasizes the effect of hydrocolloids on flavor release and perception. Nevertheless, from a practical perspective, it is just as important to understand the influence of flavor molecules on the behavior of hydrocolloids (Erdem and Ak 2021). A comparatively small number of studies have reported on how the presence of ingredients influences gelation kinetics (Jørgensen et al. 2007; Zhang et al. 2008; Can Karaca et al. 2018). Aside from conferring sensory characteristics, flavor molecules are often incorporated into alginate-based products for their antimicrobial and antioxidant properties (Pranoto et al. 2005; Rojas-Graü et al. 2007; Riquelme et al. 2017). It is reasonable to hypothesize that gelation characteristics of hydrocolloids will be altered in the presence of flavor molecules. It was reported that for high-methoxy pectin gels, the presence of limonene together with a surfactant (Tween 80) predominantly enhances the early stage of network formation, as shown by shorter gelling times (Monge et al. 2008). Furthermore, a marked reduction was observed in the storage modulus of fully developed pectin gels as a result of the addition of tutti-frutti flavor, i.e. an artificial or natural flavoring simulating the combined flavor of many different fruit (Monge et al. 2004). To study an alginate gel blank, a hydrocolloid solution with added $CaCO_3$ followed by addition of freshly prepared GDL (glucono-delta-lactone) was prepared (Can Karaca et al. 2018). The flavored alginate solution included the blank with separate addition of either 2-acetyl pyridine, benzaldehyde, eugenol, or isoamyl acetate before the carbonate addition (Erdem and Ak 2021). The number of flavor molecules added was lower than their aqueous solubility limit and gelation kinetics was studied by rheological measurements at 10, 20, 30, and 40 °C (Erdem and Ak 2021). The flavor molecules significantly changed the gelation characteristics and gelation time of alginate dispersions and the properties of the resulting alginate gels. In addition, surface-tension values of the gelation media were reduced. The most notable reduction in surface tension was observed with eugenol, which quantitatively, functioned like a conventional food surfactant. The surfactant characteristic of flavor molecules may enable innovative use of such compounds in different applications (Erdem and Ak 2021).

3.9 Demonstrating the Use of Hydrocolloids in Modifying Food Taste/Odor

The preparation of foods with taste/odor modified by hydrocolloids is demonstrated in this section using two examples. In the first recipe, the batter for fried chicken includes MC. In the second recipe, HPMC is used to produce gluten-free bread. Most polysaccharides used in food systems are highly hydrophilic in nature and have little tendency to adsorb at oil–water interfaces. Exceptions include certain tree gum exudates such as gum arabic and gum ghatti, certain pectins, and some chemically modified polysaccharides such as MC and HPMC, propylene glycol alginate, and octenyl succinylated starch (Phillips and Williams 2021). In general, the family of commercial water-soluble cellulose ethers is comprised of MC and HPMC, among other derivatives (Davidson 1980). As stated, MC (E461) and HPMC (E464) are water-soluble polymers derived from the etherification of cellulose [the E number ("E" stands for Europe) is a code for substances used as food additives, including those found naturally in many foods; https://en.wikipedia.org/wiki/E_number]. They have the useful properties of thickening, thermal gelation, acting as a surfactant, film formation, and adhesion (Davidson 1980), prompting their application in areas such as foods, agriculture, adhesives, cosmetics, construction, paints, paper, pharmaceuticals, textiles, and tobacco products. In addition, to manufacture a product for a specific aim, the properties of MC and HPMC can be modified by changing their molecular weight or relative amounts of etherifying reagents (Davidson 1980). The viscosity of both MC and HPMC increases with temperature (gelation may occur), but is not influenced by the addition of electrolytes or pH (Phillips and Williams 2021). In other words, MC and HPMC possess the rather unusual property of solubility in cold water and insolubility in hot water, so that when a solution is heated, a 3D gel structure is formed. By modifying production techniques and by altering the ratios of methyl and hydroxypropyl substitutions, it is possible to obtain products with thermal gelation temperatures ranging from 50 to 90 °C and gel textures ranging from firm to rather mushy (Davidson 1980). HPMC is surface-active, with hydrophobic (methyl) and hydrophilic (hydroxypropyl) groups distributed along the cellulose backbone, and can be adsorbed at oil–water interfaces (Camino and Pilosof 2011; Camino et al. 2011). The interfacial behavior of HPMC is strongly influenced by the chemical nature of the dispersed phase and its functionality is more similar to that of proteins than to that of other surface-active polysaccharides. MC also exhibits interfacial activity and facilitates emulsion formation, enhancing emulsion stability by forming a protective layer around the droplet; the most efficient stabilization is obtained with the lower molar mass MC (Nasatto et al. 2014).

Figure 3.12 demonstrates a recipe for fried chicken using a batter composed of cornstarch, wheat flour, vegetable oil, salt, paprika, MC, white pepper, garlic powder, and water. The marinated chicken is dipped into the batter before it is deep-fried in oil at 175 °C for five minutes. MC is an odorless and tasteless food-grade gum which has great film-forming characteristics and is resistant to fat and oil transmission, in addition to oxygen transmission. These properties result in reduced final oil uptake by the coated chicken piece as the MC forms a thermally induced gelatinous coating (http://thefriedchickenblog.blogspot.com/). Throughout the cooking or frying process, MC coagulates and forms a protective layer or barrier between the batter and breading and the meat. This coagulation effect binds the meat and the batter for the duration of frying and explains the higher percentage of moisture retained in the internal meat sample. Moreover, MC binds up to 40 times its weight in water. Consequently, this MC layer can help prevent moisture loss from the meat, as well as excess oil uptake. Thus, overall, adding MC to batter can improve yield, decrease moisture loss and decrease fat uptake (http://thefriedchickenblog.blogspot.com). The special batter (and breading) brings a completely new dimension in taste and texture to fried foods. In addition, such batters keep the final food products crispy and crunchy while they sit under heat lamps waiting to be served (http://thefriedchickenblog.blogspot.com/2013/12/the-science-behind-methylcellulose.html). MC and HPMC also improve the coating's adhesion to the food matrix and help control batter pickup and the adhesion of seasoning and flavoring agents on the food surface. Therefore, improved taste and flavor is another advantage of the MC-fortified batter (Ergun et al. 2016).

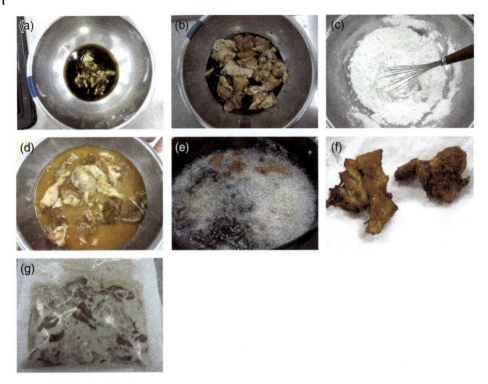

Figure 3.12 (a) Making marinade, (b) marinating cut chicken, (c) making batter, (d) dipping chicken in the batter, (e) deep-frying. (f) Serve. (g) Chicken covered with batter in a storage bag.

Figure 3.13 and the detailed recipe describe the preparation of gluten-free bread that includes HPMC. In addition to the HPMC powder, the dough contains rice flour, cornstarch, salt, white sugar, skim milk powder, shortening, instant dry yeast, and water. Gluten is a protein complex found in the Triticeae tribe of grains, which includes wheat, barley and rye. The gluten content in wheat flour provides the required organoleptic properties, such as texture and taste, to numerous bakery and other food items. Gluten also delivers the necessary processing quality for both the commercial food manufacturer and the home baker. Thus, it is very challenging to prepare bread using gluten-free flours, because gluten is considered by many to be the "heart and soul" of baked and other food products (https://patents.google.com/patent/EP3592151A1/en). The use of HPMC in a dough composition containing gluten-free flour is well known. A dough comprised of gluten-free cereal flour, a water-soluble cellulose ether, and a low-substitution cellulose ether with a molar substitution of 0.05–1.0 is provided herein. The resultant bread has a good mouthfeel and a satisfactory volume, retains softness over time, and can be consumed by anyone who has food allergies to wheat or the like (Fukasawa 2005). Another study also concluded that in addition to good mouthfeel, HPMC is favored for gluten-free baked products for the acceptable levels of volume and crumb texture that it confers, and the possibility of using the highest proportion of gluten-free flour (Osella et al. 2014).

Figure 3.13 (a) Dispersing dry yeast, (b) mixing all ingredients, (c) covering the dough with plastic wrap. (d) Dough after fermentation. (e) Putting dough into a square container. (f) Dough after the second fermentation. (g) Baked bread.

3.9.1 Fried Chicken with Methylcellulose (MC)

(Serves 5) (Figure 3.12)

600–1000 g chicken (*Hint 1*)
50 mL (⅕ cup) soy sauce
50 mL (⅕ cup) *sake*
20 g ginger
48 g (a little over ⅓ cup) cornstarch
16 g (a little under 2 Tbsp) wheat flour
5 g (a little over 1 tsp) vegetable oil
4 g (⅔ tsp) salt
2 g (1 tsp) sweet paprika
1.2 g (⅗ tsp) MC powder
1 g (½ tsp) white pepper
1 g (⅖ tsp) garlic powder
150 mL (⅗ cup) water
Vegetable oil for deep-frying
pH of batter = 4.57

For preparation, see Figure 3.12

1) To make the marinade: grind ginger, and put ground ginger, soy sauce and sake in a bowl (Figure 3.12a).
2) Cut chicken into bite-sized pieces and marinate in sauce (1) for at least 30 minutes in the refrigerator (Figure 3.12b, *Hint 2*).
3) To make the batter: mix cornstarch, wheat flour, vegetable oil, salt, paprika, MC, white pepper, and garlic powder in a bowl with a whisk (Figure 3.12c), and add water gradually with mixing (*Hint 3*).
4) Dip marinated chicken (2) in the batter (3) (Figure 3.12d, *Hint 4*).
5) Heat vegetable oil for deep-frying to 175 °C, and deep-fry the battered chicken (4) (Figure 3.12e) for five minutes (Figure 3.12f).

Preparation hints:

1) You can use any parts of the chicken.
2) Knead chicken in the sauce for better flavor.
3) Do not let lumps form.
4) You can store the batter or the chicken covered with batter in the freezer (Figure 3.12g).

3.9.2 Gluten-Free Bread with Hydroxypropyl Methylcellulose (HPMC)

(Small loaf) (Figure 3.13)

120 g (1 cup) rice flour
30 g (a little over 3 Tbsp) cornstarch
3 g (½ tsp) salt
12 g (1⅓ Tbsp) white sugar
6 g (1 Tbsp) skim milk powder
10.5 g (a little under 1 Tbsp) shortening
3 g (1½ tsp) HPMC powder
6 g (2 tsp) instant dry yeast
135 mL (a little over ½ cup) lukewarm water (40 °C)

For preparation, see Figure 3.13

1) Put dry yeast in lukewarm water and mix (Figure 3.13a).
2) Put all ingredients in a bowl and mix with an electrical hand mixer for three minutes (Figure 3.13b).
3) Cover the dough (2) with plastic wrap (Figure 3.13c) and let rise in a 40 °C oven for 20 minutes.
4) After the fermentation (Figure 3.13d), mix the dough with a spatula to remover the air.
5) Put the dough into a square container (Figure 3.13e) and let rise in a 40 °C oven for 30 minutes (Figure 3.13f).
6) Preheat the oven at 210 °C, and bake the dough for 20 minutes (Figure 3.13g).

References

Abedi, F., Sani, A.M., and Karazhiyan, H. (2014). Effect of some hydrocolloids blend on viscosity and sensory properties of raspberry juice-milk. *Journal of Food Science and Technology* 51: 2246–2250.

Andriot, I., Harrison, M., Fournier, N., and Guichard, E. (2000). Interactions between methyl ketones and β-lactoglobulin: sensory analysis, headspace analysis, and mathematical modeling. *Journal of Agricultural and Food Chemistry* 48: 4246–4251.

Arctander, S. (1969). *Perfume and Flavor Chemicals*. Montclair, NJ: Published by the author.

Axel, R. (1995). The molecular logic of smell. *Scientific American* 273: 154–159.

Azarikia, F. and Abbasi, S. (2009). On the stabilization mechanism of Doogh (Iranian yoghurt drink) by gum tragacanth. *Food Hydrocolloids* 50: 87–94.

Baek, I., Linforth, R.S.T., Blake, A., and Taylor, A.J. (1999). Sensory perception is related to the rate of change of volatile concentration in-nose during eating of model gels. *Chemical Senses* 24: 155–160.

Bakker, J. (1995). Flavor Interactions with the food matrix and their effects on perception. In: *Ingredient Interactions. Effect of Food Quality* (ed. A.G. Gaonkar), 411–439. New York: Marcel Dekker Inc.

Bascuas, S., Espert, M., Llorca, E. et al. (2021). Structural and sensory studies on chocolate spreads with hydrocolloid-based oleogels as a fat alternative. *LWT – Food Science and Technology* 135: 110228.

Beyeler, M. and Solms, J. (1974). Interaction of flavor model compounds with soy protein and bovine serum albumin. *Lebensmittel-Wissenschaft & Technologie* 7: 217–219.

Boland, A.B., Buhr, K., Giannouli, P., and van Ruth, S.M. (2004). Influence of gelatin, starch, pectin and artificial saliva on the release of 11 flavour compounds from model gel systems. *Food Chemistry* 86: 401–411.

Boland, A.B., Delahunty, C.M., and van Ruth, S.M. (2006). Influence of the texture of gelatin gels and pectin gels on strawberry flavour release and perception. *Food Chemistry* 96: 452–460.

Boutboul, A., Giampaoli, P., Feigenbaum, A., and Ducruet, V. (2001). Influence of the nature and treatment of starch on aroma retention. *Carbohydrate Polymers* 47: 73–82.

Buleon, A., Delage, M.M., Brisson, J., and Chanze, H. (1990). Single crystals of V Amylose complexed with isopropanol and acetone. *International Journal of Biological Macromolecules* 12: 25–33.

Bylaite, E., Meyer, A.S., and Adler-Nissen, J. (2003). Changes in macroscopic viscosity do not affect the release of aroma aldehydes from a pectinaceous food model system of low sucrose content. *Journal of Agricultural and Food Chemistry* 51: 8020–8026.

Bylaite, E., Ilgunaite, Z., Meyer, A.S., and Adler-Nissen, J. (2004). Influence of λ-carrageenan on the release of systematic series of volatile flavor compounds from viscous food model systems. *Journal of Agricultural and Food Chemistry* 52: 3542–3549.

Camino, N.A. and Pilosof, A.M.R. (2011). Hydroxypropylmethyl cellulose at the oil–water interface. Part II. Submicron-emulsions as affected by pH. *Food Hydrocolloids* 25: 1051–1062.

Camino, N.A., Sánchez, C.C., Rodríguez Patino, J.M., and Pilosof, A.M.R. (2011). Hydroxypropylmethyl cellulose at the oil–water interface. Part I. Bulk behavior and dynamic adsorption as affected by pH. *Food Hydrocolloids* 25: 1–11.

Can Karaca, A., Erdem, I.G., and Ak, M.M. (2018). Effects of polyols on gelation kinetics, gel hardness, and drying properties of alginates subjected to internal gelation. *Lebensmittel-Wissenschaft & Technologie* 92: 297–303.

Christensen, C.M. (1980). Effects of solution viscosity on perceived saltiness and sweetness. *Perception and Psychophysics* 28: 347–353.

Cook, D.J., Hollowood, T.A., Linforth, R.S.T., and Taylor, A.J. (2002). Perception of taste intensity in solutions of random-coil polysaccharides above and below c*. *Food Quality and Preference* 13: 473–480.

Cook, D., Linforth, R.S.T., and Taylor, A. (2003a). Effects of hydrocolloid thickeners on the perception of savory flavors. *Journal of Agricultural and Food Chemistry* 51: 3067–3072.

Cook, D.J., Hollowood, T.A., Linforth, R.S.T., and Taylor, A.J. (2003b). Oral shear stress predicts flavour perception in viscous solutions. *Chemical Senses* 28: 11–23.

Costell, E., Peyrolon, M., and Duran, L. (2000). Influence of texture and type of hydrocolloid on perception of basic tastes in carrageenan and gellan gels. *Food Science and Technology International* 6: 495–499.

Damodaran, S. and Kinsella, J.E. (1980). Flavor protein interactions. Binding of carbonyls to bovine serum albumin: thermodynamic and conformational effects. *Journal of Agricultural and Food Chemistry* 28: 567–571.

Damodaran, S. and Kinsella, J.E. (1981). Interaction of carbonyls with soy protein: thermodynamic effects. *Journal of Agricultural and Food Chemistry* 29: 1249–1253.

Davidson, R.L. (1980). *Handbook of Water-Soluble Gums and Resins*, 13–11 to 13.16. New York: McGraw-Hill.

De Roos, K.B. (2000). Physiochemical models of flavor release from foods. In: *Flavor Release* (ed. D.D. Roberts and A.J. Taylor), 126–141. Washington, DC: American Chemical Society.

Demirkesen, I. and Mert, B. (2019). Recent developments of oleogel utilizations in bakery products. *Critical Reviews in Food Science and Nutrition* 60: 2460–2479.

Dufour, C. and Bayonove, C.L. (1999). Interactions between wine polyphenols and aroma substances. An insight at the molecular level. *Journal of Agricultural and Food Chemistry* 47: 678–684.

Dufour, E. and Haertlé, T. (1990). Binding affinities of beta-ionone and related flavor compounds to beta-lactoglobulin: effects of chemical modifications. *Journal of Agricultural and Food Chemistry* 138: 1691–1695.

Ebeler, S.E., Pangborn, R.M., and Jennings, W.G. (1988). Influence of dispersion medium on aroma intensity and headspace concentration of menthone and isoamyl acetate. *Journal of Agricultural and Food Chemistry* 36: 791–796.

El Hadi, M.A., Zhang, F.J., Wu, F.F. et al. (2013). Advances in fruit aroma volatile research. *Molecules* 18: 8200–8229.

Erdem, I.G. and Ak, M.M. (2021). Gelation characteristics of sodium alginate in presence of flavor molecules. *Journal of Food Processing and Preservation* 45: e15033.

Ergun, R., Guo, J., and Huebner-Keese, B. (2016). Cellulose. In: *Encyclopedia of Food and Health* (ed. B. Caballero, P.M. Finglas and F. Toldrá), 694–702. Waltham, MA: Academic Press.

Escher, F.E., Nuessli, J., and Conde-Petit, B. (2000). Interactions of flavor compounds with starch in food processing. In: *Flavor Release* (ed. D.D. Roberts and A.J. Taylor), 230–245. Washington, DC: American Chemical Society.

Espert, M., Wiking, L., Salvador, A., and Sanz, T. (2020). Reduced-fat spreads based on anhydrous milk fat and cellulose ethers. *Food Hydrocolloids* 99: 105330.

Farkye, N.Y. (2004). Cheese technology. *The International Journal of Dairy Technology* 57: 91–98.

Fayaz, G., Goli, S.A.H., Kadivar, M. et al. (2017). Potential application of pomegranate seed oil oleogels based on monoglycerides, beeswax and propolis wax as partial substitutes of palm oil in functional chocolate spread. *Lebensmittel-Wissenschaft & Technologie* 86: 523–529.

Firoozmand, H. and Rousseau, D. (2013). Microstructure and elastic modulus of phase-separated gelatin–starch hydrogels containing dispersed oil droplets. *Food Hydrocolloids* 30: 333–342.

Fleming, A. (2013). *Umami: Why the Fifth Taste Is So Important*. London, UK: The Guardian https://www.theguardian.com/lifeandstyle/wordofmouth/2013/apr/09/umami-fifth-taste (accessed 24 February, 2022).

Fukasawa, M. (2005). Gluten-free dough composition. European Patent Application EP 1 561 380 A1.

Gardiner, M.B. (2005). The importance of being cilia. *HHMI Bulletin* 18: 32–36.

Glicksman, M. and Farkas, E. (1966). Gums in artificially sweetened foods. *Food Technology* 20: 58–61.

Godet, M.C., Bizot, H., and Buléon, A. (1995). Crystallization of amylose–fatty acid complexes. *Carbohydrate Polymers* 27: 47–52.

Godshall, M.A. (1997). How carbohydrates influence food flavor. *Food Technology* 51: 63–67.

Goubet, I., Le Quere, J.L., and Voilley, A.J. (1998). Retention of aroma compounds by carbohydrates: influence of their physicochemical characteristics and of their physical state. A review. *Journal of Agricultural and Food Chemistry* 46: 1981–1990.

Grabowska, K.J., Tekidou, S., Boom, R.M., and Goot, A.J.V.D. (2014). Shear structuring as a new method to make anisotropic structures from soy–gluten blends. *Food Research International* 64: 743–751.

Gremli, H.A. (1974). Interaction of flavour compounds with soy protein. *The Journal of the American Oil Chemists' Society* 51: 95A–97A.

Guichard, E. (2002). Interactions between flavor compounds and food ingredients and their influence on flavor perception. *Food Reviews International* 18: 49–70.

Guichard, E. and Langourieux, S. (2000). Interactions between beta-lactoglobulin and flavour compounds. *Food Chemistry* 71: 301–308.

Guinard, J.X. and Marty, C. (1995). Time-intensity measurements of flavor release from a model gel system: effect of gelling type and concentration. *Journal of Food Science* 60: 727–730.

Guo, Z., Teng, F., Huang, Z. et al. (2020). Effects of material characteristics on the structural characteristics and flavor substances retention of meat analogs. *Food Hydrocolloids* 105: 105752.

Hansen, A.P. and Booker, D.C. (1996). Flavor interaction with casein and whey protein. In: *Flavor–Food Interactions* (ed. R.G. McGorrin and J.V. Leland), 75–89. Washington, DC: American Chemical Society.

Hansen, A.P. and Heinis, J.J. (1991). Decrease of vanillin flavor perception in the presence of casein and whey proteins. *Journal of Dairy Science* 74: 2936–2940.

Hansen, A.P. and Heinis, J.J. (1992). Benzaldehyde, citral, and d-limonene flavor perception in the presence of casein and whey proteins. *Journal of Dairy Science* 75: 1211–1215.

Hansson, A., Leufvén, A., Pehrson, K., and van Ruth, S. (2003). Partition and release of 21 aroma compounds during storage of a pectin gel system. *Journal of Agricultural and Food Chemistry* 51: 2000–2005.

Harvey, B., Druaux, C., and Voilley, A. (1995). Effect of protein on the retention and transfer of aroma compounds at the lipid–water interface. In: *Food Macromolecules and Colloids* (ed. E. Dickinson and D. Lorient), 154–163. Cambridge, UK: The Royal Society of Chemistry.

Hau, M.Y.M., Gray, D.A., and Taylor, A.J. (1994). Binding of volatiles to starch. In: *Flavor– Food Interactions* (ed. R.J. McGorrin and J.V. Leland), 109–117. Washington, DC: American Chemical Society.

Hau, M.Y.M., Gray, D.A., and Taylor, A.J. (1998). Binding of volatiles to extruded starch at low water contents. *Flavour and Fragrance Journal* 13: 77–84.

Haugeneder, A., Trinkl, J., Härtl, K. et al. (2018). Answering biological questions by analysis of the strawberry metabolome. *Metabolomics* 14: 145. https://doi.org/10.1007/s11306-018-1441-x.

Hauzoukim, K.A., Martin, X., Nagalakshmi, K. et al. (2019). Development of enrobed fish products: improvement of functionality of coated materials by added aquatic polymers. *Journal of Food Process Engineering* 42: 1–9.

Hummel, J.P. and Dreyer, W.J. (1962). Measurement of protein binding phenomena by gel filtration. *Biochimica et Biophysica Acta* 63: 530–532.

Hussain, I., Grandison, A.S., and Bell, A.E. (2012). Effects of gelation temperature on Mozzarella-type curd made from buffalo and cows' milk. 1: rheology and microstructure. *Food Chemistry* 134: 1500–1508.

Ilc, T., Werck-Reichhart, D., and Navrot, N. (2016). Meta-analysis of the core aroma components of grape and wine aroma. *Frontiers in Plant Science* 7: 1472. https://doi.org/10.3389/fpls.2016.01472.

Jiang, J., Coffey, P., and Toohey, B. (2006). Improvement of odor intensity measurement using dynamic olfactometry. *Journal of the Air & Waste Management Association* 56: 675–683.

Jiménez-Colmenero, F., Cofrades, S., López-López, I. et al. (2010). Technological and sensory characteristics of reduced/ low-fat, low-salt frankfurters as affected by the addition of konjac and seaweed. *Meat Science* 84: 356–363.

Johnson, J. and Clydesdale, F.M. (1982). Perceived sweetness and redness in colored sucrose solutions. *Journal of Food Science* 47: 747–754.

Jørgensen, T.E., Sletmoen, M., Draget, K.I., and Stokke, B.T. (2007). Influence of oligoguluronates on alginate gelation, kinetics, and polymer organization. *Biomacromolecules* 8: 2388–2397.

Juteau, A., Doublier, J.L., and Guichard, E. (2004). Flavor release from *ι*-carrageenan matrices: a kinetic approach. *Journal of Agricultural and Food Chemistry* 52: 1621–1629.

Kawakishi, S. and Kaneko, T. (1987). Interaction of proteins with allyl isothiocyanate. *Journal of Agricultural and Food Chemistry* 35: 85–88.

King, B.M. and Solms, J. (1982). Interactions of volatile flavor compounds with propyl gallate and other phenols as compared with caffeine. *Journal of Agricultural and Food Chemistry* 30: 838–840.

Koca, N. and Metin, M. (2004). Textural, melting and sensory properties of low fat fresh Kashar cheese produced by using fat replacers. *International Dairy Journal* 14: 365–373.

Kolb, B. and Ettre, L. (1997). *Static Headspace-Gas Chromatography, Theory and Practice*. New York: *Wiley-VCH*.

Koliandris, A., Lee, A., Ferry, A.-L. et al. (2008). Relationship between structure of hydrocolloid gels and solutions and flavour release. *Food Hydrocolloids* 22: 623–630.

Land, D.G. (1996). Perspectives on the effects of interactions on flavor perception: an overview. In: *Flavor–Food Interactions* (ed. R.J. McGorrin and J.V. Leland), 2–11. Washington, DC: American Chemical Society.

Langourieux, S. and Crouzet, J. (1994). Study of aroma compounds–polysaccharides interactions by dynamic exponential dilution. *Lebensmittel-Wissenschaft & Technologie* 27: 544–549.

Langourieux, S. and Crouzet, J.C. (1997). Study of interactions between aroma compounds and glycopeptides by a model system. *Journal of Agricultural and Food Chemistry* 45: 1873–1877.

Lee, H.S. and Coates, G.A. (2003). Effect of thermal pasteurization on Valencia orange juice color and pigments. *Journal of Food Science and Technology* 36: 153–156.

Lindemann, B., Ogiwara, Y., and Ninomiya, Y. (2002). The discovery of umami. *Chemical Senses* 27: 843–844.

Linforth, R.S.T. and Taylor, A.J. (1999). Apparatus and methods for the analysis of trace constituents of gases. *United States Patent* 5 (869): 344.

Locker, M. (2014). Breaking Breakfast News: Fruit loops are all the same flavor, after the mandela effect now known as Froot loops. TIME.com. https://time.com/1477/breaking-breakfast-news-froot-loops-are-all-the-same-flavor/ (accessed 30 January, 2022).

Lubke, M., Guichard, E., and Le Quéré, J.L. (1999). Infrared spectroscopic study of β-lactoglobulin interactions with flavour compounds. In: *Flavor Release* (ed. D.D. Roberts and A.J. Taylor), 282–292. Washington, DC: American Chemical Society.

Luebke, W. (2017). *Hexyl acetate, 142-92-7*. www.thegoodscentscompany.com (accessed 1 April, 2022).

Maier, H.G. (1970). Volatile flavoring substances in foodstuffs. *Angewandte Chemie International Edition* 9: 917–926.

Malkki, Y., Heinio, R.L., and Autio, K. (1993). Influence of oat gum, guar gum and carboxymethyl cellulose on the perception of sweetness and flavor. *Food Hydrocolloids* 6: 525–532.

de March, C.A., Ryu, S., Sicard, G. et al. (2015). Structure–odour relationships reviewed in the postgenomic era. *Flavour and Fragrance Journal* 30: 342–361.

Mariera, L., Owuoche, J.O., and Cheserek, M. (2017). Development of sorghum-wheat composite bread and evaluation of nutritional, physical and sensory acceptability. *African Journal of Food Science and Technology* 8: 113–119.

Monge, M.E., Bulone, D., Giacomazza, D. et al. (2004). Detection of flavor release from pectin gels using electronic noses. *Sensors and Actuators B* 101: 28–38.

Monge, M.E., Negri, R.M., Giacomazza, D., and Bulone, D. (2008). Correlation between rheological properties and limonene release in pectin gels using an electronic nose. *Food Hydrocolloids* 22: 916–924.

Monsan, P., Bozonnet, S., Albenne, C. et al. (2001). Homopolysaccharides from lactic acid bacteria. *International Dairy Journal* 11: 675–685.

Morris, E.R. (1987). Organoleptic properties of food polysaccharides in thickened systems. In: *Industrial Polysaccharides: Genetic Engineering, Structure/Property Relations and Applications* (ed. M. Yalpani), 225–238. Amsterdam: Elsevier.

Morris, E.R. (1994). Rheological and organoleptic properties of food hydrocolloids. In: *Food Hydrocolloids Structures, Properties, and Functions* (ed. E. Doi), 201–210. New York: Plenum Press.

Mottram, D.S. and Nobrega, C.C. (2000). Interaction between sulfur-containing flavor compounds and proteins in foods. In: *Flavor Release* (ed. D.D. Roberts and A.J. Taylor), 274–281. Washington, DC: American Chemical Society.

Mottram, D.S., Szaumanszumski, C., and Dodson, A. (1996). Interaction of thiol and disulfide flavor compounds with food components. *Journal of Agricultural and Food Chemistry* 44: 2349–2351.

Murtaza, M.S., Sameen, A., Huma, N., and Hussain, F. (2017). Influence of hydrocolloid gums on textural, functional and sensory properties of low fat cheddar cheese from buffalo milk. *Pakistan Journal of Zoology* 49: 27–34.

Nasatto, L.P., Pignon, F., Silveira, L.J. et al. (2014). Interfacial properties of methylcelluloses: the influence of molar mass. *Polymers* 6: 2961–2973.

Nebot, M.J.G., Alegría, A., Barberá, R. et al. (2010). Addition of milk or caseinophosphopeptides to fruit beverages to improve iron bioavailability? *Food Chemistry* 119: 141–148.

Nussinovitch, A. (1997). *Gum Technology in the Food and Other Industries*. UK: Chapman and Hall.

Nussinovitch, A. (2003). *Water-Soluble Polymer Applications in Foods*. Oxford: Blackwell Publishing.

O'Keefe, S., Wilson, L.A., Resurreccion, A.P., and Murphy, P.A. (1991). Determination of the binding of hexanal to soy glycinin and beta-conglycinin in an aqueous model system using a headspace technique. *Journal of Agricultural and Food Chemistry* 39: 1022–1028.

O'Neil, T.E. (1996). Flavor binding by food proteins: an overview. In: *Flavor–Food Interactions* (ed. R.G. McGorrin and J.V. Leland), 59–74. Washington, DC: American Chemical Society.

O'Neill, T.E. and Kinsella, J.E. (1987). Binding of alkanone flavors to beta-lactoglobulin: effects of conformational and chemical modification. *Journal of Agricultural and Food Chemistry* 35: 770–774.

Oaklander, M. (2015). A new taste has been added to the human palate. TIME.com. https://time.com/3973294/fat-taste-oleogustus/ (accessed 15 March, 2022).

Omohimi, C.I., Sobukola, O.P., Sarafadeen, K.O., and Sannil, O. (2014). Effect of thermo-extrusion process parameters on selected quality attributes of meat analogue from mucuna bean seed flour. *Nigerian Food Journal* 32: 21–30.

Osella, C., Torre, M., and Sanchez, H. (2014). Safe foods for celiac people. *Food and Nutritional Science* 5: 787–800.

Osen, R., Toelstede, S., Wild, F. et al. (2014). High moisture extrusion cooking of pea protein isolates: raw material characteristics, extruder responses, and texture properties. *Journal of Food Engineering* 127: 67–74.

Overbosch, P., Afterof, W.G.M., and Haring, P.G.M. (1991). Flavor release in the mouth. *Food Reviews International* 7: 137–184.

Palanisamy, M., Topfl, S., Aganovic, K., and Berger, R.G. (2018). Influence of iota carrageenan addition on the properties of soya protein meat analogues. *LWT – Food Science and Technology* 87: 546–552.

Pangborn, R.M. and Szczesniak, A.S. (1974). Effect of hydrocolloids and viscosity on flavor and odor intensities of aromatic flavor compounds. *Journal of Texture Studies* 4: 467–482.

Pangborn, R.M., Trabue, I.M., and Szczesniak, A.S. (1973). Effect of hydrocolloids on oral viscosity and basic taste intensities. *Journal of Texture Studies* 4: 224–241.

Pangborn, R.M., Gibbs, Z.M., and Tassan, C. (1978). Effect of hydrocolloids on apparent viscosity and sensory properties of selected beverages. *Journal of Texture Studies* 9: 415–436.

Patel, A.R., Rajarethinem, P.S., Grędowska, A. et al. (2014). Edible applications of shellac oleogels: spreads, chocolate paste and cakes. *Food and Function* 5: 645–652.

Pelletier, E., Sostmann, K., and Guichard, E. (1998). Measurement of interactions between beta-lactoglobulin and flavor compounds (esters, acids, and pyrazines) by affinity and exclusion size chromatography. *Journal of Agricultural and Food Chemistry* 46: 1506–1509.

Phillips, G.O. and Williams, P.A. (2021). *Handbook of Hydrocolloids*, 3e, Chapter 15, 481–507. Sawston, UK: Woodhead Publishing.

Poll, L. and Flink, J.M. (1984). Aroma analysis of apple juice: influence of salt addition on headspace volatile composition as measured by gas chromatography and corresponding sensory evaluations. *Food Chemistry* 13: 193–207.

Pranoto, Y., Salokhe, V.M., and Rakshit, S.K. (2005). Physical and antibacterial properties of alginate-based edible film incorporated with garlic oil. *Food Research International* 38: 267–272.

Pszczola, D.E. (2005). Ingredients. Making fortification. *Food Technology* 59: 44–61.

Rahimi, J., Khosrowshahi, A., Maddadlou, A., and Aziznia, S. (2007). Texture of low-fat Iranian white cheese as influenced by gum tragacanth as a fat replacer. *Journal of Dairy Science* 90: 4058–4070.

Rankin, S.A. and Bodyfelt, F.W. (1996). Headspace diacetyl as affected by stabilizers and emulsifiers in a model dairy system. *Journal of Food Science* 61: 921–923.

Reineccius, G. (2005). *Flavor Chemistry and Technology*, 2e. Boca Raton, FL: CRC Press.

Reiners, J., Nicklaus, S., and Guichard, E. (2000). Interactions between β-lactoglobulin and flavor compounds of different chemical classes. Impact of the protein on the odour perception of vanillin and eugenol. *Le Lait* 80: 347–360.

Allied Market Research (2021). https://www.globenewswire.com/news-release/2021/12/14/2351820/0/en/Global-Food-Flavors-Market-To-Reach-19-22-Billion-By-2030-Allied-Market-Research.html (accessed 20 March, 2022).

Riquelme, N., Herrera, M.L., and Matiacevich, S. (2017). Active films based on alginate containing lemongrass essential oil encapsulated: effect of process and storage conditions. *Food and Bioproducts Processing* 104: 94–103.

Roberts, D.D. and Acree, T.E. (1995). Simulation of retronasal aroma using a modified headspace technique: investigating the effects of saliva, temperature, shearing, and oil on flavour release. *Journal of Agricultural and Food Chemistry* 43: 2179–2186.

Roberts, D.D., Elmore, J.S., Langley, K.R., and Bakker, J. (1996). Effects of sucrose, guar gum, and carboxymethylcellulose on the release of volatile flavor compounds under dynamic conditions. *Journal of Agricultural and Food Chemistry* 44: 1321–1326.

Rogacheva, S., Espinosa-Diaz, M.A., and Voilley, A. (1999). Transfer of aroma compounds in water–lipid systems: binding tendency of β-lactoglobulin. *Journal of Agricultural and Food Chemistry* 47: 259–263.

Rogers, N.R., Mcmahon, D.J., Daubert, C.R. et al. (2010). Rheological properties and microstructure of cheddar cheese made with different fat contents. *Journal of Dairy Science* 93: 4565–4576.

Rojas-Graü, M.A., Avena-Bustillos, R.J., Olsen, C. et al. (2007). Effects of plant essential oils and oil compounds on mechanical, barrier and antimicrobial properties of alginate-apple puree edible films. *Journal of Food Engineering* 81: 634–641.

Rothe, M. and Specht, M. (1976). Notes about molecular weights of aroma compounds. *Nahrung* 20: 281–286.

Ruggiero, M.T., Sibik, J., Axel Zeitler, J., and Korter, T.M. (2016). Examination of L-glutamic acid polymorphs by solid-state density functional theory and terahertz spectroscopy. *Journal of Physical Chemistry A* 120: 7490–7495.

Rühmkorf, C., Rübsam, H., Becker, T. et al. (2012). Effect of structurally different microbial homoexopolysaccharides on the quality of gluten-free bread. *European Food Research and Technology* 235: 139–146.

van Ruth, S.M. and King, C. (2003). Effect of starch and amylopectin concentrations on volatile flavour release from aqueous model food systems. *Flavor and Fragrance Journal* 18: 407–416.

Rutschmann, M.A. and Solms, J. (1990a). Formation of inclusion complexes of starch with different organic compounds. II. Study of ligand binding in binary model systems with decanal, 1-naphthol, monostearate and monopalmitate. *Lebensmittel-Wissenschaft & Technologie* 23: 70–79.

Rutschmann, M.A. and Solms, J. (1990b). Formation of inclusion complexes of starch with different organic compounds. III. Study of ligand binding in binary model systems with (-) limonene. *Lebensmittel-Wissenschaft & Technologie* 23: 80–83.

Rutschmann, M.A. and Solms, J. (1990c). Formation of inclusion complexes of starch with different organic compounds. IV. Ligand binding and variability in helical conformations of V amylose complexes. *Lebensmittel-Wissenschaft & Technologie* 23: 84–87.

Saha, D. and Bhattacharya, S. (2010). Hydrocolloids as thickening and gelling agents in food: a critical review. *Journal of Food Science and Technology* 47: 587–597.

Sakhale, B.K., Badgujar, J.B., Pawar, V.D., and Sananse, S.L. (2011). Effect of hydrocolloids incorporation in casing of samosa on reduction of oil uptake. *Journal of Food Science and Technology* 48: 769–772.

Santana, P., Huda, N., and Yang, T.A. (2013). The addition of hydrocolloids (carboxymethylcellulose, alginate and konjac) to improve the physicochemical properties and sensory characteristics of fish sausage formulated with surimi powder. *Turkish Journal of Fisheries and Aquatic Sciences* 13: 561–569.

Sanyal, M.K., Pal, S.C., Gangopadhyay, S.K. et al. (2011). Influence of stabilizers on quality of sandesh from buffalo milk. *Journal of Food Science and Technology* 48: 740–744.

Schirle-Keller, J.P., Chang, H.H., and Reineccius, G.A. (1992). Interaction of flavor compounds with microparticulated proteins. *Journal of Food Science* 57: 1448–1451.

Seal, C.J., Jones, A.R., and Whitney, A.D. (2006). Whole grains uncovered. *Nutrition Bulletin* 31: 129–137.

Sébille, B., Thuaud, N., Piquion, J., and Behar, N. (1987). Determination of association constants of β-cyclodextrin and β-cyclodextrin-bearing polymers with drugs by high-performance liquid chromatography. *Journal of Chromatography* 409: 61–69.

Sébille, B., Zini, R., Vidal-Majdar, C. et al. (1990). Separation procedures used to reveal and follow drug-protein binding. *Journal of Chromatography* 531: 51–77.

Shankar, M.U., Levitan, C.A., and Spence, C. (2010). Grape expectations: the role of cognitive influences in color–flavor interactions. *Consciousness and Cognition* 19: 380–390.

Shaviklo, A.R. and Fahim, A. (2014). Quality improvement of silver carp fingers by optimizing the level of major elements influencing texture. *International Food Research Journal* 21: 283–290.

Small, D.M. and Green, B.G. (2012). A proposed model of a flavor modality. In: *The Neural Bases of Multisensory Processes* (ed. M.M. Murray and M.T. Wallace), Chapter 36, 717–739. Boca Raton, FL: CRC Press/Taylor & Francis.

Soukoulis, C., Chandrinos, I., and Tzia, C. (2008). Study of the functionality of selected hydrocolloids and their blends with κ-carrageenan on storage quality of vanilla ice cream. *LWT – Food Science and Technology* 41: 1816–1827.

Stevens, A. (2018). Are gummy bear flavors just fooling our brains? *NPR*. https://www.npr.org/sections/thesalt/2018/01/08/575406711/are-gummy-bear-flavors-just-fooling-our-brains (accessed 24 March 2022).

Su, K., Brunet, M., Festring, D. et al. (2021). Flavour distribution and release from gelatine-starch matrices. *Food Hydrocolloids* 112: 106273.

Talab, A.S. and Abou-Taleb, M. (2021). Effect of different hydrocolloids on the quality criteria of fish fingers during frozen storage. *Egyptian Journal of Aquatic Biology and Fisheries* 25: 323–335.

Taylor, A.J. (1998). Physical chemistry of flavor. *International Journal of Food Science & Technology* 33: 53–62.

Taylor, A.J. (1999). Flavour matrix interactions. In: *Current Topics in Flavours and Fragrances* (ed. K.A.D. Swift), 123–138. Dordrecht: Kluwer Academic Publishers.

Taylor, A.J., Linforth, R.S.T., Harvey, B.A., and Blake, A. (2000). Atmospheric pressure chemical ionization mass spectrometry for in vivo analysis of volatile flavour release. *Food Chemistry* 71: 327–338.

Taylor, A.J., Besnard, S., Puaud, M., and Linforth, R.S.T. (2001). in vivo measurement of flavour release from mixed phase gels. *Biomolecular Engineering* 17: 143–150.

Torii, K., Uneyama, H., and Nakamura, E. (2013). Physiological roles of dietary glutamate signaling via gut-brain axis due to efficient digestion and absorption. *Journal of Gastroenterology* 48: 442–451.

Trygg, E. (1991). *Odor Sensation and Memory*. New York: Praeger.

Tuinier, R., Rolin, C., and de Kruif, C.G. (2002). Electrosorption of pectin onto casein micelles. *Biomacromolecules* 3: 623–638.

Vaisey, M., Brunon, R., and Cooper, J. (1969). Some sensory effects of hydrocolloid sols on sweetness. *Journal Food Science* 34: 397–400.

van de Velde, F., de Hoog, E.H.A., Oosterveld, A., and Tromp, R.H. (2015). Protein-polysaccharide interactions to alter texture. *Annual Review of Food Science and Technology* 6: 371–388.

Voilley, A., Beghin, V., Charpentier, C., and Peyron, D. (1991). Interactions between aroma substances and macromolecules in a model wine. *Lebensmittel-Wissenschaft & Technologie* 24: 469–472.

Voilley, A., Espinoza-Diaz, M., Druaux, C., and Landy, P. (2000). Flavor release from emulsion and complex media. In: *Flavor Release* (ed. D.D. Roberts and A.J. Taylor), 142–152. Washington, DC: American Chemical Society.

Wagoner, B., Çakır-Fuller, E., Shingleton, R. et al. (2020). Viscosity drives texture perception of protein beverages more than hydrocolloid type. *Journal of Texture Studies* 51: 78–91.

Wang, Y., Sorvali, P., Laitila, A. et al. (2018). Dextran produced *in situ* as a tool to improve the quality of wheat-faba bean composite bread. *Food Hydrocolloids* 84: 396–405.

Wang, Y., Compaoré-Sérémé, D., Sawadogo-Lingani, H. et al. (2019). Influence of dextran synthesized *in situ* on the rheological, technological and nutritional properties of whole grain pearl millet bread. *Food Chemistry* 285: 221–230.

Wang, Y., Trani, A., Knaapila, A. et al. (2020). The effect of *in situ* produced dextran on flavour and texture perception of wholegrain sorghum bread. *Food Hydrocolloids* 106: 105913.

Weel, K.G.C., Boelrijk, A.E.M., Alting, A.C. et al. (2002). Flavor release and perception of flavored whey protein gels: perception is determined by texture rather than by release. *Journal of Agricultural and Food Chemistry* 50: 5149–5155.

Wilke, C.R. and Chang, P. (1995). Correlation of diffusion coefficients in dilute solutions. *AIChE Journal* 1: 264–270.

Wolfe, J., Kluender, K., and Levi, D. (2012). *Sensation & Perception*, 3e, 7. Sunderland, MA: Sinauer Associates.

Wu, S.Y., Pérez, M.D., Puyol, P., and Sawyer, L. (1999). β-Lactoglobulin binds palmitate within its central cavity. *Journal of Biological Chemistry* 274: 170–174.

Yven, C., Guichard, E., Giboreau, A., and Roberts, D.D. (1998). Assessment of interactions between hydrocolloids and flavor compounds by sensory, headspace, and binding methodologies. *Journal of Agricultural and Food Chemistry* 46: 1510–1514.

Zhang, J., Daubert, C.R., Mulligan, J.H., and Foegeding, E.A. (2008). Additive effect on the rheological behavior of alginate gels. *Journal of Texture Studies* 39: 582–603.

4

Use of Hydrocolloids to Control Food Viscosity

In contrast to aroma, color, and taste, which can be formulated and introduced into compound or processed foods, texture cannot be added "from a bottle." The viscosity of a medium is enhanced through immobilization of water by macromolecules with some potential intermolecular bonding. Viscosity is defined as the internal friction of a fluid or its tendency to resist flow. The International Handbook of Food Additives mentions numerous substances that can be added to foods. Over 10% of these constituents are described as texturizers, thickeners, viscosity modifiers, and bodying agents. They give the food technologist a large selection of aids to develop desired textures. This chapter includes sections on flow and viscosity, including necessary definitions, flow equations and models, a description of the thickening abilities of hydrocolloids and their use as viscosity formers in foods, time dependency of hydrocolloid solutions, fluid gels, and a demonstration of the use and performance of hydrocolloids to control viscosity of foods in recipes for Italian and French dressings that include typical hydrocolloids.

4.1 Viscosity of Fluids

4.1.1 The Field of Flow and Viscosity

The tendency of a fluid to flow effortlessly or with difficulty has unlimited practical importance (Bourne 2002). Sir Isaac Newton (Figure 4.1), an English astronomer, alchemist, theologian, and one of the greatest mathematicians and physicists of all time, was one of the first scientists to study the flow of fluids. He hypothesized that a fluid's flow is directly proportional to the applied force. This principle defines the class of "Newtonian fluids," and water (the main constituent of the Earth's hydrosphere and of the fluids in all known living organisms) is the best-known Newtonian fluid (Bourne 2002). Incompressible fluids were studied by the Irish English physicist and mathematician Sir George Gabriel Stokes (Figure 4.2) (Stokes 1842), and he is considered the founder of the efflux-type viscometer. The French physicist and physiologist Jean Léonard Marie Poiseuille studied flow in capillary tubes and is regarded as one of the originators of modern viscometry (Poiseuille 1846).

Use of Hydrocolloids to Control Food Appearance, Flavor, Texture, and Nutrition, First Edition.
Amos Nussinovitch and Madoka Hirashima.
© 2023 John Wiley & Sons Ltd. Published 2023 by John Wiley & Sons Ltd.

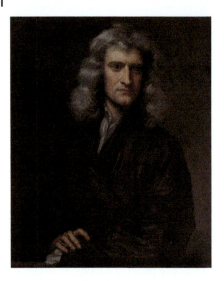

Figure 4.1 Sir Isaac Newton (1643–1727), author of the *Principia*. *Source:* University of Cambridge.

Figure 4.2 Sir George Gabriel Stokes, an Irish English physicist and mathematician (1819–1903). *Source:* Unknown author/Wikipedia Commons/Public Domain.

4.1.2 Laminar Flow and Turbulent Flow

Laminar flow and turbulent flow differ in their behavior (Figure 4.3). Laminar flow is streamlined flow in a fluid. Turbulent flow is fluid flow in which the velocity varies erratically in magnitude and direction. Laminar flow is slow and turbulent flow occurs at high rates (Bourne 2002). Osborne Reynolds (Figure 4.4) (1842–1912) was a British engineer, physicist, and educator best known for his work in hydraulics and hydrodynamics (https://www.britannica.com/biography/Osborne-Reynolds). The difference

Figure 4.3 Laminar and turbulent water flow over the hull of a submarine. As the relative velocity of the water increases, turbulence occurs. The USS *Los Angeles* (SNN 688). *Source:* US Navy/Wikipedia Commons/Public Domain.

Figure 4.4 Prof. Osborne Reynolds (1842–1912). *Source:* John Collier/Wikipedia Commons/Public Domain.

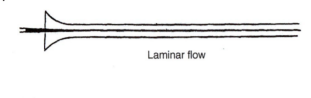
Laminar flow

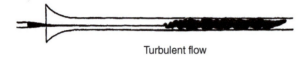
Turbulent flow

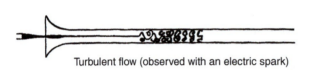
Turbulent flow (observed with an electric spark)

Figure 4.5 Reynolds' observations of the nature of flow in his best-known experiment on fluid dynamics in pipes. Drawn by Osborne Reynolds. *Source:* Osborne Reynolds/Wikimedia Commons/ Public domain.

between laminar and turbulent flow is demonstrated in Figure 4.5, a drawing that was published in Reynolds' influential 1883 paper "An experimental investigation of the circumstances which determine whether the motion of water in parallel channels shall be direct or sinuous and of the law of resistance in parallel channels" (Reynolds 1883). In the figure, water flows from left to right in a transparent tube, and dye (in black) flows in the middle. The nature of the flow (either laminar or turbulent) can be easily observed. Under laminar flow, the dye is seen to move straight down the tube. Under turbulent flow, there are numerous eddies, currents and vortices as the dye moves down the pipe (Bourne 2002).

The Reynolds number is a dimensionless number that can be defined by an equation that can take several forms. One of these forms is:

$$R_e = \frac{2\rho Q}{\pi r \eta} \quad (4.1)$$

where R_e is the Reynolds number; ρ is the density of liquid; Q is the rate of flow; r is the radius of the pipe and η is the viscosity (Bourne 2002). The critical Reynolds number R_c characterizes the rate at which the flow changes from laminar to turbulent. For pipe flow, it takes place at about 2200, a number that determines the lowest velocity at which turbulent flow occurs but does not determine the highest velocity at which laminar flow occurs. If a pipe is very smooth and the fluid is free of colloidal or suspended material, laminar flow can occur beyond the critical Reynolds number. Note that a Newtonian fluid at very high shear rate looks non-Newtonian (Bourne 2002).

4.2 Important and Useful Definitions

4.2.1 Dynamic Viscosity and Fluidity

Dynamic viscosity, termed "absolute viscosity" or simply "viscosity," is the tendency of a fluid to resist flow or the internal friction of a liquid (Bourne 2002). It is defined by the equation

$$\eta = \frac{\sigma}{\dot{\gamma}} \tag{4.2}$$

where η is the viscosity; σ is the shear stress, and $\dot{\gamma}$ is the shear rate. In the older metric system, the unit of measurement of viscosity was the poise (P), named after the French scientist Poiseuille. This is a large unit of measurement and therefore the centipoise cP was used, $100\,cP = 1\,P$. The newer International System of Units (SI), adopted in 1960, defines dynamic viscosity in Pascal seconds (Pa . s) and $1000\,mPa\cdot s = 1\,Pa\cdot s$

The reciprocal of dynamic viscosity is fluidity (ϕ). It is denoted by:

$$\phi = \frac{\dot{\gamma}}{\sigma} \tag{4.3}$$

4.2.2 Kinematic Viscosity

The kinematic viscosity is defined as the absolute viscosity of a liquid divided by its density at the same temperature (Shashi Menon 2015):

$$v = \frac{\eta}{\rho} = \frac{\sigma}{\rho\dot{\gamma}} \tag{4.4}$$

where v is the kinematic viscosity, η is the absolute viscosity, and ρ is the density in g/cm³. The units of kinematic viscosity are ft²/s in English units and m²/s in SI units. Other commonly used units for kinematic viscosity are Stokes and centistokes (cS) (Shashi Menon 2015).

In the petroleum industry, two other units for kinematic viscosity are also used. Expression in these units characterizes the time taken for a fixed volume of liquid to flow through an orifice of defined size. Both absolute and kinematic viscosities change with temperature. As the temperature increases, liquid viscosity decreases and vice versa. Nevertheless, unlike specific gravity, the viscosity–temperature relationship is not linear (Shashi Menon 2015).

4.2.3 Relative Viscosity

Relative viscosity is defined by the ratio between the viscosity of a solution and the viscosity of the pure solvent:

$$\eta_{rel} = \frac{\eta}{\eta_s} \tag{4.5}$$

where η_{rel} is the relative viscosity, η is the viscosity of the solution, and η_s is the viscosity of the solvent (Bourne 2002).

4.3 Flow Equations

4.3.1 Definitions of Apparent Viscosity, Shear Stress, and Shear Rate

Apparent viscosity is the viscosity of a non-Newtonian fluid expressed as though it were a Newtonian fluid. It is a coefficient calculated from empirical data as if the fluid complied with Newton's law. Apparent viscosity is denoted by the symbol η_a (Bourne 2002). *Shear stress* is the stress component applied tangential to the plane on which the force acts. It is expressed in units of force per unit area. The SI unit for shear stress is Pascal (Pa) with units of Newtons per square meter (N/m^2) (Bourne 2002). *Shear rate* is the velocity gradient established in a fluid as a result of an applied shear stress. It is expressed in units of reciprocal seconds (s^{-1}) (Bourne 2002).

4.3.2 The General Equation for Viscosity

Viscous performance can be Newtonian or non-Newtonian. *Newtonian flow* means that the shear rate is directly proportional to the shear stress and the viscosity is independent of the shear rate within the laminar flow range. The viscosity is given by the slope of the shear stress–shear rate curve. The viscosity of a Newtonian fluid does not change with different shear rates (Bourne 2002). Typical Newtonian fluids are water, and watery beverages such as beer, coffee, tea and carbonated beverages, sugar syrups, most honeys, edible oils, filtered juices, and milk. However, most fluid and semifluid foods fall into one of quite a few classes of *non-Newtonian fluids*. For *plastic* (or Bingham) fluid, a minimum shear stress (i.e. yield stress) must be exceeded before flow begins. Furthermore, the apparent viscosity of plastic fluid is different at each shear rate. Values of plastic yield stress of some foods have been published (Rha 1980). An example of a true Bingham plastic is a concentrated meat extract. Its shear stress–shear rate plot shows a linear relationship all the way down to zero shear rate (Halmos and Tiu 1981). In *pseudoplastic* flow, increasing shear force gives a more than proportional increase in shear rate and the curve (i.e. convex line) begins at the origin. In this case, the apparent viscosity decreases with increasing shear rate. A wide variety of foods, such as butter, margarine, applesauce, tomato catsup, mayonnaise, peanut butter, and many puddings are either plastic or pseudoplastic in nature. The shear stress–shear rate of *dilatant flow* begins at the origin but is categorized by equal increments in the shear stress causing less than equal increments in the shear rate. In this case, for a shear stress–shear rate plot, a concave line begins at the origin and apparent viscosity increases with increasing shear rate (Bourne 2002). Instances of foods exhibiting dilatant flow are high-solids raw starch suspensions, and some chocolate syrups. High concentrations of suspended matter usually render the product non-Newtonian and can lead to plastic flow or dilatant flow. The latter property only appears in suspensions with ca. 40–70% solids concentration. Dilatant flow is relatively uncommon in the food industry and quite rare in finished food products (Bourne 2002). All of these flow types can be described by the following equation:

$$\sigma = b\dot{\gamma}^s + C \tag{4.6}$$

where σ is the shear stress, b is a proportionality factor (this factor is the viscosity for a Newtonian fluid), C is the yield stress, and s is the the pseudoplasticity constant, which is

Table 4.1 The relationship between type of flow and general viscosity equation.

Type of flow	s	C	Equation form[a]
Newtonian	1	0	$\sigma = b\dot{\gamma} = \eta\dot{\gamma}$
True plastic	1	>0	$\sigma = b\dot{\gamma} + C$
Pseudoplastic	$0 < s < 1$	0	$\sigma = b\dot{\gamma}^s$
Dilatant	$1 < s < \infty$	0	$\sigma = b\dot{\gamma}^s$
Pseudoplastic with yield value	$0 < s < 1$	>0	$\sigma = b\dot{\gamma}^s + C$
Dilatant with a yield value	$1 < s < \infty$	>0	$\sigma = b\dot{\gamma}^s + C$

Source: From Bourne (2002)/With permission from Elsevier. The general viscosity equation is $\sigma = b\dot{\gamma}^s + C$.
[a] The term b is a proportionality factor and for a Newtonian fluid, this factor is the viscosity η; C is the yield stress and s the pseudoplasticity constant.

an index of the degree of nonlinearity of the shear stress–shear rate curve, and $\dot{\gamma}$ is the shear rate. Table 4.1 presents the relationship between flow type and the general viscosity equation.

4.3.3 The Power Equation

The power equation is also known as the Ostwald–de Waele model. Armand Michel A. de Waele (1887–1966) was a British chemist, noted for his contributions to rheology, and after whom the Ostwald–de Waele relationship for non-Newtonian fluids is named (Reiner and Schoenfeld-Reiner 1933; Saramito 2016). A power law is a functional relationship between two quantities, where a relative change in one quantity results in a proportional relative change in the other quantity, independent of the initial size of those quantities: one quantity varies as the power of another (https://en.wikipedia.org/wiki/Power_law):

$$\sigma = K\dot{\gamma}^n \tag{4.7}$$

where σ is the shear stress, K is a consistency index, $\dot{\gamma}$ is the shear rate, and n is a dimensionless number that designates closeness to Newtonian flow. For Newtonian flow, $n = 1$; for dilatant flow, $n > 1$; and for pseudoplastic flow, $n < 1$. The logarithmic form of Eq. (4.7) is:

$$\log \sigma = \log K + n \log \dot{\gamma} \tag{4.8}$$

Plotting log shear stress versus log shear rate, for fluids that obey the power equation, produces a linear relationship with a slope equal to n.

4.3.4 The Herschel-Bulkley Model

This non-Newtonian fluid model was introduced by Winslow Herschel and Ronald Bulkley (Herschel and Bulkley 1926; Tang and Kalyon 2004):

$$\sigma = \sigma_0 + K\dot{\gamma}^n \tag{4.9}$$

In this model, the strain experienced by the fluid is related to the stress in a complicated, nonlinear way. σ_0 is the yield stress and the rest is in the same form as the power equation. Although the occurrence of true yield stress has been debated, it has been recognized as an engineering reality. Today, yield stress is regularly measured and used in the food industry not merely for basic process calculations and manufacturing practices, but also as a test for sensory and quality indices and to define the effect of composition and manufacturing procedures on structural and functional properties (Sun and Gunasekaran 2009). Extrapolation of the shear stress versus shear rate data gained from conventional rheometers is the most common indirect technique for measuring yield stress. The experimental data, also denoted as equilibrium flow curves, can be interpreted with or without a rheological model, for example, the Bingham, Herchel-Bulkley or Casson models (Sun and Gunasekaran 2009).

4.3.5 Casson Equation

A popular model of non-Newtonian fluid is known as the Casson model (Casson 1959):

$$\sqrt{\sigma} = \sqrt{\sigma_0} + \eta_a \sqrt{\dot{\gamma}} \tag{4.10}$$

where σ is the shear stress, σ_0 is the yield stress, and η_a is the apparent viscosity and $\dot{\gamma}$ is the shear rate.

The Casson model is used in pigment oil suspensions to estimate the behavior of fluid flows. This model has been widely configured by numerous researchers under distinct conditions of fluid flow. A vertical exponentially stretching cylinder was used for boundary layer flow investigations of Casson fluid (Malik et al. 2014). The flow of blood through a narrow artery with bell-shaped stenosis was investigated in a mathematical study treating blood as a Casson fluid (Venkatesan et al. 2013). Moreover, heat-transfer analysis of thin film flows of a third-grade fluid in the presence of magnetohydrodynamic drive (MHD) on a vertical moving belt with slip boundary conditions was performed (Gul et al. 2013).

As concerns food systems, the International Office of Cocoa, Chocolate, and Confectionary (IOCCC) has adopted the Casson model as a standard to describe the flow behavior of chocolate (Steffe 1996). A modified Casson equation for chocolate was also suggested and instead of using the square root of shear stress, $\sigma^{0.6}$ was found to give more consistent results (Chevalley 1991). Peanut butter is a defined viscoplastic food, which contains a thick concentrated suspension of small non-colloidal peanut particles in peanut oil (Corradini and Peleg 2005). Additional vegetable oils and stabilizers may be added to the oil matrix to scatter the solid particles. The rheological and physical characteristics of two types of peanut butter were studied (Citerne et al. 2001). One product was a non-stabilized suspension with solids and particles in peanut oil, and the second was stabilized with vegetable oil and contained additives such as salt and sugar. The "apparent yield stress" of both types was obtained by fitting the flow curves, which were measured with parallel plate geometry attached to a stress-controlled rheometer, to Casson and Bingham models. The presence of additives was clearly reflected in the measured yield stress data. The "100% peanut butter" exhibited much smaller yield values, 21.5 and 27 Pa, than the "smooth" type peanut butter, 363 and 374 Pa, based on Casson and Bingham models, respectively (Citerne et al. 2001; Sun and Gunasekaran 2009).

4.4 Thickening and Viscosity-Forming Abilities of Hydrocolloids – A General Approach

The main hydrocolloid thickeners used in food products are guar gum, locust bean gum (LBG), starches, xanthan gum, methyl cellulose (MC), carboxymethylcellulose (CMC), hydroxymethyl propyl cellulose (HPMC), and microcrystalline cellulose, to name a few. The selection of hydrocolloid is dependent on its functional features, availability and obviously, price. For these reasons, starches are the most frequently used thickening agents. Nevertheless, due to its unique rheological features, xanthan is the thickener of choice in many applications (Phillips and Williams 2021).

The viscosity of polymer solutions demonstrates a noticeable increase above a critical polymer concentration, usually denoted as C^*, analogous to the shift from the so-called "dilute region," where the polymer molecules are free to move independently without interpenetration, to the "semi-dilute region," where molecule accumulation gives rise to overlapping polymer coils and interpenetration occurs (Phillips and Williams 2021). Customarily, polysaccharide solutions display Newtonian behavior at concentrations well below C^*, that is, their viscosity is independent of the shear rate. However, above C^*, non-Newtonian behavior is regularly observed. A characteristic viscosity–shear rate side view for a polymer solution above C^* demonstrates three separate regions: (i) a low-shear Newtonian plateau, (ii) a shear-thinning region, and (iii) a high-shear Newtonian plateau (Phillips and Williams 2021). At low shear rates, the rate of disruption of entanglements is less than the rate of re-entanglement, and hence viscosity is independent of shear. Above a critical shear rate, disentanglement predominates and the viscosity drops to a minimum plateau value at infinite shear rate (Phillips and Williams 2021).

A few empirical mathematical models refer to flow characteristics. The power law model adequately describes the behavior of some fluids over a limited range of shear rates. However, most non-Newtonian fluids exhibit more complex behavior and are better characterized by other models (Cross 1965; Subbaraman et al. 1971; Boger 1977; Gastone et al. 2014). The rheological model based on the Cross equation is one of the most popular in use today. The Cross model provides a simple way of quantifying the "full" viscosity/shear rate profile for a shear-thinning fluid. It can be located in more or less every research rheometer software package, and can be used to extract some significant facts from the "full" viscosity vs. shear rate profile (http://www.rheologyschool.com/advice/rheology-tips/25-making-use-of-models-the-cross-model). Here is the Cross equation, giving viscosity η as a function of shear rate $\dot{\gamma}$:

$$\eta = \eta_\infty + \frac{\eta_0 - \eta_\infty}{1 + \left(C\dot{\gamma}\right)^m} \tag{4.11}$$

where η_0 is the zero shear viscosity, i.e. the magnitude of the viscosity at the lower Newtonian plateau. It is a critical material property and can prove valuable in assessing suspension and emulsion stability, estimating comparative polymer molecular weight and tracking changes due to process or formulation variables etc. (http://www.rheologyschool.com/advice/rheology-tips/25-making-use-of-models-the-cross-model); η_∞ is the infinite shear viscosity. This tells us how our product is likely to behave in very-high-shear processing situations such as blade, knife, and roller coating; m is the (Cross) rate constant. It is dimensionless and is a measure of the degree of dependence of viscosity on shear rate in the shear-thinning region. A value of zero for m indicates Newtonian behavior with m tending to unity for increasing shear-thinning behavior. C is the Cross time constant (or sometimes consistency) and has dimensions of time. The reciprocal, $1/C$, gives us a critical shear rate that is a useful indicator of the onset shear rate for shear thinning; m and $1/C$ can be related to texture, application properties, pumping, mixing and pouring characteristics, and many other everyday flow processes that often occur in the shear-thinning region of the fluid's flow behavior (http://www.rheologyschool.com/advice/rheology-tips/25-making-use-of-models-the-cross-model).

4.5 Hydrocolloids as Viscosity Formers in Foods

Hydrocolloids are used in food processing to improve thickening, gelling, structuring, stabilization, encapsulation and water binding (Lapasin 2012; Torres et al. 2012). Key hydrocolloid thickeners are xanthan gum, cellulose gum, galactomannan and glucomannan. Xanthan gums carry very high low-shear viscosity (yield stress), high shear thinning, and maintain viscosity in the presence of electrolytes over a broad pH range and at high temperatures. CMC is characterized by high viscosity that is reduced by the addition of electrolyte and at low pH. MC and HPMC increase viscosity with temperature (gelation may occur) but are not influenced by the addition of electrolytes or pH. Galactomannans (guar gum and LBG) and glucomannan (konjac) have very high low-shear viscosity and strong shear thinning. They are not influenced by the presence of electrolytes but can degrade and lose viscosity at high and low pH and when subjected to high temperatures (Phillips and Williams 2021).

Many plants naturally exude viscous, gummy fluids (Nussinovitch 2010). The most important of these to the food industry are gum arabic, gum ghatti, gum karaya, and gum tragacanth (Glicksman 1969). Gum arabic is the oldest known of all natural gums. It is extremely soluble and is not very viscous at low concentrations. To achieve high viscosities, a concentration of 40–50% must be used (Glicksman 1969). At concentrations of up to 40%, gum arabic solutions present typical Newtonian behavior. Above 40%, the solution presents pseudoplastic characteristics as denoted by a decrease in viscosity with increasing shear stress (Araujo 1966). Gum ghatti is used as a stabilizer for oil-in-water emulsions (Glicksman 1969). Gum karaya is used in textile printing operations as a thickening agent for the dye in direct color printing on cotton fabrics (Verbeken et al. 2003). The viscosity of a 1% solution of gum karaya at pH 4.0–4.7 is 3300 cP. The viscosity is maximal at about pH 8.5 (Glicksman 1969). Gum tragacanth has been used as a stabilizer, emulsifier, and thickener in food products. Its outstanding water-absorbing abilities make it an exceptional thickening agent. Gum tragacanth is used in numerous ordinary commercial products with low viscosity, such as jellies and pourable dressings. It is also used in liquid and semisolid foods, for example, syrups, mayonnaise, sauces, liqueurs, dairy drinks, candy, ice cream, desserts, and popsicles (Phillips and Williams 2021). Viscosity is gum tragacanth's most important property. The viscosity of a 1% solution of the highest grade of gum is about 3400 cP. A 2–4% concentration gives a thick paste when properly hydrated. Gum tragacanth's maximum initial viscosity is at pH 8, but its maximum stable viscosity is near pH 5 (Glicksman 1969).

Different galactomannans, consisting of the hydrocolloids fenugreek, guar gum, tara gum, and LBG, are generally used for thickening purposes (Nussinovitch and Hirashima 2014). A LBG solution is extremely viscous and sticky: a 1% concentration of good-quality gum may have a viscosity of about 3500 cP. Its viscosity is little affected by pH within the range of 3–11 (Glicksman 1969). Guar gum hydrates rapidly in cold-water systems to produce highly viscous solutions. Dilute solutions of guar gum (<1%) are less thixotropic than solutions of 1% or higher; commercial products vary widely in their viscosity. A 1% dispersion of good-quality guar gum has a viscosity of 3000–6000 cP (Glicksman 1969). Hydroxypropylation of guar galactomannan has been used to improve its thickening properties (Zhan et al. 2013). High-galactosyl and high-molecular-weight galactomannans such as guar (M/G 2 : 1) and Prosopis gum (M/G 2 : 1) display good solubility in the cold and due to the large molecular sizes, they contribute impressively to viscosity augmentation, even at low concentrations. High-M/G ratio galactomannans such as LBG (M/G 4 : 1), *Delonix regia* (5 : 1), and *Sophora japonica* (5.7 : 1) need to be heated for full solubilization. The association between mannan chains heightens viscosity and could encourage gel formation at high concentration and low temperature (Phillips and Williams 2021). High-galactosyl galactomannans such as fenugreek (M/G 1 : 1) and *Leucaena* (M/G 1.3 : 1) are highly soluble in water, even cold water. Moreover, high-galactosyl galactomannans show improved emulsification properties (Garti et al. 1997; Wu et al. 2009). Galactomannan- and glucomannan-based synergistic combinations have already been widely explored in the food and cosmetic industries for the purposes of thickening and gelation (Phillips and Williams 2021). The addition of corn fiber gum demonstrated a notable viscosifying action for aqueous solutions of non-gelling polysaccharides, probably due to hydrogen-bonding interactions (Zhang et al. 2015). In addition, such synergistic combinations between okra polysaccharide and κ-carrageenan or high-methoxy pectin were also detected (Chen et al. 2019; Li et al. 2019).

Psyllium seed gum comes from a group of plants belonging to the genus *Plantago* (Figure 4.6). Like quince gum, psyllium seed gum hydrates slowly in water to form a clear solution or highly viscous dispersion (at concentrations of up to 1%). Quince seed gum,

Figure 4.6 *Plantago major.* Image taken on 20 March 2004 in the outdoor botanical garden of the Technion, Haifa, Israel. *Source:* Iorsh/Wikipedia Commons/Public Domain.

which comes from the botanical tree species *Cydonia vulgaris* Pers. or *Cydonia oblonga* Miller (Figure 4.7), creates highly viscous dispersions at concentrations of up to 1.5% (Glicksman 1969).

Pectin has a complex heterogeneous structure. The properties of pectic substances depend greatly on their molecular weight and degree of substitution. The viscosity of pectin solutions depends on the degree of methylation of the pectin, concentration, temperature, pH and presence of salts and their composition (Glicksman 1969). Commercial pectins can be used for gelation and thickening purposes. For any food application, the galacturonic content of the pectin needs to be at least 65% by weight (Endress 2011). In the case of a fruit preparation for yogurts, pectins are mostly utilized for their thickening effect on the dairy matrix following mixing or production of a layered dessert (Phillips and Williams 2021).

Arabinogalactan (Figure 4.8) is a water-soluble gum found principally in larch trees. It is unique in that it confers high viscosity at low concentration and in that a highly concentrated solution of over 60% can be prepared. Solutions with 40% arabinogalactan exhibit Newtonian-flow characteristics (Glicksman 1969). Due to their thickening properties, in the digestive system, water-soluble arabinoxylans are thought to enhance the viscosity of the digesta in the small intestine and to impair dispersion and mixing of the food mass with the fluid layer adjacent to the mucosal surface (Mendis and Simsek 2014).

Agar, carrageenan, furcellaran, and alginate are important gelling agents. Agar can be used for a wide range of functionalities, such as binding, stabilizing, thickening and gelling, and as a water-retention agent, emulsifier, water-soluble dietary fiber, and satiety agent (Phillips and Williams 2021). Agar was the first hydrocolloid isolated, extracted from seaweed. Its viscosity depends on the source of the raw material and varies widely.

Figure 4.7 Painting of quince fruit (*Cydonia oblonga* fruit, tree and foliage). Dated before 1835. *Source:* Pancrace Bessa/Wikipedia Commons/Public Domain.

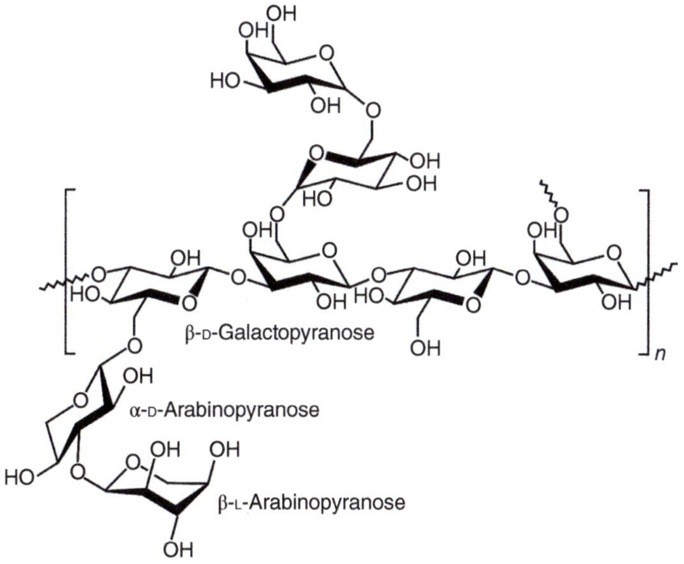

Figure 4.8 Structure of arabinogalactan. *Source:* Adapted from Mano et al. (2007); Nothingserious.

At temperatures above its gelation point, its viscosity is relatively constant from pH 4.5–9.0 and is only minorly affected by age or ionic strength within the limits of pH 6.0–8.0 (Glicksman 1969). The practical applications of carrageenans are based on their gel-forming, texturizing, thickening, emulsifying, stabilizing, mouthfeel-providing, and moisture-binding properties (Phillips and Williams 2021). Furcellaran and kappa carrageenan are commonly used at 0.5–1.0% for gelling purposes and at 0.1–0.5% for stabilization and thickening. Iota and lambda carrageenans are also often used in amounts of 0.1–0.5% (Schorsch et al. 2000). Commercial alginates are offered in a comprehensive range of viscosity types, ranging from 10 cP for a 1% solution of low viscosity, to 2000 cP for a high-viscosity solution of the same concentration. Viscosity is dependent on concentration, temperature, molecular weight, pH, and the presence of polyvalent metal cations (Glicksman 1969). Important properties related to the gelling ability of *gelatins* are gel formation, texture, thickening, and water binding (Phillips and Williams 2021). The viscosity conferred by gelatin in a water solution is one of its most important properties. Commercial gelatins vary in viscosity between 20 and 70 mP. In dilute solutions, the effect of changes in shear rate on the apparent viscosity of gelatin solutions is small. Viscosity increases more or less exponentially with increasing gelatin concentration. Viscosity also increases exponentially with decreasing temperature (Glicksman 1969).

The most important textural property of starch is its viscosity or consistency when it is heated and cooled (Glicksman 1969). Of the two components of starch, amylose has the most valuable task as a hydrocolloid. The extended conformation of amylose generates high viscosity in aqueous dispersions that differs very little with temperature. Increasing amylose concentration may decrease the stickiness of hydrogels but might increase the modulus of the resultant gels (Eggleston et al. 2018). Starches can be modified by gelatinization followed by rapid drying, upon which the starch loses its native semi-crystalline structure. Such starches are termed pregelatinized or instantized and because of the process, they have cold water-thickening properties, except when they have been held under conditions that promote retrogradation of the amylose or amylopectin before drying. In addition, the original botanical source of the starch and the gelatinization regime (i.e. duration, shear and temperature) will affect its viscosity and performance (Phillips and Williams 2021).

Dextran is a neutral polysaccharide, readily solubilizing in hot or cold water to give a clear and viscous solution (Glicksman 1969). Dextrans are used as emulsifying and thickening agents, high-viscosity gums, explosives, deflocculants in paper products, for the secondary recovery of petroleum from oil-drilling mud, soil conditioners, and surgical sutures (Penczek 2018).

Xanthan gum dissolves in hot or cold water to form a viscous solution (Glicksman 1969). The resultant solutions are very viscous at very low concentrations, and are strongly shear thinning. The viscosity is not influenced to any great extent by variations in pH or the presence of salts. Xanthan gum's high viscosity at low shear enables it to inhibit particle sedimentation and droplet creaming. Its shear-thinning feature guarantees that the product flows freely from a container after shaking. Therefore, it is extensively used in sauces and salad dressings (Phillips and Williams 2021). In addition, xanthan gum is used as a thickener for people suffering from dysphagia (i.e. swallowing disorders) (Rofes et al. 2014; Torres et al. 2019). Hydrocolloids can be utilized to yield stable emulsions, thus preventing creaming, droplet aggregation, and coalescence. Creaming can be inhibited by increasing the viscosity of the continuous phase by means of polymer thickeners or a low concentration of gelling agent; and causing the oil droplets to weakly flocculate and form a 3D structure. This can be promoted by the attendance of non-adsorbed high-molar-mass polymers through the so-called depletion flocculation mechanism. High-molecular-mass charged polymers such as xanthan gum enhance emulsion droplet-creaming rates due to depletion flocculation at concentrations as low as 0.075%. At higher concentrations, xanthan gum will inhibit creaming due to the increased viscosity of the continuous phase (Phillips and Williams 2021). There are two kinds of double emulsions: oil-in-water-in-oil (O/W/O) and water-in-oil-in-water (W/O/W), the latter being the most common. W/O/W emulsions are usually formed by first dispersing a water phase in an oil phase, and then dispersing this system in another water phase (Lamba et al. 2015). Hydrocolloids such as gelatin (Zhu et al. 2018), pectin (Oppermann et al. 2018), xanthan gum (Oppermann et al. 2018), gum arabic (Niu et al. 2017) and alginate (Sato et al. 2018) are added to either the inner or outer water phases as gelling or thickening agents to enhance the stability of the double emulsions (Phillips and Williams 2021).

Sodium CMC has typical non-Newtonian properties when in solution. Its behavior is primarily pseudoplastic. At very low shear rate, solutions of all CMC gums approach Newtonian flow. Standard commercial products can range from 10 cP (13 000 molecular weight) to 19 000 cP (140 000 molecular weight) for a 2% CMC solution (Glicksman 1969). CMC stabilizes emulsions by thickening their water phase. It has a broad variety of applications, including acidic beverages, dairy beverages, wine, processed meat, fillings, bakery products, ice creams, and frozen desserts (Phillips and Williams 2021). The thickening effect of CMC is particularly useful for soups, sauces, and ketchup. Aerated frozen products are stabilized by thickening the water phase. In addition, CMC prevents particles from sedimenting during storage. The European Union allows use of CMC as an emulsifying, stabilizing, thickening, and gelling agent with the status "quantum satis" – i.e. the amount necessary to offer a desired technological advantage (Phillips and Williams 2021). Plant-derived cellulose contains both hemicellulose and lignin, but cellulose produced by bacteria, also termed bacterial cellulose, is cellulose in its pure form. Cellulose can also be produced by other microorganisms (Nussinovitch and Hirashima 2019). Bacterial cellulose is a biopolymer synthesized mainly by *Acetobacter xylinum* and *Acetobacter hansenii*. Its structure and applications have been reviewed extensively, but less attention has been paid to its role as a food hydrocolloid. Early work showed that when dispersed in water, a very high viscosity can be achieved (Phillips and Williams 2021). After dispersion in water and after swelling, bacterial cellulose forms a continuous 3D-network structure that holds water and develops viscosity. The 3D-network structure is little affected by heat, acids or salts. In addition to pseudoplastic viscosity, these fluids present good suspension and dispersion-stabilizing capabilities (Nussinovitch and Hirashima 2019).

4.6 Time Dependence of Hydrocolloid Solutions

Newtonian fluids are time-independent. In several fluids, the shear stress is a function of both shear rate and the length of time that it is subjected to shearing force. There are four main types of time dependency. In *thixotropic* solutions, the apparent viscosity decreases with time of shearing, implying a progressive breakdown of the structure. Nevertheless, the change is reversible; in other words, the fluid reverts to its original state and rebuilds itself (Bourne 2002; Razavi and Karazhiyan 2009). Modeling of the thixotropic behavior of food products is based on equations, such as the Weltman model (Weltman 1943), first-order stress decay models (Steffe 1996) and the structural kinetic model (Nguyen et al. 1998). Experimentally, the flow's time dependence is demonstrated by running a hysteresis cycle. The area of hysteresis enables estimating the magnitude of the product's thixotropy (Longrée et al. 1966; Barbosa-Canovas and Peleg 1983; Carbonell et al. 1991; Thebaudin et al. 1998). The flow curves and time-dependent rheological behavior of two hydrocolloids obtained from the tubers of wild terrestrial orchids (salep), and balangu seeds (*Lallemantia royleana*) were studied. All samples exhibited thixotropic behavior at the concentrations used, and both forward and backward measurements were characterized by the power law model (Razavi and Karazhiyan 2009). Four time-dependent models – the second-order structural kinetic model, Weltman model, and first-order stress decay models with zero and non-zero equilibrium stress values – were used to describe the thixotropic behavior and dissimilar parameters of these models were analyzed. The rate of the thixotropic breakdown was found to increase with increasing shear rate for salep, whereas the extent of thixotropy decreased for balangu (Razavi and Karazhiyan 2009). Another report dealt with the effect of shear rate on apparent viscosity of gum solutions, and time dependency was found to be a significant feature in some foods. For instance, aqueous solutions of gums exhibiting a high degree of shear thinning or thixotropy had no sliminess (Szczesniak and Farkas 1962). The correlation between organoleptic sliminess and the degree of viscosity decrease with increasing rate of shear is logical, and follows the definition of sliminess (i.e. a thick material that coats the mouth and is difficult to swallow) (Szczesniak and

Farkas 1962). The more rapidly the solution decreases in viscosity under tongue motion, the faster and easier it can be swallowed. The slower the change in viscosity in the mouth, the more difficult it is to swallow. The reason for the diverse behavior of different gums lies in the degree of particle dispersion, molecular weight, molecular shape, strength of intermolecular bonds, and other characteristics (Szczesniak and Farkas 1962). This finding was also confirmed by others (Stone and Oliver 1966).

In the thixotropic class, we can also find some starch pastes. Starch pastes and gels are constituents whose rheological properties are determined by the contributions and interactions of both the dispersed phase (swollen starch granules) and the continuous viscous matrix (Doublier et al. 1987; Carnali and Zhou 1996). An increase in starch content leads to a marked intensification of paste viscosity, but altering starch concentration does not seem to significantly affect the thixotropic characteristics of the paste. Temperature has a major effect on the viscosity and thixotropic behavior of maize starch pastes (Nguyen et al. 1998). The rheological behavior of aqueous starch pastes of two types of maize starch were characterized as strongly thixotropic and shear-thinning (pseudoplastic). The time-dependent rheology of the studied starches was satisfactorily modeled by postulating that the internal gel structure formed by the polysaccharide polymers and the swollen starch granules breaks down irreversibly under shear following third-order kinetics (Nguyen et al. 1998). At equilibrium, the starch pastes displayed pseudoplastic behavior, which could be characterized by the power law. The type of starch, viz. the amylose/amylopectin proportion, starch concentration, and temperature all affected the rheological behavior of the studied starch pastes (Nguyen et al. 1998).

Mesquite (Prosopis) gum is an exudate of the mesquite tree (*Prosopis* sp.) (Mudgil 2020). An aqueous solution of mesquite gum (*Prosopis juliflora*) reveals low viscosity compared to other carbohydrate gums. The viscosity of complex carbohydrates makes them suitable for use as a dietary fiber source (Mudgil 2018), i.e. they are resistant to enzymatic secretions of the human gastrointestinal tract, and they are beneficial for human health (Barak and Mudgil 2020). Gum solutions (<15% concentration) showed "shear-thickening behavior" at shear rates $>100\,s^{-1}$. This behavior may have resulted from the experimental conditions, which produced turbulent flow in the geometry (coaxial cylinders) used in the study. Another reason for shear-thickening performance of gum solution is high shear rates which lead to changes in molecular shape or conformation (Vernon-Carter and Sherman 1980). However, some other researchers did not detect any such performance in commercial mesquite gum. They reported that an aqueous solution of mesquite gum (20% concentration, w/w) in 0.1 M sodium chloride at 20 °C demonstrates "Newtonian behavior" with viscosity values of 25–30 mPa·s. They also reported "shear-thinning behavior" of mesquite gum at concentrations higher than 20% (w/w) (Goycoolea et al. 1995). This shear-thinning behavior did not resemble that of entangled linear polysaccharides (Mudgil et al. 2016) as no increase in slope (log η versus log $\dot{\gamma}$) or "Newtonian plateau" was observed.

In the *shear-thinning* class, the apparent viscosity decreases with time and the change is irreversible. In other words, the product/gum solution stays in a thinner state when the shear stress is removed (Bourne 2002). Examples can be found in some gum solutions and starch pastes. For instance, the effects of two thickening agents on the rheology and microstructure of soy protein desserts was studied (Arancibia et al. 2015). The dessert samples were prepared with two concentrations of soy protein isolate (6 and 8% w/w), each with either modified starch or CMC. The flow curves of all systems demonstrated typical shear-thinning behavior and observable hysteresis loops (Arancibia et al. 2015). In another publication, CMC, xanthan gum, and guar gum were reported to have different shear-thinning behaviors (Phillips and Williams 2009), as well as resistance to degradation by salivary alpha-amylase (Ferry et al. 2006; Weber et al. 2009; Newman et al. 2016). In time-dependent and non-Newtonian shear-thinning products, perceived thickness is challenging to predict with rheological parameter values since flow in the mouth may be a combination of shear and elongational flow (van Vliet 2002). Some researchers have suggested a relationship between slipperiness and the shear-thinning behavior of gums (Szczesniak and Farkas 1962; Richardson et al. 1989; Vickers et al. 2015; Ong et al. 2018).

It is important to note that a fluid may exhibit both thixotropic and shear-thinning properties. Another type of behavior is rheopexy, where the absolute value of viscosity increases after switching from a high shear rate to a low one (Yusof et al. 2020). This type of behavior is rare in food systems (Bourne 2002). The structure-recovery capacity of highly concentrated emulsions under shear flow was studied via their rheopexy (Zhao and Zhang 2018), and another study dealt with "rheopexy" by looking at high-molecular-weight polyacrylamide (Al-Hashmi et al. 2013). Tomato juice containing 1% soy protein demonstrated rheopectic behavior at shear rates $<250\,s^{-1}$ due to enhanced aggregation between pectins and soy protein (Tiziani and Vodovotz 2005). A synovial fluid and bovine serum albumin solution revealed rheopectic behavior at shear rates $<10^{-1}\,s^{-1}$ attributable to protein aggregation (Oates et al. 2005). These manuscripts concluded that rheological behavior is complex and that numerous features affect it, e.g. associations or aggregation of macromolecules via their deformation and reorientation caused by the shear stress; nevertheless, its source is currently not well understood.

An interesting manuscript dealt with exceptional phenomena stemming from the feature of super megamolecules. Sacran is a supergiant cyanobacterial polysaccharide extracted from the jelly extracellular matrix of the river plant *Aphanothece sacrum* (Okajima et al. 2007), which has been mass cultivated in Japan for a many years. Sacran is famous for its particularly high absolute molecular weight of $1.0–2.2\times10^7$ g/mol (Okajima et al. 2008; Okeyoshi et al. 2018). Sacran has irregular rheological properties (Yusof et al. 2020). Rheopexy was detected at low shear rates, where the shear viscosity increased with elapsed time. It became constant within 900 seconds. Comparable behavior has been reported in, e.g. polyacrylamide (Al-Hashmi et al. 2013), with an increase in viscosity of ∼6 Pa·s, although the viscosity increase for the sacran solution was particularly large. The shear viscosity for sacran aqueous solutions increased with time at low shear rates. Upon cessation of shearing, it almost recovered its initial viscosity. The storage modulus was also observed to increase after the shear was applied. These results indicated that the rheopectic behavior initiates from the weak and transient crosslinking of sacran chains (Yusof et al. 2020). The concentration dependence of the changes in shear viscosity, storage modulus, and zeta potential demonstrated a crossover concentration of 0.15% (w/w), i.e. the rheopectic mechanism changed at this crossover concentration; above it, the number of crosslinking points per unit chain might increase in proportion to the sacran concentration (Yusof et al. 2020).

In the *shear-thickening* class, the apparent viscosity increases with time of shearing and the change is irreversible. Pastes were prepared from regular maize starch, waxy maize

Figure 4.9 Andrey Nikolayevich Tikhonov, leading Soviet mathematician and geophysicist. *Source:* Konrad Jacobs/Wikipedia Commons/CC BY-SA 2.0.

starch, potato starch, and the commercial starches E1412 (distarch phosphate) and E1420 (acetylated starch) and their rheological properties were evaluated (Ptaszek 2010). At a temperature of 20 °C, all starch pastes and mixtures of starch and hydrocolloids behaved like shear-thickening fluids. The same was true at 50 °C, except for potato starch paste whose viscosity decreased when $\dot{\gamma}$ increased (Ptaszek 2010). A modified equation of state was developed for the shear-thickening systems. The model was applied to describe both the rheological properties of various starch pastes and pastes with added hydrocolloid. The parameters of the model were estimated using the Tikhonov regularization method. To compare the results, some other models were also fitted to the experimental data (Ptaszek 2010). Named after Andrey Tikhonov (Figure 4.9), the Tikhonov regularization method is used to solve ill-posed problems. It is particularly useful in mitigating the problem of multicollinearity in linear regression, which commonly occurs in models with large numbers of parameters (Kennedy 2003; Ptaszek 2010). Another manuscript dealt with aqueous solutions of mesquite gum which exhibited shear thickening when the shear rate exceeded $100 \, s^{-1}$ (Vernon-Carter and Sherman 1980).

4.7 Fluid Gels

Fluid gels are particulate gel suspensions that are formed by applying a shear field to a biopolymer solution while it gels (Norton et al. 1999; Gabriele et al. 2009; Farres et al. 2014). The immersion of abundant elastic gel particles in a water phase gives fluid gels unique rheological properties (Norton et al. 1999). The creation of materials based on edible hydrocolloid gels with controlled microstructure has enabled a broader range of food applications, e.g. as a fat replacer in low-energy products. They also serve in pharmaceutical applications for controlled drug release (Farres et al. 2014; Mahdi et al. 2014). In addition, being able to produce different textures from identical biopolymers merely by governing the processing conditions, for instance shear, is a most important benefit (Norton et al. 2006).

Agarose is the main gelling agent of agar–agar (Nussinovitch 1997; Nussinovitch and Hirashima 2014). Agarose solution undergoes a sol–gel transition upon cooling and forms a 3D network because of its molecular structure and physicochemical properties. The gelation of agarose was shown to follow two steps upon cooling, which could be addressed to the formation of helices and their aggregation (Nordqvist and Vilgis 2011; Russ et al. 2013). Moreover, a coil-to-helix transition takes place throughout cooling and has been described by mean field Zimm–Bragg methodology (Zimm and Bragg 1959; Nowak and Vilgis 2004; Vilgis 2015). Fluid gels with different concentrations of agarose (0.5, 1.0, and 2.0% w/w) were examined for their rheological properties and microstructure (Ghebremedhin et al. 2021). Rheological measurements of the microgel particles demonstrated an increase in storage and loss modulus with growing concentration. The fluid gel (1% w/w) displayed the lowest viscosity in the low-shear range and the shortest linear-viscoelastic range. In addition, the effect on the microstructure and size of the gel particles was studied by light microscopy and particle-size analysis (Ghebremedhin et al. 2021). As the concentration of agarose increased, the particle size and number of unordered chains present at the particle surface decreased. Specific models suggested impacts of particle size, concentration and "hairy" projections on the rheological and tribological properties of the formed fluid gels (Ghebremedhin et al. 2021).

Hydrocolloids, including gellan gum, are used to prepare structured liquids, sometimes loosely termed "fluid gels." As already noted, in such applications, shear is applied to a hydrocolloid system throughout the gelation process. Fluid gel establishment is employed, to a large extent, in the manufacturing of custards and gravies or during high-temperature short-time, or ultra-high-temperature processing. Fluid gels are occasionally used to suspend herbs and spices in pourable dressings (Sworn et al. 1995). Gellan gum is an extracellular polysaccharide produced through fermentation of the organism *Sphingomonas elodea* (Moorhouse et al. 1981). Aside from the production of fluid gels, gellan gum has very wide possible uses in foods, for instance, in jellification, texturizing, film-forming, suspension, and structuring (Sanderson 1990; Gibson 1992; Kelco 1993). Brittle gellan gels are prepared by hydrating the gum in hot water, adding gelling ions to the hot solution, then cooling to below the setting temperature of the system. A wide range of ions, including sodium, potassium, calcium, and hydrogen, induce gelation (Grasdalen and Smidsrod 1987). The properties of gellan gels rely on the types of ions present and their concentrations (Sanderson et al. 1988). To manufacture smooth, consistent gellan gum fluid gels, the systems must be formulated to provide weak gelation, through ion type, and concentration or gellan gum concentration. Shear should be applied either at or after the transition point. Even though there may be differences at a molecular level between fluid gels produced by shearing at the transition point and those sheared afterwards, qualitatively, they both form smooth homogeneous fluids in which a wide variety of materials can be suspended (Sworn et al. 1995).

4.8 Demonstrating the Use of Hydrocolloids to Control Viscosity in Foods

Salad dressings have become quite popular in recent years. Consequently, the food industry is facing the challenge of manufacturing a wide variety of dressings, including low-fat dressings, to satisfy consumer demand (McIlveen-Farley and Armstrong 1995). In developing such products, special concern should be directed to their perception, which is influenced by the availability, nature, and quality of the included flavor components (Overbosch et al. 1991). In other words, the matrix plays an important role in controlling flavor release at each step of these food products' preparation and consumption (Druaux and Voilley 1997). Food dressings are oil-in-water emulsions. They are frequently used to season food and enhance its taste, color or aroma. Emulsifiers are important for the emulsion's formation and stabilization (Turgeon et al. 1996; Ford et al. 2004). The complete flavor of oil-in-water dressing stems from a combination of aroma, taste, and mouthfeel (McClements and Demetriades 1998). In addition, its lipid content is of marked significance not only for the perceived intensity, but also for the temporal profile of the flavors (Druaux and Voilley 1997). Moreover, fat is vital for numerous supplementary properties, such as texture, color, emulsification, and lubricity (Vafiadis 1996).

Here we present the preparation of creamy Italian dressing and French dressing (Figures 4.10 and 4.11, respectively). Italian dressing is a vinaigrette-type salad dressing. In American cuisine, it consists of water, vinegar or lemon juice, vegetable oil, chopped bell peppers, sugar or corn syrup, herbs, and species (including oregano, fennel, dill and salt, and sometimes onion and garlic) (https://en.wikipedia.org/wiki/Italian_dressing). In the presented recipe (Figure 4.10), mayonnaise is the lipid source. It increases the fat content and provides increased fattiness to the formulation. Xanthan gum was chosen as a stabilizer for the dressing, as it is often used as a stabilizer for oil-in-water emulsions (Nussinovitch 1997). Ideally, the product should have high yield value (permitting the suspension of spices, herbs, and vegetables and enabling the dressing to cling to the salad as well as to have body) and strong pseudoplasticity. These requirements make xanthan favorable for this application (Nussinovitch 1997), where product stability will not be influenced by low pH, high salt or thermal treatment (McNeely and Kang 1973). In the range of 10–90 °C, xanthan's viscosity is almost unaffected by the presence of salts (Urlacher and Dalbe 1992). The viscosity of xanthan from 5 to 70 °C also helps yield a uniform texture and good stability. The amount of xanthan required depends on the oil content, with lower oil content requiring a higher amount of xanthan in the product. In fact, the same flow properties can be achieved with different oil levels by adjusting the level of xanthan (Nussinovitch 1997). Furthermore, xanthan solutions are stable over a wide pH range (2.5–11) (Pettitt 1982). Albeit not in the recipe presented here, xanthan gum is often used in a mixture with other thickeners because it provides the desired rheological behavior for sauces and dressings (Ma and Barbosa-Canovas 1995). It is also interesting to note that several researchers have reported cases in which xanthan decreases flavor intensity (Pangborn and Szczesniak 1974; O'Carroll 1997).

French dressing is a creamy dressing with a pale orange to bright red color. It is made of oil, vinegar, sugar, and other flavorings. The color is derived from tomato and often

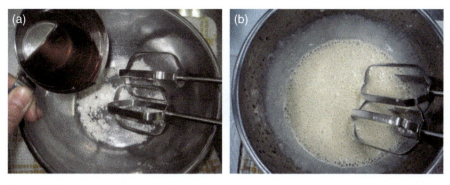

Figure 4.10 Creamy Italian dressing. (a) Adding red wine vinegar, (b) mixing all ingredients.

Figure 4.11 French dressing. (a) Mixing white sugar, salt, PGA and spices, (b) mixing all ingredients. (c) Serve.

paprika. In the 19th century, *French dressing* was synonymous with vinaigrette, which is still the terminology used by the American professional culinary industry (https://en.wikipedia.org/wiki/French_dressing). Beginning in the early 20th century, American recipes for French dressing commonly added new flavorings to the vinaigrette, including Worcestershire sauce, onion juice, ketchup, sugar, and Tabasco sauce, but the name was retained (Norton 1917). In our recipe (Figure 4.11), olive oil serves as the lipid source and propylene glycol alginate (PGA) is used as a thickener and stabilizer. PGA was developed and manufactured commercially in 1944 (Nussinovitch and Hirashima 2014). Both alginate and PGA slow the separation of oil and water phases, thus stabilizing the dressing or sauce. The influence of pH on alginate solutions varies, depending on the type of alginate used. The viscosity of the solution is related to the gum's molecular weight, but it is also influenced by the level of residual calcium from the manufacturing process and/or other components in the recipe. Another reason for using PGA in this recipe is its stability at acidic pH, providing resistance to acid degradation (Nussinovitch and Hirashima 2014).

Figure 4.12 presents some commercial thickening agents for dysphagia (and see the discussion in Chapter 1). Many commercial thickening agents are sold in Japan. Starch, guar gum, and xanthan gum have been used as thickeners. Recently, xanthan gum has become Japan's main thickener. Thickeners are easy to use; they are simply added to a liquid, such as water, tea, juice, and soup.

Figure 4.12 (a) Commercial thickening agents for dysphagia, added to water (b) or green tea (c) to thicken.

4.8.1 Creamy Italian Dressing

(makes 300 g) (Figure 4.10)

190 g (a little under ¾ cup) mayonnaise
45 mL (3 Tbsp) red wine vinegar
9 g (1 Tbsp) white sugar
2 g (⅓ tsp) salt
0.7 g (⅓ tsp) xanthan gum powder
60 mL (4 Tbsp) water
1 g (½ tsp) garlic powder
0.5 g (⅕ tsp) black pepper
0.5 g (⅙ tsp) onion powder
pH = 3.50

For preparation, see Figure 4.10.

1) Mix white sugar, salt, and xanthan gum powder in a bowl, and stir in water, mayonnaise and red wine vinegar (Figure 4.10a) with a hand mixer.
2) Add garlic powder, black pepper and onion powder (*Hint 1*) to the dressing and mix well (Figure 4.10b).

Preparation hint:

1) You can use any spices that you like.

4.8.2 French Dressing

(makes 100 g) (Figure 4.11)

25 g (a little over 2 Tbsp) olive oil
30 mL (3 Tbsp) water
20 mL (1 ⅓ Tbsp) apple cider vinegar
5 g (½ Tbsp) white sugar
2.5 g (½ tsp) salt
0.5 g propylene glycol alginate (PGA)
Pepper, mustard (optional)
pH = 2.69

For preparation, see Figure 4.11

1) Mix white sugar, salt, PGA, pepper and mustard in a bowl (Figure 4.11a) (*Hint 1*), and stir in water and apple cider vinegar with a hand mixer.
2) Add olive oil gradually while stirring, and mix well until the solution emulsifies (Figure 4.11b and c).

Preparation hint:

1) You can use any spices that you like.

References

Al-Hashmi, A.R., Luckham, P.F., and Grattoni, C.A. (2013). Flow-induced-microgel adsorption of high-molecular weight polyacrylamides. *Journal of Petroleum Science and Engineering* 112: 1–6.

Arancibia, C., Bayarri, S., and Costell, E. (2015). Effect of hydrocolloid on rheology and microstructure of high-protein soy desserts. *Journal of Food Science and Technology* 52: 6435–6444.

Araujo, O.E. (1966). Effects of certain preservatives on the aging characteristics of *Acacia*. *Journal of Pharmaceutical Sciences* 55: 636–639.

Barak, S. and Mudgil, D. (2020). Effect of guar fiber on physicochemical, textural and sensory properties of sweetened strained yoghurt. *Biointerface Research in Applied Chemistry* 10: 5564–5568.

Barbosa-Canovas, G.V. and Peleg, M. (1983). Flow parameters of selected commercial semi-liquid food products. *Journal of Texture Studies* 14: 213–234.

Boger, D.V. (1977). Demonstration of upper and lower Newtonian fluid behavior in a pseudoplastic fluid. *Nature* 265: 126–128.

Bourne, M. (2002). *Food Texture and Viscosity, Concept and Measurement*, 2e. Cambridge, MA: Academic Press.

Carbonell, E., Costell, E., and Dursan, L. (1991). Rheological behavior of sheared jams: relation with fruit content. *Journal of Texture Studies* 22: 33–43.

Carnali, J.O. and Zhou, Y. (1996). An examination of the composite model for starch gels. *Journal of Rheology* 40: 221–234.

Casson, N. (1959). A flow equation for pigment-oil suspensions of the printing ink type. In: *Rheology of Disperse Systems* (ed. C.C. Mill), 84–104. Oxford: Pergamon Press.

Chen, J., Chen, W., Duan, F. et al. (2019). The synergistic gelation of okra polysaccharides with kappa-carrageenan and its influence on gel rheology, texture behavior and microstructures. *Food Hydrocolloids* 87: 425–435.

Chevalley, J. (1991). An adaptation of the Casson equation for the rheology of chocolate. *Journal of Texture Studies* 22: 219–229.

Citerne, G.P., Carreau, P.J., and Moan, M. (2001). Rheological properties of peanut butter. *Rehological Acta* 40: 86–96.

Corradini, M.G. and Peleg, M. (2005). Consistency of dispersed food systems and its evaluation by squeezing flow viscometry. *Journal of Texture Studies* 36: 605–629.

Cross, M.M. (1965). Rheology of non-Newtonian fluids: a new flow equation for pseudoplastic systems. *Journal of Colloid Science* 20: 417–437.

Doublier, J.L., Llamas, G., and Paton, D. (1987). A rheological investigation of oat starch pastes. *Cereal Chemistry* 64: 21–26.

Druaux, C. and Voilley, A. (1997). Effect of food composition and microstructure on volatile flavour release. *Trends in Food Science and Technology* 8: 364–368.

Eggleston, G., Finley, J.W., and DeMan, J.M. (2018). Carbohydrates. In: *Principles of Food Chemistry* (ed. J.M. DeMan), 165–229. Cham, Switzerland: Springer International Publishing.

Endress, H.U. (2011). Pectins: Production, properties and applications. In: *Renewable Resources for Functional Polymers and Biomaterials: Polysaccharides, Proteins and Polyesters* (ed. P.A. Williams), Chapter 8, 210–260. Cambridge, UK: The Royal Society of Chemistry.

Farres, I.F., Moakes, R.J.A., and Norton, I.T. (2014). Food hydrocolloids designing biopolymer fluid gels: a microstructural approach. *Food Hydrocolloids* 42: 362–372.

Ferry, A.L.S., Hort, J., Mitchell, J.R. et al. (2006). Viscosity and flavour perception: why is starch different from hydrocolloids? *Food Hydrocolloids* 20: 855–862.

Ford, L.D., Borwankar, R.P., Pechak, D., and Schwimmer, B. (2004). Dressings and sauces. In: *Food Emulsions* (ed. S.E. Friberg, K. Larsson and J. Sjoblom), 525–572. New York: Marcel Dekker Inc.

Gabriele, A., Spyropoulos, F., and Norton, I.T. (2009). Kinetic study of fluid gel formation and viscoelastic response with kappa-carrageenan. *Food Hydrocolloids* 23: 2054–2061.

Garti, N., Madar, Z., Aserin, A., and Sternhemein, B. (1997). Fenugreek galactomannan as food emulsifiers. *LWT – Food Science and Technology* 30: 305–311.

Gastone, F., Tosco, T., and Sethi, R. (2014). Green stabilization of microscale iron particles using guar gum: bulk rheology, sedimentation rate and enzymatic degradation. *Journal of Colloid Interface Science* 421: 33–43.

Ghebremedhin, M., Seiffert, S., and Vilgis, T.A. (2021). Physics of agarose fluid gels: rheological properties and microstructure. *Current Research in Food Science* 4: 436–448.

Gibson, W.G. (1992). Gellan gum. In: *Thickening and Gelling Agents for Foods* (ed. A.P. Imeson), 227–249. London: Blackie Academic and Professional.

Glicksman, M. (1969). *Gum Technology in the Food Industry*. New York: Academic Press.

Goycoolea, F.M., Morris, E.R., Richardson, R.K., and Bell, A.E. (1995). Solution rheology of mesquite gum in comparison with gum arabic. *Carbohydrate Polymers* 27: 37–45.

Grasdalen, H. and Smidsrod, O. (1987). Gelation of gellan gum. *Carbohydrate Polymers* 7: 371–393.

Gul, T., Shah, R.A., Islam, S., and Arif, M. (2013). MHD thin film flows of a third grade fluid on a vertical belt with slip boundary conditions. *Journal of Applied Mathematics* 2013: 707286.

Halmos, A.L. and Tiu, C. (1981). Liquid foodstuffs exhibiting yield stress and shear degradability. *Journal of Texture Studies* 12: 39–46.

Herschel, W.H. and Bulkley, R. (1926). Konsistenzmessungen von Gummi-Benzollösungen. *Kolloid Zeitschrift* 39: 291–300.

Kelco (1993). *Gellan Gum, Multifunctional Polysaccharide for Gelling and Texturizing*. San Diego, CA.

Kennedy, P. (2003). *A Guide to Econometrics*, 5e, 205–206. Cambridge, MA: The MIT Press.

Lamba, H., Sathish, K., and Sabikhi, L. (2015). Double emulsions: emerging delivery system for plant bioactives. *Food and Bioprocess Technology* 8: 709–728.

Lapasin, R. (2012). *Rheology of Industrial Polysaccharides: Theory and Applications*. Berlin: Springer Science & Business Media.

Li, X., Dong, Y., Guo, Y. et al. (2019). Okra polysaccharides reduced the gelling-required sucrose content in its synergistic gel with high-methoxyl pectin by microphase separation effect. *Food Hydrocolloids* 95: 506–516.

Longrée, K., Behaver, S., Buck, P., and Nowrey, J.E. (1966). Viscous behavior of custard systems. *Journal of Agriculture and Food Chemistry* 14: 653–659.

Ma, L. and Barbosa-Canovas, G.V. (1995). Rheological characterization of mayonnaise. Part II: Flow and viscoelastic properties at different oil and xanthan gum concentrations. *Journal of Food Engineering* 25: 409–425.

Mahdi, M.H., Conway, B.R., and Smith, A.M. (2014). Evaluation of gellan gum fluid gels as modified release oral liquids. *International Journal of Pharmaceutics* 475: 335–343.

Malik, M.Y., Naseer, M., Nadeem, S., and Rehman, A. (2014). The boundary layer flow of Casson nanofluid over a vertical exponentially stretching cylinder. *Applied Nanoscience* 4: 869–873.

Mano, J.F., Silva, G.A., Azevedo, H.S. et al. (2007). Natural origin biodegradable systems in tissue engineering and regenerative medicine: present status and some moving trends. *Journal of the Royal Society* 4: 999–1030.

McClements, D. and Demetriades, K. (1998). An integrated approach to the development of reduced-fat food emulsions. *Critical Reviews in Food Science and Nutrition* 38: 511–536.

McIlveen-Farley, H. and Armstrong, G. (1995). Consumer acceptance of low fat and fat-substituted dairy products. *Journal of Consumer Studies and Home Economics* 19: 277–287.

McNeely, W.H. and Kang, K.S. (1973). Xanthan and some other biosynthetic gums. In: *Industrial Gums* (ed. R.L. Whistler and N.J. BeMiller), 473–521. New York: Academic Press.

Mendis, M. and Simsek, S. (2014). Arabinoxylans and human health. *Food Hydrocolloids* 42: 239–243.

Moorhouse, R., Colegrove, G.T., Sandford, P.A. et al. (1981). PS60: a new gel forming polysaccharide. In: *Solution Properties of Polysaccharides*, ACS Symp. Ser., vol. 150 (ed. D.A. Brant), 111–124. Washington DC: ACS.

Mudgil, D. (2018). Influence of partially hydrolyzed guar gum as soluble fiber on physicochemical, textural and sensory characteristics of yoghurt. *Journal of Microbiology, Biotechnology and Food Sciences* 8: 794–797.

Mudgil, D. (2020). Mesquite gum (Prosopis gum): structure, properties & applications – a review. *International Journal of Biological Macromolecules* 159: 1094–1102.

Mudgil, D., Barak, S., and Khatkar, B.S. (2016). Effect of partially hydrolyzed guar gum on pasting, thermo-mechanical and rheological properties of wheat dough. *International Journal of Biological Macromolecules* 93: 131–135.

Newman, R., Vilardell, N., Clavé, P., and Speyer, R. (2016). Effect of bolus viscosity on the safety and efficacy of swallowing and the kinematics of the swallow response in patients with oropharyngeal dysphagia: white paper by the European Society for Swallowing Disorders (ESSD). *Dysphagia* 31: 232–249.

Nguyen, Q.D., Jensen, C.T.B., and Kristensen, P.G. (1998). Experimental and modeling studies of the flow properties of maize and waxy maize starch pastes. *Chemical Engineering Journal* 70: 165–171.

Niu, F.G., Zhang, Y.T., Chang, C.H. et al. (2017). Influence of the preparation method on the structure formed by ovalbumin/gum arabic to observe the stability of oil-in-water emulsion. *Food Hydrocolloids* 63: 602–610.

Nordqvist, D. and Vilgis, T.A. (2011). Rheological study of the gelation process of agarose based solutions. *Food Biophysics* 6: 450–460.

Norton, J.Y. (1917). Heavy French dressing. In: *Mrs. Norton's Cook-Book: Selecting, Cooking, and Serving for the Home Table*, 354. Whitefish, MT: Kessinger Publishing, LLC.

Norton, I.T., Jarvis, D.A., and Foster, T.J. (1999). A molecular model for the formation and properties of fluid gels. *International Journal of Biological Macromolecules* 26: 255–261.

Norton, I.T., Frith, W.J., and Ablett, S. (2006). Fluid gels, mixed fluid gels and satiety. *Food Hydrocolloids* 20: 229–239.

Nowak, C. and Vilgis, T.A. (2004). Rod-coil multiblock copolymers: structure and stability. *Europhysics Letters* 68: 44–50.

Nussinovitch, A. (1997). *Hydrocolloid Applications, Gum Technology in the Food and Other Industries*, 154–168. London: Chapman and Hall, Blackie Academic and Professional.

Nussinovitch, A. (2010). *Plant Gum Exudates of the World, Sources, Distribution, Properties and Applications*. Boca Raton, FL: CRC Press, Taylor & Francis Group.

Nussinovitch, A. and Hirashima, M. (2014). *Cooking Innovations, Using Hydrocolloids for Thickening, Gelling and Emulsification*. Boca Raton, FL: CRC Press.

Nussinovitch, A. and Hirashima, M. (2019). *More Cooking Innovations, Novel Hydrocolloids for Special Dishes*. Boca Raton, FL: CRC Press.

Oates, K.M.N., Krause, W.E., Jones, R.L., and Colby, R.H. (2005). Rheopexy of synovial fluid and protein aggregation. *The Journal of the Royal Society Interface* 3: 167–174.

O'Carroll, P. (1997). Making it work. *The World of Ingredients: The journal of the Practicing Food Technologist* 6: 16–18.

Okajima, M.K., Ono, M., Kabata, K., and Kaneko, T. (2007). Extraction of novel sulfated polysaccharides from *Aphanothece sacrum* (Sur.) Okada, and its spectroscopic characterization. *Pure and Applied Chemistry* 79: 2039–2046.

Okajima, M.K., Bamba, T., Kaneko, Y. et al. (2008). Supergiant ampholytic sugar chains with imbalanced charge ratio form saline ultra-absorbent hydrogels. *Macromolecules* 41: 4061–4064.

Okeyoshi, K., Shinhama, T., Budpud, K. et al. (2018). Micelle-mediated self-assembly of microfibers bridging millimeter-scale gap to form three-dimensional-ordered polysaccharide membranes. *Langmuir* 34: 13965–13970.

Ong, J.J., Steele, C.M., and Duizer, L.M. (2018). Sensory characteristics of liquids thickened with commercial thickeners to levels specified in the International Dysphagia Diet Standardization Initiative (IDDSI) framework. *Food Hydrocolloids* 79: 208–217.

Oppermann, A.K.L., Noppers, J.M.E., Stieger, M., and Scholten, E. (2018). Effect of outer water phase composition on oil droplet size and yield of (w_1/o/w_2) double emulsions. *Food Research International* 107: 148–157.

Overbosch, P., Agterof, W.G.M., and Haring, P.G.M. (1991). Flavor release in the mouth. *Food Reviews International* 7: 137–184.

Pangborn, R.M. and Szczesniak, A.S. (1974). Effect of hydrocolloids and viscosity on flavor and odor intensities of aromatic flavor compounds. *Journal of Texture Studies* 4: 467–482.

Penczek, S. (2018). *Models of Biopolymers by Ring-Opening Polymerization*. Boca Raton, FL: CRC Press.

Pettitt, D.J. (1982). Xanthan gum. In: *Food Hydrocolloids*, vol. 1 (ed. M. Glicksman), 127–149. Boca Raton, FL: CRC Press.

Phillips, G.O. and Williams, P.A. (2009). *Handbook of Hydrocolloids*, 2e. Boca Raton, FL: CRC Press.

Phillips, G.O. and Williams, P.A. (2021). *Handbook of Hydrocolloids*, 3e. Kidlington, UK: Woodhead Publishing Series in Food Science, Technology and Nutrition.

Poiseuille, J.L.M. (1846). Recherches expérimentales sur le mouvement des liquides dans les tubes de très petits diamètres. *Mémoires présentés par Divers Savans à l'Académie Royale des Sciences de l'Institut de France* 9: 433–544.

Ptaszek, A. (2010). Rheological equation of state for shear-thickening food systems. *Journal of Food Engineering* 100: 322–328.

Razavi, S.M.A. and Karazhiyan, H. (2009). Flow properties and thixotropy of selected hydrocolloids: experimental and modeling studies. *Food Hydrocolloids* 23: 908–912.

Reiner, M. and Schoenfeld-Reiner, R. (1933). Viskosimetrische Untersuchungen an Lösungen hochmolekularer Naturstoffe. I. Mitteilung. Kautschuk in Toluol. *Kolloid Zeitschrift* 65: 44–62.

Reynolds, O. (1883). An experimental investigation of the circumstances which determine whether the motion of water shall be direct or sinuous, and of the law of resistance in parallel channels. *Proceedings of the Royal Society of London* 35: 84–99.

Rha, C. (1980). *Food Rheology Principles and Practice*, Lecture Notes, Food Materials Science Fabrication Lab. Cambridge, MA: MIT Press.

Richardson, R.K., Morris, E.R., Ross-Murphy, S.B. et al. (1989). Characterization of the perceived texture of thickened systems by dynamic viscosity measurements. *Food Hydrocolloids* 3: 175–191.

Rofes, L., Arreola, V., Mukherjee, R. et al. (2014). The effects of a xanthan gum-based thickener on the swallowing function of patients with dysphagia. *Alimentary Pharmacology and Therapeutics* 39: 1169–1179.

Russ, N., Zielbauer, B.I., Koynov, K., and Vilgis, T.A. (2013). Influence of non-gelling hydrocolloids on the gelation of agarose. *Biomacromolecules* 14: 4116–4124.

Sanderson, G.R. (1990). Gellan gum. In: *Food Gels* (ed. P. Harris), 201–233. London: Elsevier Science.

Sanderson, G.R., Bell, V.L., Clark, R.C., and Ortega, D. (1988). The texture of gellan gum gels. In: *Gums and Stabilizers for the Food Industry*, vol. 4 (ed. G.O. Phillips, D.J. Wedlock and P.A. Williams), 219–229. Oxford: IRL Press.

Saramito, P. (2016). *Complex Fluids: Modeling and Algorithms*, 65. Cham, Switzerland: Springer International Publishing.

Sato, A.C.K., Polastro, M.Z., Furtado, G.D., and Cunha, R.L. (2018). Gelled double-layered emulsions for protection of flaxseed oil. *Food Biophysics* 13: 316–323.

Schorsch, C., Jones, M.G., and Norton, I.T. (2000). Phase behaviour of pure micellar casein/κ-carrageenan systems in milk salt ultrafiltrate. *Food Hydrocolloids* 14: 347–358.

Shashi Menon, E. (2015). *Transmission Pipeline Calculations and Simulations Manual*. Chapter 3 – Physical Properties, 29–82. Amsterdam: Elsevier.

Steffe, J.F. (1996). *Rheological Methods in Food Process Engineering*. East Lansing, MI: Freeman Press.

Stokes, G.G. (1842). On the steady motion of incompressible fluids. *Transactions of the Cambridge Philosophical Society* 7: 439–453.

Stone, H. and Oliver, S. (1966). Effect of viscosity on the detection of relative sweetness intensity of sucrose solutions. *Journal of Food Science* 31: 129–134.

Subbaraman, V., Mashelkar, R.A., and Ulbrecht, J. (1971). Extrapolation procedures for zero shear viscosity with a falling sphere viscometer. *Rheologica Acta* 10: 429–433.

Sun, A. and Gunasekaran, S. (2009). Yield stress in foods: measurements and applications. *International Journal of Food Properties* 12: 70–101.

Sworn, G., Sanderson, R.G., and Gibson, W. (1995). Gellan gum fluid gels. *Food Hydrocolloids* 9: 265–271.

Szczesniak, A.S. and Farkas, E. (1962). Objective characterization of the mouthfeel of gum solutions. *Journal of Food Science* 27: 381–385.

Tang, H.S. and Kalyon, D.M. (2004). Estimation of the parameters of Herschel–Bulkley fluid under wall slip using a combination of capillary and squeeze flow viscometers. *Rheologica Acta* 43: 80–88.

Thebaudin, J.Y., Lefebvre, A.C., and Doublier, J.L. (1998). Rheology of starch pastes from starches of different origins: applications to starch-based sauces. *LWT – Food Science and Technology* 31: 354–360.

Tiziani, S. and Vodovotz, Y. (2005). Rheological effects of soy protein addition to tomato juice. *Food Hydrocolloids* 19: 45–52.

Torres, M.D., Moreira, R., Chenlo, F., and Vazquez, M.J. (2012). Water adsorption isotherms of carboxymethyl cellulose, guar, locust bean, tragacanth and xanthan gums. *Carbohydrate Polymers* 89: 592–598.

Torres, O., Yamada, A., Rigby, N.M. et al. (2019). Gellan gum: a new member in the dysphagia thickener family. *Biotribology* 17: 8–18.

Turgeon, S.L., Sanchez, C., Gauthier, S.F., and Paquin, P. (1996). Stability and rheological properties of salad dressing containing peptidic fractions of whey proteins. *International Dairy Journal* 6: 645–658.

Urlacher, B. and Dalbe, B. (1992). Xanthan gum. In: *Thickening and Gelling Agents for Food* (ed. A. Imeson), 206–226. Glasgow: Blackie Academic and Professional.

Vafiadis, D.K. (1996). Culture clash. *Dairy Field* 179: 47–49.

Venkatesan, J., Sankar, D.S., Hemalatha, K., and Yatim, Y. (2013). Mathematical and numerical modeling of flow and transport. *Journal of Applied Mathematics* 2013: 583809.

Verbeken, D., Dierckx, S., and Dewettinck, K. (2003). Exudate gums: occurrence, production, and applications. *Applied Microbiology and Biotechnology* 63: 10–21.

Vernon-Carter, E.J. and Sherman, P. (1980). Rheological properties and applications of mesquite tree (*Prosopis juliflora*) gum. 1. Rheological properties of aqueous mesquite gum solutions. *Journal of Texture Studies* 11: 339–349.

Vickers, Z., Damodhar, H., Grummer, C. et al. (2015). Relationships among rheological, sensory texture, and swallowing pressure measurements of hydrocolloid-thickened fluids. *Dysphagia* 30: 702–713.

Vilgis, T.A. (2015). Gels: model systems for soft matter food physics. *Current Opinion in Food Science* 3: 71–84.

van Vliet, T. (2002). On the relation between texture perception and fundamental mechanical parameters for liquids and time dependent solids. *Food Quality and Preference* 13: 227–236.

Weber, F.H., Clerici, M.T.P.S., Collares-Queiroz, F.P., and Chang, Y.K. (2009). Interaction of guar and xanthan gums with starch in the gels obtained from normal, waxy and high-amylose corn starches. *Starch* 61: 28–34.

Weltman, R.N. (1943). Breakdown of thixotropic structure as a function of time. *Journal of Applied Physics* 14: 343–350.

Wu, Y., Cui, S.W., Eskin, M., and Doff, H.D. (2009). An investigation of four commercial galactomannans on their emulsion and rheological properties. *Food Research International* 42: 1141–1146.

Yusof, F.A.A., Yamaki, M., Kawai, M. et al. (2020). Rheopectic behavior for aqueous solutions of megamolecular polysaccharide sacran. *Biomolecules* 10: 155. https://doi.org/10.3390/biom10010155.

Zhan, Z., Du, B., Peng, S. et al. (2013). Homogeneous synthesis of hydroxypropyl guar gum in an ionic liquid 1-butyl-3-methylimidazolium chloride. *Carbohydrate Polymers* 93: 686–690.

Zhang, F., Luan, T., Kang, D. et al. (2015). Viscosifying properties of corn fiber gum with various polysaccharides. *Food Hydrocolloids* 43: 218–227.

Zhao, H.R. and Zhang, K.M. (2018). The structure recovery capacity of highly concentrated emulsions under shear flow via studying their rheopexy. *Journal of Dispersion Science and Technology* 39: 970–976.

Zhu, Q., Qiu, S., Zhang, H. et al. (2018). Physical stability, microstructure and micro-rheological properties of water-in-oil-in-water (W/O/W) emulsions stabilized by porcine gelatin. *Food Chemistry* 253: 63–70.

Zimm, B.H. and Bragg, J.K. (1959). Theory of the phase transition between helix and random coil in polypeptide chains. *The Journal of Chemical Physics* 31: 526–535.

5

Use of Hydrocolloids to Improve the Texture of Crispy, Crunchy, and Crackly Foods

5.1 Introduction

Food texture is primarily judged by the tactile senses' detection of physical stimuli resulting from contact between some part of the body and the food. Touch is the main means of sensing texture but kinesthetics (sense of movement and position), and sometimes sight (degree of slump, rate of flow) and sound (associated with crisp, crunchy, and crackly textures) also come into play. This chapter starts with definitions of crispness and crunchiness and their dependence on a food's moisture and oil content. The mechanical, acoustical, and temporal aspects of crunchiness and crispness are described and analyzed. An entire section is devoted to the definition of crackly foods and how to influence their properties. Methodologies for improving the texture of crispy and crunchy foods using hydrocolloids are described in detail, with sections on coatings and batters for vacuum frying. A special section is devoted to those foods that are treated for the enhancement of acoustic properties by various hydrocolloids. Cases such as the contribution of inulin to crispness of biscuits, pizza, and wafers; crispness of banana chips; specialty starches as functional ingredients and in snack foods; and protein-rich extruded snacks are also described. A recipe for baked tortilla chips is provided, as well as photographs of typical crisp and crunchy products.

5.2 Definitions of Crispness and Crunchiness

Crispness and crunchiness are defined differently among consumers, dictionaries, and scholars. Acoustic, mechanical, and sensorial approaches have been used to provide information on these parameters. Sensory measurements include sound intensity and biting force. Mechanical techniques mimic mastication and include compression, stretch, and shear. Acoustical practices quantify frequency, intensity, and sound measures. Crispness and crunchiness also have temporal characteristics (Tunick et al. 2013). Crispness and crunchiness are textural qualities that are frequently related to the firmness and freshness of natural crops and manufactured foods (Tunick et al. 2013). Crunchy and crispy foods are attractive and pleasurable. Crispness has been labeled the most adopted textural parameter of a product because it is generally liked, and this obvious textural characteristic is linked to top-quality cooking (Szczesniak and

---

Use of Hydrocolloids to Control Food Appearance, Flavor, Texture, and Nutrition, First Edition.
Amos Nussinovitch and Madoka Hirashima.
© 2023 John Wiley & Sons Ltd. Published 2023 by John Wiley & Sons Ltd.

Kahn 1971, 1985). Crunchiness is often put on display and has been linked with pleasure and fun (Szczesniak and Kahn 1971, 1985). Sounds made while eating can modulate people's perceptions of numerous aspects of foods, and may affect taste awareness (Zampini and Spence 2010). Crispness and crunchiness are vital when addressing food texture, but academics agree that these qualities are not well defined (Chauvin et al. 2008; Varela et al. 2008c, 2008d). In a study of auditory rating scales for foods, crispness and crunch demonstrated the largest positive correlations with enjoyableness (Chauvin et al. 2008). From time to time, the term "crispy" is used to illustrate qualities defined by others as "crunchy" (Chauvin et al. 2008). Some researchers have considered these terms to be interchangeable (Chen et al. 2005). A strong correlation has been established between crispness and crunchiness by some (Vickers 1984b, 1985; Seymour and Hamann 1988; Ioannides et al. 2007), whereas others consider them to be dissimilar factors (Seymour et al. 1991; Dacremont 1995; Roudaut et al. 2002).

The term *crisp* is derived from the Latin *crispus*, meaning, "curled." This adjective meant "wrinkled" or "rippled" in 14th century English and the meaning was altered to "brittle but hard or firm" in the 16th century (Onions 1982). Dictionary characterizations of crispness consist of "firm, dry, and brittle, especially in a way considered pleasing" (Concise Oxford English Dictionary 2008) "firm but easily broken or crumbled; brittle" (American Heritage Dictionary of the English Language 2006) and "easily crumbled, brittle, desirably firm and crunchy" (Merriam-Webster's Collegiate Dictionary 2008). Finding a middle ground classification might be "desirably firm and brittle, and easily crumbled" (Tunick et al. 2013). The term *crunch* became an English verb in the 19th century, modified from *cranch*, a 17th century word that was most likely commonplace (Onions 1982; Tunick et al. 2013). Meanings include: "crush (a hard or brittle foodstuff) with the teeth, making a loud grinding sound" (Concise Oxford English Dictionary 2008); "chew with a noisy crackling sound" (American Heritage Dictionary of the English Language 2006), and "chew or press with a crushing noise" (Merriam-Webster's Collegiate Dictionary 2008). A compromise for these classifications could be "chew with a crushing noise" (Tunick et al. 2013). Textural terminologies diverge because of variations in language and culture. Individuals repeatedly use the terms crunchy, crispy, and brittle interchangeably and yet they are not identical (Peleg 1983). For instance, the Japanese and Chinese each have about a dozen words referring to these three attributes, including words translating as "sprinkling" and "rustling" (Szczesniak 1988). Quantifiable descriptors are required to delimit that which is crunchy or crispy (Tunick et al. 2013). Sensory-panel intensity data and values for crispness and crunchiness, adjusted to a 100-point scale for comparison, are given in Table 5.1. These data provide an idea of what are considered to be crunchy and crispy foods, but they are not statistically valid due to the different techniques used to determine the values and/or unreported specimen size (Tunick et al. 2013).

5.3 Dependence of Crunchiness and Crispness on Moisture and Oil Content

Foods can be divided into wet crispy and dry crispy based on their moisture content (Mohamed et al. 1982; Chauvin et al. 2008). Raw fruit and vegetables are typical examples of wet crispy foods. Their texture depends on many factors, among them size and some

Table 5.1 Intensity data for crispness and crunchiness, converted to a scale of 0–100 and rounded to the nearest 5.

Product	Crispness value[a]	Crunchiness value[a]	Size	References
Banana	0	0	1.3 cm slice	Chauvin et al. (2008)
Cereal marshmallow bar	15	0	⅙ bar	Chauvin et al. (2008)
Gala apple	25	25	1.3 cm slice	Chauvin et al. (2008)
Cucumber	30	40	1.0 cm slice	Vincent et al. (2002)
Carrot	45	70	7.0 mm slice	Vincent et al. (2002)
Peanut	65	30	One peanut	Vickers (1985)
Celery	65	60	1.0 cm stalk segment	Vincent et al. (2002)
Graham cracker	75	65	0.4×3.0 cm	Vickers (1985)
Ginger snap cookie	75	70	0.5×1.2 cm	Vickers (1985)
Peanut brittle	90	80	0.7×1.0 cm	Vickers (1985)
Melba toast	100	100	½ cracker	Chauvin et al. (2008) and Meilgaard et al. (2007)

Source: Adapted with changes from Tunick et al. (2013).
[a] Data from first bite are used for crispness scores and from chewing with molars for crunchiness scores.

mechanical properties of their water-filled cellular structures during consumption (Varela et al. 2008c). Water activity (a_w) is defined as: $a_w = p/p^*$ where p is the partial water vapor pressure in equilibrium with the solution, and p^* is the (partial) vapor pressure of pure water at the same temperature. Water activity is a key characteristic for food product design and food safety (https://en.wikipedia.org/wiki/Water_activity); a_w values of fresh fruit and vegetables range between 0.960 and 0.999 (Chirife and Ferro Fontan 1982). Force is required to bite into and chew the fresh produce, and the sounds of these actions evidence the product's crispness (Vickers 1985, 1987). Dry crisp products are comprised of cells filled with air instead of water. Crackers, potato chips (Figure 5.1), and popcorn (Figure 5.2) are dry crisp products with $a_w < 0.1$ (Katz and Labuza 1981). On the one hand, researchers claim that the sensations of wet crispness and dry crispness do not differ, and that they are based on the same auditory cues. On the other hand, panelists were able to distinguish between dry and wet food crispness. Thus, it was concluded that the difference is related to the means of sound propagation in water or air (Vickers and Christensen 1980; Chauvin et al. 2008). Crispness and crunchiness are related to a_w. The sensory crispness scores of crackers, extruded snacks, and potato chips decreased with moisture content in a straight-line relationship with $r^2 > 0.9$ (Katz and Labuza 1981; Srisawas and Jindal 2003). Force-deformation curves and acoustic spectra of crunchy dry foods lose their jaggedness when moisture is added (Rohde et al. 1993). Data for breakfast cereals (Sauveageot and Blond 1991) were re-deduced and considerable alterations were observed in the range of a_w corresponding to most of the crispness and crunchiness (Peleg 1994). The speed of deformation in conjunction with a_w was also related to crispness. A critical a_w at which crispness is lost was defined

Figure 5.1 Grandma Utz's Kettle-Style Potato Chips. *Source:* Evan-Amos/Wikipedia Commons/Public Domain.

Figure 5.2 Popcorn. *Source:* ElinorD/Wikipedia Commons/Public Domain.

(Castro-Prada et al. 2009). As the speed of deformation increased from 10 to 40 mm/s, the critical a_w increased from 0.40 to 0.5–0.6 (Castro-Prada et al. 2009).

Crispness and crunchiness of fried foods depend on, among other issues, their oil content. For example, the crispness of cooked frozen French fries was significantly correlated with frying time. Oil uptake increased as moisture was lost and leveled off at about 29% as the moisture content dropped below 35% (Du Pont et al. 1992). Whey protein isolate was used as a post-breading dip to decrease oil absorption, resulting in higher crunchiness due to increased moisture retention (Mah and Brannan 2009). Voids in fried foods are created by water evaporation during frying. Condensation of water vapor when the product was cooled reduced the internal pressure, essentially sucking oil adhering to the food surface into the product (Dana and Saguy 2006). Oil uptake does not lead to substantial changes in the mechanical properties of crispy cellular foods; nevertheless, the number of acoustic events and the acoustic energy are greatly reduced upon fracture (van Vliet et al. 2007). This consequence seems to be due to the reflection of sound at the oil–air interface and rises within 20 minutes of frying, resulting in loss of crispness (van Vliet et al. 2007).

5.4 Mechanical, Acoustical, and Temporal Aspects of Crunchiness and Crispness

Mastication fractures crispy or crunchy foods. Mechanical assessments are used to determine the crisp and crunch properties of rigid foods, involving bending, shear, and compression methods. Representative mechanical techniques for determining crispness are summarized in Table 5.2. Compression may involve puncture, and is directly characterized by peak force. Squeezing a sample that is sandwiched between parallel plates or using a piston inside a cylinder does not produce a break point with rigid structures (Pons and Fiszman 1996). A jagged force-deformation curve is correlated to sensory crunchiness and crispness (Peleg 1994) and the latter is associated with big fracture events (Vincent 1998). Once a food is fractured, the resultant pieces can be crunchy or crispy. Marcona almonds are sweet, gourmet almonds from Spain that are very popular in Mediterranean countries. They have a delicate aroma and taste (https://www.thespruceeats.com/what-is-a-marcona-almond-1375433) that is reminiscent of the almond essence used in baked goods. Image analysis was used to quantify the fracture properties and microstructural features of roasted Marcona almonds (Varela et al. 2008a, 2008b). This method proved to be a very useful tool for analyzing the crispness/crunchiness behavior of the samples. Disruption of the inner parenchyma in samples roasted for longer times proved to be the principal reason for their increased brittleness. The heterogeneity produced by heat degradation contributed to the

Table 5.2 Representative mechanical techniques for determining crispness.

Product	Technique	Conditions	Measurement	References
Apples	Puncture	Penetrate 8 mm at 4 mm/s	Puncture force	Harker et al. (2002)
Biscuits	Constant loading rate	1.23 kg/s for 5 s	Fracture force and rate, work done	Vickers and Christensen (1980)
Extruded corn puffs	Compression	50 mm/s, 67% compression	Time to reach maximum force	Duizer et al. (1998)
Hazelnuts	Compression	⅙ mm/s	Force and deformation through second fracture point	Saklar et al. (1999)
Pickles	Texture profile analysis	⅝ mm/s	Brittleness, hardness, total work	Jeon et al. (1975)

Source: Adapted with changes from Tunick et al. (2013).

failure of the material under compression, changing it from a deformable hard solid (raw sample) to a brittle one with enhanced crispy/crunchy characteristics (Varela et al. 2008a).

Crispness or crunchiness is fundamentally auditory perception (Tunick et al. 2013). Acceptability of a product depends on its anticipated sound level (Duizer 2001). The sound formed by eating or chewing a solid food can be identified through conduction of air to the ear, through the soft tissues in the mouth, and via conduction through the jawbone (Duizer et al. 1998). For a strong awareness of the nature and intensity of bone-mediated sounds released while food is being chewed, the noises need to be propagated at a frequency of 160 Hz. Air-mediated noises must be propagated at 160 Hz and amplified at 3.5 kHz. The input of air and bone conductance to the establishment of sounds perceived during the sensory evaluation of foods has been discussed (Dacremont et al. 1991). The two sounds need to be joined and equalized to experience precise acoustic sensations when the food is consumed (Dacremont et al. 1991). Air conduction is the main phenomenon when biting into the food with the incisors with an open mouth, whereas bone conduction dominates when chewing with molars and with a closed mouth. Bone-conducted sounds are characteristically of lower frequencies because some of the sound is absorbed by soft mouth tissue and by the jaw (Duizer 2001).

A ground-breaking study on noises made when food is chewed was conducted in the 1960s (Drake 1963, 1965). Volunteers chewed a number of foods and the sounds were amplified by microphones and recorded. The first of this series of studies dealt with food-crushing sounds (Drake 1963), and the second compared these sounds' subjective and objective data (Drake 1965). The crispness of food (Vickers and Bourne 1976a) and a psycho-acoustical theory of crispness were later reviewed and evaluated (Vickers and Bourne 1976b). The relationship between chewing sounds and perception of a food's crispness was further evaluated and found to be highly correlated (Christensen and Vickers 1981). Acoustic emission, fracture behavior, and morphology of dry crispy foods were further discussed by Luyten and van Vliet (2006). It was concluded that prerequisites for crispy and crunchy foods are high crack speed and the sound that accompanies it (Luyten and van Vliet 2006). Disintegration of crunchy and crispy foods yielded jagged acoustic outputs, with large fluctuations in peak number and amplitude. Fast Fourier transform (FFT) was used to interpret the data (Barrett et al. 1994). The mouth detects the force and the results are viewed in terms of stress (Vincent 1998). Fractal analysis can be used to analyze acoustic signatures (Tesch et al. 1995). Acoustic tests of cheese balls and croutons during compression produced a sigmoidal relationship between apparent fractal dimension and water activity (Tesch et al. 1995). Different parameters derived from acoustic data include the amplitude of the produced sound (Drake 1965; Al Chakra et al. 1996), maximum sound level (Chen et al. 2005), mean peak height (Vickers 1987; Srisawas and Jindal 2003), number of sound events (Vickers and Bourne 1976a; Vickers 1987; Srisawas and Jindal 2003), sound duration and energy (Al Chakra et al. 1996), and sound pressure and intensity (Seymour and Hamann 1988).

Crunchiness is associated with the elapsed time of chewing. Such assessments fall under compression or multiple break forces. It was noted that analyzing a series of chews provides more information than that from only the first chew (Lee et al. 1998). Nevertheless, the total time to be considered when determining crunchiness is subject to disagreement. Crunchiness has been defined as perceived intensity of repeated incremental failure during a single complete bite with the molars (Barrett et al. 1994). Another definition is "observed accumulative intensity of force necessary for repeated incremental failures of the product by chewing up to five times with molars" (Guraya and Toledo 1996), whereas other researchers distinguish a relationship between crunchiness and chewing effort from the fifth to tenth chew. This latter study indicated the importance of the temporal changes taking place in the sample throughout mastication to awareness of textural features (Brown et al. 1998).

5.5 Crackly Foods

We can learn a lot about the texture of a food, whether it is crispy, crunchy, or crackly, from the chewing sounds perceived while biting and masticating. In other words, what we hear can help us categorize the textural properties of what is being eaten (Spence 2015). In reality, with the understanding of just how significant the sound is to inclusive multisensory familiarity comes the realization of why food salespersons devote much of their time to highlighting its crispy, crunchy, and crackly sounds in their commercials (Vickers 1977). Crispness, crunchiness, and crackliness are not only valuable descriptors of food texture; they are also desirable textural qualities for many foods (Chauvin et al. 2008). Nevertheless, crackly perceptions have not received much research attention. Table 5.3 lists literature citations for the crackliness of foods. Crackly foods are characteristically recognized by the sharp sudden and repeated bursts of noise that they make upon chewing (Vickers 1984a)). Masking these sounds leads to a reduction in perceived crackliness. The number of sounds produced offers a practical measure of this parameter (Chauvin et al. 2008). Multidimensional scaling output demonstrated that crispness, crunchiness, and crackliness are different sensory texture parameters with distinct concepts and can be analyzed by auditory cues alone. In addition, it was demonstrated that there is a perceptual difference between texture attributes for dry and wet foods and that differentiating between them is essential for descriptive analysis training (Chauvin et al. 2008).

Table 5.3 Literature citations and definitions of crackliness.

Definition	Technique	References
To make small, sharp, sudden and repeated noises.	At first bite or during chewing	Vickers (1984a)
Products that produce low-pitched sounds with a high level of bone conduction. Discrimination between crackly and crunchy foods could be due to vibrations propagated by bone conduction that also generate vibrotactile sensations.	Only incisors	Dacremont (1995)
Mixture of sound and bite force. Breaks upon biting with incisors. Crackling is harder than crispness: it snaps.	At first bite	Dijksterhuis et al. (2005)
Audible for a long period during chewing, bigger pieces, mainly of the crust.	During chewing	Dijksterhuis et al. (2005)

Source: Adapted with changes from Chauvin et al. (2008).

Foods that make a crackly sound include pork scratchings, or the appropriately termed pork crackling (Spence 2015). Pork rind is the culinary term for the skin of a pig. It can be rendered, fried in fat, baked or roasted to produce a kind of pork crackling (US) or scratching (UK) which is served in small pieces as a snack or side dish (https://en.wikipedia.org/wiki/Pork; https://web.archive.org/web/20121209035852/http:/freshersfoods.com/history-of-pork-scratchings/).

The crispness, crunchiness, and hardness of foods are quite strongly correlated. Different languages use different terms – or have no terms at all – to capture some of these textural distinctions (Spence 2015). The French *croustillant* and *croquant* can be translated into English as crispness and crunchiness, respectively (Drake 1989), and *craquant* as crackliness. However, the meanings of these corresponding French and English terms are not exactly equivalent. Szczesniak (1988) found that lettuce is an example of a food that is commonly described as crispy by Americans. The French would be more likely describe lettuce as *craquante* (crackly) or *croquante* (crunchy) but not *croustillante* (crispy). However, most results obtained with English descriptors can be transposed to French descriptors (Dacremont 2007).

Texture is often monitored quantitatively: subjects are asked to review the crispness intensity of foods (from not crispy to very crispy); then, physical characteristics of the food are sought that vary with crispness intensity. Dacremont (2007) took a different, qualitative approach to texture assessment; subjects were asked whether the tested foods fit with their concept of crispness (crunchiness or crackliness), and the spectral characteristics of the biting sounds used by the subjects to decide whether a food was crispy, crunchy, or crackly were determined. Subjects judged texture by both eating the foods and then listening to the reconstructed sounds, making it possible to assess how much information was lost when sensations other than the auditory ones were removed. Then, both air- and bone-conducted sounds were analyzed by a FFT-signal analyzer that calculates the spectral power density of biting sounds (Dacremont 2007). The spectral composition of the eating sounds generated by crispy, crunchy, and crackly foods was studied by examining eight food products that were selected according to three criteria: (i) they produce a sound when eaten, (ii) they can be easily cut into standard sizes and shapes, and (iii) they have a uniform texture (e.g. without a filling). The selected foods were roasted almonds, *bricelet* (dry biscuits), raw carrot, *cracotte* (extruded bread), *feuillete* (flaky pastry), *katimini* (flaky pastry), *langue de chat* (dry biscuit), and *speculoos* (dry biscuit) (Dacremont 2007). The separate air- and bone-conducted food sounds generated by six subjects biting into these eight foods were recorded and analyzed by FFT-signal analyzer. A panel of 60 subjects categorized the eight foods according to their texture – crispy, crunchy and crackly – and these textural characteristics were defined by spectral characteristics of biting sounds. Varela et al. (2006) reported that the sensory crispness of roasted almonds is highly correlated with the size of the acoustic peaks emitted by crackling. Moreover, Varela et al. (2008c) found that number of sound peaks (NSP) was best at discriminating precooked chicken nuggets and was directly related to food crispness.

Every sensation is multisensory – a combination of all senses working together. Touch is no different. The parchment-skin illusion is a famous study that showed how even the feel of our own skin can be completely swayed by something other than touch – in this case, sound. What you feel – such as a food's texture – is not just about touch. The sound it makes when you touch it greatly affects how you will perceive the sensation (https://www.sensebook.co.uk/the-parchment-skin-illusion). Using the parchment-skin illusion, researchers found that the crackliness perceived when chewing chips is significantly affected by modifying the overall sounds recorded near the subject's mouth and then playing these modified sounds back to them via headphones (Guest et al. 2002). Crispy foods (such as extruded flat breads) produce high-pitched sounds that display a high level of frequencies >5 kHz, particularly through air conduction (Dacremont 2007). Crunchy foods (such as raw carrot) create low-pitched sounds with a characteristic peak in the frequency range of 1.25–2 kHz for air conduction. Crackly foods (such as dry biscuits) (Figure 5.3) generate low-pitched sounds that are strongly conducted through the bone. It was hypothesized that discrimination between crunchy and crackly foods could be due to vibrations transmitted by bone conduction that also generate vibrotactile sensations (Dacremont 2007). Cellular brittle foods possess crispness, crunchiness, and crackliness and textural characteristics that are frequently connected to the freshness and firmness of natural produce and mass-produced foods (Wang et al. 2020). Crackliness is the sudden low-pitched sound event that such brittle cellular products produce when pressure is applied and evaluated with the molars and closed mouth (Michael et al. 2013). The effect of the structure formed by different types of hydrocolloids [low-methoxyl pectin, a mixture of xanthan gum (Figure 5.4) and locust bean gum, and a mixture of xanthan gum and guar gum] and aeration times (three, five, seven, and nine minutes) on textural properties of freeze-dried gels was investigated (Ciurzynska et al. 2017).

Freeze-drying is considered the best technique for obtaining a high-quality dried gel product. Nevertheless, several studies have shown that freeze-dried gels are characterized by a collapse of their internal structure, due to collapsed pore structure. Such low-density gels have an open cellular structure (Rassis et al. 2002; Sundaram and Durance 2008). It was claimed that the mechanical signature of brittle and crumby materials with a cellular structure is irregular (jagged) and irreproducible. The "noise" is a record of internal

Figure 5.3 Peek, Frean & Co's Biscuits, London. *Source:* Victoria and Albert Museum/Wikipedia Commons/Public Domain.

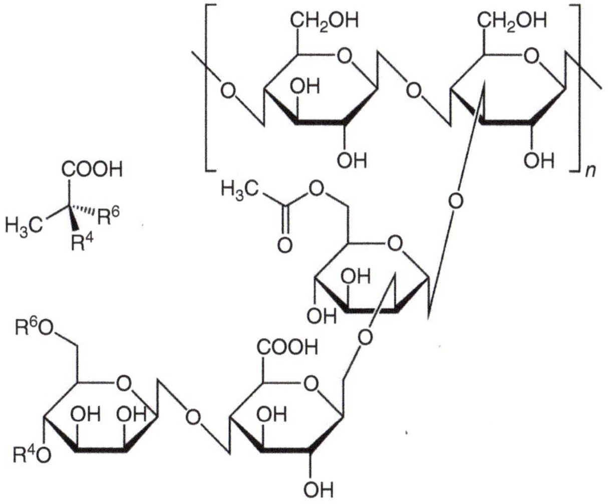

Figure 5.4 Structure of xanthan gum. *Source:* NEUROtiker/Wikimedia Commons/Public domain.

rupture events that should be monitored and analyzed but should not be smoothed or discarded (Peleg and Normand 1993). The total acoustic energy generated throughout deformation of raw materials and food products closely correlates with their mechanical parameters (Jakubczyk and Kaminska 2007; Marzec and Ziołkowski 2007). In other words, intensification in the values of mechanical parameters in freeze-dried gels contributes to a stronger acoustic emission. Looking at foods other than freeze-dried gels, it was observed that cookies have large, thin-walled pores that provide greater energy of acoustic events and amplitude than cakes with small pores (Marzec and Ziołkowski 2007). The degree of force increases and decreases throughout compression differed for different gel compositions. Freeze-dried low-methoxyl pectin gels were characterized by a large number of peaks, which may indicate fracture toughness and may be related to the presence of calcium ions in the solid walls (Nussinovitch et al. 2001).

Relationships among mechanical and acoustic properties and porosity, in addition to carefully selected parameters of the structure of freeze-dried gels, may bring us closer to explaining them (Ciurzynska et al. 2017). In one experiment, all freeze-dried gels were characterized by high porosity, in the range of 97–99%. Freeze-dried gels with low-methoxyl pectin had the lowest porosity of all samples tested. The hydrocolloid type in the gel had a notable impact on the average Feret diameter in the samples. Gels with low-methoxyl pectin were characterized by bigger pore size than those with a mixture of xanthan gum and guar gum. Thinner walls and pore edges were associated with reduced hardness of freeze-dried gels (Nussinovitch et al. 2004). It is likely that the smaller, but much more numerous pores in gels with a mixture of xanthan gum and guar gum affect the notable increase in porosity for those samples (Ciurzynska et al. 2017). The porosity of freeze-dried gels with a mixture of xanthan gum and guar gum (Figure 5.5) increased inconsequentially with an increase in aeration time. Curves of the relationship between force and distance demonstrated variations in deformation, connected to the differential structures of freeze-dried gels built from different hydrocolloids. Freeze-dried gels with low-methoxyl pectin had markedly higher water activity and the highest acoustic amplitude (Ciurzynska et al. 2017).

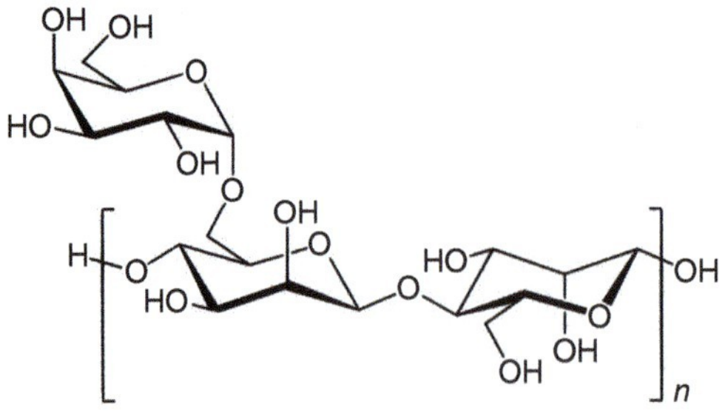

Figure 5.5 Structural formula of a guar gum unit. *Source:* Yikrazuul/Wikimedia Commons/Public domain.

5.6 Methods for Improving the Texture of Crispy and Crunchy Foods Using Hydrocolloids

5.6.1 Vacuum Frying

Vacuum frying is a novel way of increasing the value of fried foods through minimization of the final oil uptake by the product. The product is heated at low pressure, and the boiling points of the frying oil and the water in the product are decreased. Moreover, the absence of air during vacuum frying might inhibit lipid oxidation and enzymatic browning, consequently retaining most of the product's color and nutrients (Andres-Bello et al. 2011). Frying mainly changes the microstructure of raw vegetables, which defines their final physical and sensory properties. The most significant textural attribute of potato chips and French fries is crispness, reflecting high quality and freshness. A potato chip should be firm and snap without difficulty when bent, emitting a crunchy sound (Krokida et al. 2001). Numerous studies in the snack food industry have attempted to understand oil uptake during the vacuum-frying process to reduce the fat content of such products while maintaining their sought-after organoleptic characteristics (Andres-Bello et al. 2011). Vacuum frying is carried out under pressures that are well below atmospheric levels, preferably below 50 Torr (6.65 kPa). Several factors strongly affect oil absorption in fried products, including oil quality and composition, product shape, temperature and frying time, moisture content, initial porosity, and pre- and post-treatments (Saguy and Pinthus 1995; Bouchon and Aguilera 2001). Many studies have been carried out with different pretreatments combined with traditional frying or vacuum frying. Table 5.4 summarizes the pretreatments used before vacuum frying. The solutes in these treatments include dextrin (Figure 5.6) and maltodextrin (Figure 5.7). Common pretreatments in vacuum frying include blanching, osmotic dehydration, predrying in a microwave, freezing, or their combinations. The major objective of pretreatments is to minimize the final oil content in the product, as well as improve its texture by eliminating strong shrinkage (Andres-Bello et al. 2011). In addition, much consideration has been given to the use of hydrocolloids, e.g. methylcellulose (MC), hydroxypropyl methylcellulose (HPMC), long fiber cellulose, and corn zein, to decrease surface permeability and hinder oil uptake (Bouchon and Aguilera 2001). MC was employed to increase stability, modify the surface and control the moisture content of fried products (Chaisawang and Suphantharika 2005). Coating bananas chips with 1% (w/v) pectin solution reduced oil content in the fried product by approximately 23% (Singthong and Thongkaew 2009). Banana chips coated with guar gum and subjected to a higher-speed oil centrifugation step in vacuum frying preserved good quality

Table 5.4 Pretreatments carried out before vacuum frying.

Product	Treatment	Solute	Concentration (%)	Temperature (°C)	Time	References
Carrot chips	Blanched + immersed in fructose	NaCl/malt syrup:dextrin	2/2:1	100/25	2 min/1 h/ overnight	Fan et al. (2005)
Green beans Mango snacks	Immersed in citric acid and maltodextrin solution	Citric acid/ maltodextrin	0.15/50	25	1 h	Da Silva and Moreira (2008)
Mango chips	Osmotic dehydration	Citric acid/ maltodextrin	c. 0.15/40, 50, 65 w/v	22/40	45, 60 and 70 min	Munes and Moreira (2009)

Source: Adapted with changes from Andres-Bello et al. (2011).

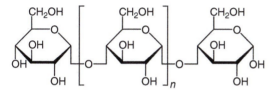

Figure 5.6 Structure of poly-(1→4)-alpha-D-glucose (e.g.dextrin). Source: NEUROtiker/Wikimedia Commons/Public domain.

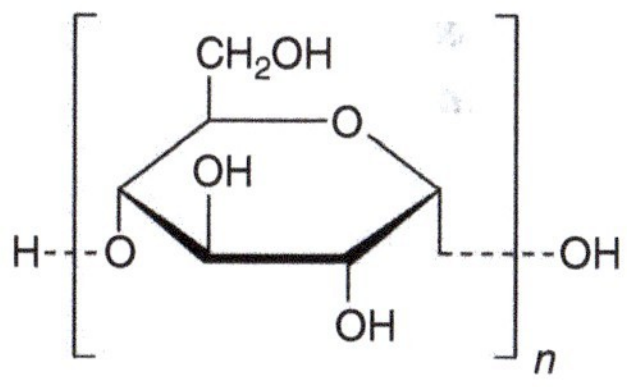

with low oil content (Sothornvit 2011). French fries soaked in a sugar solution (40% w/w) and then fried in the traditional manner showed a 60% decrease in fat content, whereas immersing them in NaCl (20% w/w) and maltodextrin (20% w/w) solutions for the same treatment duration resulted in lower reductions in oil content of 35% and 15%, respectively (Krokida et al. 2001). As evidenced by these results, numerous products can be subjected to vacuum frying. An interesting case is the treatment of peas by microwave vacuum predrying to achieve the crispiest product. The effect of coating with sodium carboxymethyl cellulose (CMC), and hot-air or vacuum-microwave predrying on the physicochemical properties and sensory scores of vacuum-fried peas was studied (Zhu et al. 2015). Throughout, more irregular pores resulted in better crispness of the products. Peas treated by hot-air predrying were the hardest, because their microstructure was characterized by tight packing and a strong connection between the cells, which reduced the quantity of pores during vacuum frying. The break force of fried peas as a control was not as high as that of the treated products, with the exception of the vacuum microwave pretreatment,

Figure 5.7 Chemical structure of maltodextrin. Source: Edgar 181/ Wikimedia Commons/Public domain.

$\alpha\text{-}1,4$

$2 < n < 20$

due to the highest free water contained in the controls, and more pores developing when the water evaporated during frying (Zhu et al. 2015).

5.6.2 Coating and Batter

Deep frying is a cooking technique in which food is submerged in hot fat, usually oil (The Culinary Institute of America 2007). Frying cooks foods rapidly due to the high temperatures involved, and produces an attractive flavor and texture. However, these foods are higher in calories than when cooked in water or by other methods. Consequently, fried foods with lower calorie contents are in demand and can be achieved, to a certain extent, by the use of appropriate coatings (Gates 1987). Such coatings can further act as viscosity-control agents; they can improve adhesion, pickup control, and freeze-thaw stability, and assist in retaining the crispness of the battered/breaded fried food (Varela and Fiszman 2011). In fact, crispness is a significant aspect of most fried food products (Sothornvit 2011), with breaking force or hardness indicating its extent. Peas coated with CMC displayed a higher breaking force than control peas. Coated peas had a stiff, resistant film on their surface that protected the structure throughout frying against mechanical damage (Zhu et al. 2015).

Deep-frying batter is defined as a liquid mixture comprised of water, flour, starch and seasonings in which food products are dipped prior to frying (Suderman and Cunningham 1983). This technique is customarily designed to give the food a crispy texture and characteristic flavor. The production of such foods with low oil content and high textural characteristics, such as crispness, is desirable (Primo-Martín et al. 2009, 2010; Primo-Martín and van Deventer 2011). The pasting properties of crosslinked starches demonstrated that the higher the crosslinking, the more resistant the starch is to gelatinization and granule disintegration. In batters with a constant solid-to-water ratio, batter with high crosslinked starch showed additional water loss during frying. As a result of this, crosslinked starches had lower moisture content following storage and less oil was retained after frying. Crispness, measured instrumentally as sound intensity, was highest for the high-crosslinked starch at 1 and 20 minutes after frying (Primo-Martín 2012). The emitted sound was characterized by number of sound pulses and pulse intensity, which are related to the perception of crispness (Drake 1963; Vickers and Bourne 1976b); Castro-Prada et al. 2009). Overall, high crosslinking of wheat starch enhanced the crispness perception of deep-fried battered food (Primo-Martín 2012).

Adhesion of breaded coatings to baked or fried meat or fish is greatly improved with the inclusion of lightly inhibited and bleached cornstarch in the batter (Langan 1987). One such batter mix consisted of 210 parts water and 90 parts cornstarch which had been treated for 30 minutes with 5% (on a per weight basis) chlorine (Lachmann 1969). Coating crispness is further enhanced with the inclusion of modified high-amylose starches. Modified high-amylose starch used in batters for breaded products forms a film when cooked, retarding moisture movement (Anonymous 1994). Rice flour-based batter formulations with 15% oxidized cornstarch and methylcellulose can be used as an alternative to the traditional wheat flour-based batter for chicken drumstick coating. This results in a healthier product due to lower fat absorption (Mukjprasirt et al. 2001). French fries coated with batter containing 60–90% wheat flour and 10–30% high-amylose starch have a very smooth appearance with firm crispness (Anonymous 1994).

5.7 Enhancement of Food Acoustic Properties Using Various Hydrocolloids

5.7.1 Contribution of Inulin to Crispness of Biscuits, Pizza, and Wafers

Inulins (see Figure 1.9) are polydisperse mixtures of fructan chains with varying degrees of polymerization (DPs), the latter depending on the crop from which they originate (Nussinovitch and Hirashima 2019). Inulin is composed of 2–60 fructose units, with one terminal glucose molecule. Inulin is a generic term describing all β-(2,1) fructans (Nussinovitch and Hirashima 2019). In the long list of inulin applications in foods, a few are related to crispness. The texture of biscuits varies widely, from crispy to soft. In general, the ability to keep the product crispy is an important quality factor. For instance, to improve biscuit crispness and fiber enrichment, 2–10% native inulin (average DP 8–13) and short-chain inulin/oligofructose (average DP 7–9/max DP 10) are added. To achieve crispness enhancement in pizza, 1–3% long-chain and native inulins are added (Franck 2002). The principal characteristic of wafers is their crispness, as it affects the perception of freshness. The crispness of a product is characterized by the sound throughout breakage. It is essential that wafers retain their crispness during storage and preferably, when they are filled or in contact with moist ingredients, such as when used as a sandwich wafer. When the wafer product includes inulin, it is crispier. The degree of crispness depends on the wafer density, the moisture content and the absorption profile of the included starch (Phillips and Williams 2021). Absorption profile is defined as the relationship between water content and water activity. A product retains its crispness up to $a_w = 0.1–0.2$. When the absorption profile line is steep, the product remains crispy only to a water content of around 3%. With a less steep absorption profile, the product stays crisp to higher water content. Inulin is able to bind water, so there will be less free water in the product. The conclusion is that a wafer with inulin can contain more moisture and stay crisp (Schaller-Povolny et al. 2000). In fact, native inulins contribute to both crispness enhancement and dough/batter viscosity. The above examples are all of bakery products; nevertheless, other families of products can also benefit from the inclusion of inulins. Inclusion of inulin in extruded cereals, cereal clusters and coated cereals improves their functionality and increases their crispness and bowl life.

5.7.2 Crispness of Banana Chips

Hydrocolloids have good barrier properties against oxygen, carbon dioxide and lipids, which could decrease oil absorption during deep-fat frying (Mallikarjunan et al. 1997; Williams and Mittal 1999; Albert and Mittal 2002; Rimac-Brncic et al. 2004). The influence of the hydrocolloids alginate, CMC and pectin on oil absorption and crispness was studied with banana chips (Singthong and Thongkaew 2009). The control banana chips (no hydrocolloid treatment) had oil contents as high as 40 g/100 g sample, whereas the sample blanched in 0.5 g $CaCl_2$/100 mL water followed by immersion in 1 g alginate/100 mL water exhibited a small but significant decrease in oil uptake to 38 g/100 g sample. The other samples, treated with 0.5 g $CaCl_2$ and 1 g pectin per 100 mL water, and with 0.25 g $CaCl_2$ and 1 g CMC per 100 mL water absorbed much less oil, approximately 23 g/100 g sample

(Singthong and Thongkaew 2009). Crispness of banana chips was evaluated in terms of hardness using a texture analyzer. A low hardness value indicated high crispness and vice versa. Crispness (hardness) of banana chips was significantly affected by hydrocolloid treatments. Banana chips coated with $CaCl_2$ and hydrocolloids (alginate, CMC and pectin) were significantly less crisp than the control sample. The structure of banana is governed by the presence of pectic substances, which are part of the intercellular material (Singthong and Thongkaew 2009). The lower hardness value (more crispness) of the control sample could be attributed to the degradation of pectic substances during cooking, resulting in a weak structure of the middle lamella matrix and concomitant loss of intercellular adhesion, along with weakening of the cell wall, which could play a vital role in cell rupture throughout the handling of banana chips. Conversely, the higher hardness value (less crispness) of the coated samples could be due to the combined effects of $CaCl_2$ and hydrocolloids (Singthong and Thongkaew 2009). A special case is the preparation of banana peel chips. Banana peel contains between 13.5 and 21.3% pectin, depending on the variety. In general, pectin improves the sensory attributes of common chips. Banana peel chips were produced from those varieties that had the highest percentage yield of pectin content (Rashid et al. 2016). A preparation method for the banana peel chips was developed. The banana peels were thoroughly washed with water and cut into strips; the soft inner parts of the peels were discarded; the peels were soaked in brine followed by immersion in a sugar solution before frying. The sugar-coated peels were fried at 170 °C for three minutes. After cooling and removal of the excess oil, the banana peel chips were kept in an airtight container to retain their crispness (Rashid et al. 2016). The study demonstrated banana peel chips as a novel food product (Rashid et al. 2016).

5.7.3 Specialty Starches as Functional Ingredients

Pregelatinized starches that retain their granular shape advance the use and forming properties of dough, resulting in expanded products (Huang and Rooney 2001). Use of 2–14% high-amylose starch improved dough for leavened pastries with reduced thickness and proofing time, and contributed a tender stable crust with more uniform thickness and less tendency to shrink (Radley 1976). High-amylose starch was used to build up extruded pasta, preventing its disintegration when retorted in sauce and enhancing the crispness of extruded half-products, which are expanded in hot oil well ahead of consumption (Dias et al. 1997). High-amylose cornstarch can be used in extruded and fried snack foodstuffs to produce a crisp, consistently browned product. Snacks have a high degree of mouth melt, a lesser amount of waxiness, better-quality texture and better crispness when a modified starch and/or high-amylopectin flour or starch is included in the dough composition, and when the water absorption index of the starch-based materials is taken into account (Villagran et al. 2002). Low-calorie farinaceous snacks with high-amylopectin flour or starch in the dough have a high degree of mouth melt and enhanced crispness (Gutcho 1973a; Martines et al. 2002).

Food products such as baked goods (cones, cookies, wafers) containing either a high-molecular-weight starch hydrolysate of DE (dextrose equivalent) 5–30 or a crystallite hydrate-producing sugar such as raffinose or trehalose, or their mixture, showed enhanced crispness at higher moisture levels (Huang et al. 2002). Adding modified waxy cornstarch to oat flour increased the value of extruded snack foods (Karam et al. 2001), and improved the texture and crispness of sheeted snack foods and crackers (Anonymous 2001). Fried crackers were made from jackfruit (30% w/w) and a blend of sago and cassava starches (1 : 1) and had acceptable physicochemical characteristics (Mustapha et al. 2015). The type and level of starch played an important role in cracker production as they influenced the textural and sensory acceptability of the final product. The amylose to amylopectin ratio, size, shapes and rigidity of the starch granules have been proposed as important factors affecting the expansion and textural properties of crackers (Dreher et al. 1984; Wang 1997; Cheow et al. 2004).

Starch esters such as hydroxypropylated distarch phosphates can confer firmness and textural properties to dairy foodstuffs (Dorp 1996). The physical properties and taste of cooked potato products could be markedly improved by placing raw, peeled potatoes in an aqueous medium containing a suitable crosslinking agent to form crosslinks between the labile hydrogen atoms on the alcohol moieties of the starch and sugar molecules on the surface of the raw potatoes. These potato products had markedly superior and longer-lasting crispness, as well as greatly improved color and flavor (Gutcho 1973b). Snack foods prepared from potato starch and granules, and pregelatinized starch and salt, are manufactured by extrusion followed by drying, frying, flavoring and packaging. Effects of inclusion of maltodextrins from different origins (corn, waxy corn and potato) on the texture of these cocktail foods demonstrated that waxy cornstarch gives a harder, crisper and denser product. Incorporation of potato starch gave a softer and less dense texture. Using corn maltodextrins, the extrusion capacity, crispness, chewability and brittleness increased with an increased DE value of the maltodextrins. Crosslinking starch ether (crosslinking degree 0.01–1) with the dough noticeably improved the reconstitution and texture of dried pasta (Sajilata and Singhal 2005).

5.7.4 Specialty Starches in Snack Foods

Specialty starches in snack foods can contribute to the required characteristics for such foods, including: augmented expansion, enhanced crispness, reduced oil absorption and enhanced inclusive eating value (Sajilata and Singhal 2005). A snack is defined as a light meal consumed between regular meals. This definition includes a comprehensive variety of products that can take numerous forms. When choosing a specialty starch for a specific application, both production and marketing requirements are considered. The market-associated properties are structure, aesthetics, organoleptic aspects and shelf stability (Dias et al. 1997). Corn starch is versatile, easily modified, and finds many uses in the food industry. Advertisement by the US Food Administration, 1918, indicated corn starch as "wholesome" and "nutritious" (Figure 5.8) https://hmn.wiki/nn/Corn_starch. In addition to their textural and viscosity benefits, specialty food starches frequently decrease the cost of well-known food products (Langan 1987). Modified food starch is a food additive. Restrictions for its modification, use and labeling are openly defined in the US Code of Federal Regulations (21 CFR 172.892) (Moore et al. 1984). Modified starches contain mainly starch with low to very low levels of substituent groups. Far-reaching safety studies examined by independent regulatory bodies guarantee their safety (Wurzburg and Vogel 1984).

Figure 5.8 Advertisement by the US Food Administration, 1918, indicating corn starch as "wholesome" and "nutritious." *Source:* The Library of Congress.

5.7.5 Protein-Rich Extruded Snack

Extrusion cooking is an industrial process in today's food industry used to produce breakfast cereals and snacks, textured vegetable proteins, animal feed, etc. (Patel et al. 2016). Advantages of extrusion cooking include time saving, energy efficiency, and freedom from effluent generation (Riaz 2000). The density, volume, and expansion index of the puffed products are important physical properties which are related to their expansion, crispness, and sensory properties (De Silva et al. 2014). Textural properties of extrudates are highly influenced by the degree of expansion. These parameters are affected by feed composition, extent of cooking, and melt flow in the die (Igual et al. 2020). The most common grains used in the extrusion of snacks and breakfast cereals are corn, wheat, rice, and oats. Sorghum is not a major ingredient in extruded snacks or breakfast cereals (Riaz 2000). Starch is a main ingredient in puffed snacks. It makes an outstanding contribution to the expansion and crispness of the final product. Therefore, starchy snacks are commonly high-calorie foods with a poor nutritional profile. Consequently, improvement in their nutritional quality is requested by consumers, to advance public health and save the products in a competitive market (Alam et al. 2014). Rennet casein and tapioca starch had a positive effect on the texture score of a protein-rich extruded snack, while sorghum flour had a negative effect. In addition, the interactive effects of these three factors on texture score were significant. However, rennet casein had no significant effect on the sensory texture score of wheat-flour extrudates (Voort and Stanley 1984). Favorable interactions between casein and sorghum flour levels and starch and sorghum flour levels with respect to texture and overall acceptability score of the snack base made it possible to strike a good balance between the ingredients to obtain an optimum combination. The optimized base could potentially be converted into a highly palatable expanded snack (Patel et al. 2016).

Increasing the protein content in snack foods can enhance their ability to satiate, in line with numerous health-based trends presently seen in the food industry. Understanding the effects of adding high levels of protein in a food matrix is indispensable for product development (Kreger et al. 2012). To increase the low protein content of cereal-based snacks, they can be enriched with various types of plant- and animal-based proteins such as milk protein (Brncic et al. 2008), fish protein (Pandi et al. 2019), spirulina (Lucas et al. 2018), protein-rich insect powder (Garcia-Segovia et al. 2020; Igual et al. 2020), and high-protein grains such as pulses (Shah et al. 2017). Pulse flours (e.g. bean, lentil, chickpea, and soybean flour) are rich, low-cost, and available sources of plant proteins (Alonso et al. 2000; Pardeshi and Chattopadhyay 2010; Klüver and Meyer 2015; Marengo et al. 2016). One study discussed the types and levels of proteins included in extruded snacks. The effects of changing protein type and level on the sensory aspects of a model extruded snack food were examined. Independent variables were the level of total protein and protein type in the preparation. The protein level was changed from 28 to 43% (w/w) in 5% increments. The protein type varied in the ratio of whey to soy protein, ranging from 0 : 100 to 100 : 0, in 25% increments. Descriptive analysis was conducted to profile the samples' sensory characteristics (Kreger et al. 2012). Protein type was observed to be the predominant variable in differentiating the sensory characteristics of the samples. Soy protein conveyed nutty, grainy aromas by mouth, and improved expansion throughout processing, resulting in a lighter, crispier texture. Whey protein conveyed dairy-related aromas by mouth and inhibited expansion during processing, resulting in a denser, crunchier texture (Kreger et al. 2012).

A method of increasing the nutritional aspects while preserving the quality of extruded maize snacks was described (Sharifi et al. 2021). It involved the inclusion of soybean flour (up to 20% w/w) as a source of plant protein, fiber and bioactive compounds, and changing the feed moisture content. Extruded samples were manufactured using a single-screw extruder (Sharifi et al. 2021). The inclusion of soybean flour as the central component in the extrusion process decreased the value of the puffed snacks by reducing sensory qualities, expansion and crispness. Therefore, a combination of soybean flour with cereal flour was recommended (Li et al. 2005; Veronica et al. 2006). For expanded products, low moisture contents (often <6%) are necessary to attain crispness (Guy 2001). Consequently, foods with 15 and 20% soybean flour and feed moisture contents of more than 15% might necessitate additional drying to generate a crispy texture (Sharifi et al. 2021). Overall, increasing the soybean flour and feed moisture content induced numerous changes in the physicochemical properties of the snacks, including higher moisture content, lower expansion and volume, reduced crispness, and formation of more wrinkly and air bubbles with thicker cell walls. The addition of <20% soybean flour and increased feed moisture content resulted in improved nutritional value and physical properties of the snacks (Sharifi et al. 2021).

5.8 Demonstrating the Preparation of Crunchy Products

The preparation of crunchy foods is demonstrated in this section using an example recipe for the preparation of baked tortillas (Figure 5.9), and supporting photographs of two commercial products (Figures 5.10 and 5.11). Tortilla is a Mexican round, thin, flat bread prepared from unleavened cornmeal or, less commonly, wheat flour. Customarily, the corn (maize) for tortillas is boiled with unslaked lime to soften the kernels and loosen the hulls. The grains are ground on a stone saddle quern. Small pieces of dough are flattened by hand into thin disks, a task that requires substantial skill. The tortilla is then baked (https://www.britannica.com/topic/tortilla). A simple dough manufactured from ground, dried hominy, salt, and water can also be formed into flat discs and cooked on a very hot surface (https://en.wikipedia.org/wiki/Corn_tortilla). During the processing of baking and

Figure 5.9 Baked tortilla chips. (a) Corn tortillas. (b) Cutting tortillas into wedges, (c) spreading oil on wedges, (d) sprinkling salt and seasonings, (e, f) baking and rotating, (g) baking until brown.

Figure 5.10 Commercial fabricated potato chips with starch.

Figure 5.11 Commercial fabricated fried potato.

cooking tortilla chips, numerous complex physicochemical changes take place. These changes are also typical for ready-to-eat snack foods and include: starch gelatinization, water evaporation, crust formation, and achievement of a golden color (Bouchon and Aguilera 2001; Kocchar and Certz 2004); these changes ultimately affect the product's value in terms of expansion and crispness (Mustapha et al. 2015). The crunchiness of the tortilla chips and an increase in product hardness result from the drying. Potato chips are thin slices of potato, fried quickly in oil and then salted (https://www.encyclopedia.com/manufacturing/news-wires-white-papers-and-books/potato-chip). The potato chip was invented in 1853 by a chef named George Crum at a restaurant called Moon's Lake House in Saratoga Springs, New York. Two advances paved the way for its mass production. In 1925, the automatic potato-peeling machine was invented and a year later, at Laura

Scudder's potato chip company, sheets of waxed paper were ironed into bags. The chips were hand-packed into the bags, which were then ironed shut. In 1969, General Mills and Proctor & Gamble presented fabricated potato chips – Chipos and Pringles®. These were manufactured from potatoes that had been cooked, mashed, dehydrated, reconstituted into a dough, and cut into uniform pieces. They further diverged from earlier chips in that they were packaged in break-proof, oxygen-free canisters. Figures 5.10 and 5.11 show fabricated potato chips and fabricated fried potatoes (mashed and hash browns, respectively). These products frequently contain added hydrocolloids, modified starch, potato starch, guar gum, curdlan, and HPMC. While potato chips consist of a packed array of potato cells and are thus microstructurally comparable, upon hydration, the potato cells in fabricated products are detached and dispersed in a suspension, whereas the tissue structure of the traditional chips remains essentially intact. It was observed that a significant proportion of the fat in fried chips is associated as a thin film with the potato cell walls, both at the surface and within the structure, whereas the fat is present as droplets on the fabricated chip surface (Dhital et al. 2018). During the processing of dehydrated potato products, the cellular structure of the potato is disrupted (Villagran et al. 2001). This causes snack food products manufactured from dehydrated potato products, such as fabricated chips, to have a considerably lower level of crispness compared to equivalent products prepared from fresh, raw, or cooked whole potatoes (Villagran et al. 2001). Efforts to increase the crispness of snack food products made from dehydrated potato products have included the addition of fibrous cellulosic material to the snack food dough (Saunders and McLaughlin 1980; Feeney et al. 1989). In the former study, modified food starch was added to the potato-based dough to increase the crispness of French fries made from it (Saunders and McLaughlin 1980).

5.8.1 Baked Tortilla Chips

(Serves 3–5)

3 (17 cm diameter, 90 g) packages corn tortillas (Figure 5.9a)
9 g (1 Tbsp) vegetable oil
6 g (1 tsp) salt
3 g seasonings (optional, *Hint 1*)

For preparation, see Figure 5.9

1) Preheat oven to 175 °C.
2) Cut each tortilla into eight triangles (wedges) (Figure 5.9b).
3) Spread vegetable oil on each tortilla wedge (2) (Figure 5.9c, *Hint 2*).
4) Sprinkle salt and seasonings (optional) (Figure 5.9d) on the tortilla wedges (3).
5) Bake for eight minutes in the oven (1), rotate the tortilla chips (Figure 5.9e), and bake for another five to eight minutes until the chips become brown (Figure 5.9f).

Preparation hints:

1) You can use lime juice, pepper, cumin powder, garlic powder, chili powder, and so on.
2) You don't need a lot of vegetable oil, but be sure to spread it well to the outer edges of the tortilla. If lime juice is used, mix vegetable oil and lime juice well, and then spread.

5.8.2 Commercial Fabricated Potato Chips

The potato chip is a thin slice of potato fried in oil or baked in an oven until crisp (https://kids.britannica.com/students/article/potato-chip/483662). It may be salted or flavored after cooking. Because the fabricated potato chips introduced by General Mills and Proctor & Gamble were prepared from cooked, mashed, dehydrated potatoes that had been reconstituted into a dough and then cut into uniform pieces, the USDA stipulated that the new variety must be labeled "potato chips made from dried potatoes." There are many shapes of fabricated potato chips on the market (https://www.encyclopedia.com/manufacturing/news-wires-white-papers-and-books/potato-chip). Figure 5.10 shows a few of these. They are generally prepared by extrusion and starch is added to their composition.

5.8.3 Commercial Fabricated Fried Potato

Fabricated mashed fried potato products come in numerous shapes. Figure 5.11 demonstrates two of these. They are generally formed by extrusion and sold frozen. They are prepared by either frying or baking. The addition of cellulosic ethers, specifically MC and HPMC, to the mashed potatoes before extrusion prevents oil absorption. When the extruded potato slice is fried, the formed MC film prevents oil pickup and reduces the greasiness of the final product (Glicksman 1986).

References

Alam, S.A., Jrvinen, J., Kirjoranta, S. et al. (2014). Influence of particle size reduction on structural and mechanical properties of extruded rye bran. *Food and Bioprocess Technology* 7: 2121–2133.

Albert, S. and Mittal, G.S. (2002). Comparative evaluation of edible coatings to reduce fat uptake in a deep-fried cereal product. *Food Research International* 35: 445–458.

Al Chakra, W., Allaf, K., and Jemai, A.B. (1996). Characterization of brittle food products: application of the acoustical emission method. *Journal of Texture Studies* 27: 327–348.

Alonso, R., Orue, E., Zabalza, M.J. et al. (2000). Effect of extrusion cooking on structure and functional properties of pea and kidney bean proteins. *Journal of the Science of Food and Agriculture* 80: 397–403.

American Heritage Dictionary of the English Language (2006). *American Heritage Dictionary of the English Language*, 4e. Boston: Houghton Mifflin.

Andres-Bello, A., Garcıa-Segovia, P., and Martınez-Monzo, J. (2011). Vacuum frying: an alternative to obtain high-quality dried products. *Food Engineering Reviews* 3: 63–78.

Anonymous (1994). Starch innovation in batter systems. *Food Marketing and Technology* 8: 16–18.

Anonymous (2001). Light cakes and crunchy snacks-manipulating the moisture in bakery products. *Innovations in Food Technology* 10: 16–17.

Barrett, A.H., Cardello, A.V., Lesher, L.L., and Taub, I.A. (1994). Cellularity, mechanical failure, and textural perception of corn meal extrudates. *Journal of Texture Studies* 25: 77–95.

Bouchon, P. and Aguilera, J.M. (2001). Microstructural analysis of frying of potatoes. *International Journal of Food Science and Technology* 36: 669–676.

Brncic, M., Sven, K., Bosiljkov, T. et al. (2008). Enrichment of extruded snack products with whey protein. *Dairy* 53: 275–295.

Brown, W.E., Langley, K.R., and Braxton, D. (1998). Insight into consumers' assessments of biscuit texture based on mastication analysis—hardness versus crunchiness. *Journal of Texture Studies* 29: 481–497.

Castro-Prada, E.M., Primo-Martín, C., Meinders, M.B.J. et al. (2009). Relationship between water activity, deformation speed, and crispness characterization. *Journal of Texture Studies* 40: 127–156.

Chaisawang, M. and Suphantharika, M. (2005). Effects of guar gum and xanthan gum additions on physical and rheological properties of cationic tapioca starch. *Carbohydrate Polymers* 61: 288–295.

Chauvin, M.A., Younce, F., Ross, C., and Swanson, B. (2008). Standard scales for crispness, crackliness and crunchiness in dry and wet foods: relationship with acoustical determinations. *Journal of Texture Studies* 39: 345–368.

Chen, J., Karlsson, C., and Povey, M. (2005). Acoustic envelope detector for crispness assessment of biscuits. *Journal of Texture Studies* 36: 139–156.

Cheow, C.S., Kyaw, Z.Y., Howell, N.K., and Dzulkifly, M.H. (2004). Relationship between physicochemical properties of starches and expansion of fish cracker 'keropok'. *Journal of Food Quality* 27: 1–12.

Chirife, J. and Ferro Fontan, C. (1982). Water activity of fresh foods. *Journal of Food Science* 47: 661–663.

Christensen, C.M. and Vickers, Z.M. (1981). Relationships of chewing sounds to judgments of food crispness. *Journal of Food Science* 46: 574–578.

Ciurzynska, A., Marzec, A., Mieszkowska, A., and Lenart, A. (2017). Structure influence on mechanical and acoustic properties of freeze-dried gels obtained with the use of hydrocolloids. *Journal of Texture Studies* 48: 131–142.

Concise Oxford English Dictionary (2008). *Concise Oxford English Dictionary*, 11e. New York: Oxford University Press.

Dacremont, C. (1995). Spectral composition of eating sounds generated by crispy, crunchy and crackly foods. *Journal of Texture Studies* 26: 27–43.

Dacremont, C. (2007). Spectral composition of eating sounds generated by crispy, crunchy and crackly foods. *Journal of Texture Studies* 26: 27–43.

Dacremont, C., Colas, B., and Sauvageot, F. (1991). Contribution of air- and bone-conduction to the creation of sounds perceived during sensory evaluation of foods. *Journal of Texture Studies* 22: 443–456.

Dana, D. and Saguy, I.S. (2006). Mechanism of oil uptake during deep-fat frying and the surfactant effect—theory and myth. *Advances in Colloid and Interfacial Science* 128–130: 267–272.

Da Silva, P. and Moreira, R. (2008). Vacuum frying of high quality fruit and vegetable based snacks. *LWT – Food Science and Technology* 41: 1758–1767.

De Silva, E.M.M., Ramırez Ascheri, J.L., Piler de Carvalho, C.W. et al. (2014). Physical characteristics of extrudates from corn flour and dehulled carica bean flour blend. *LWT – Food Science and Technology* 58: 620–626.

Dhital, S., Baier, S.K., Gidley, M.J., and Stokes, J.R. (2018). Microstructural properties of potato chips. *Food Structure* 16: 17–26.

Dias, F.F., Tekchandani, H.K., and Mehta, D. (1997). Modified starches and their use by the food industry. *Indian Food Industry* 16: 33–39.

Dijksterhuis, G., Hannemieke, L., De Wijk, R., and Mojet, J. (2005). A new sensory vocabulary for crisp and crunchy dry model foods. *Food Quality and Preference* 18: 37–50.

Dorp, V.M. (1996). Starch ether for milk products. *Lebensmitteltechnik* 28: 34–35.

Drake, B.K. (1963). Food crushing sounds. An introductory study. *Journal of Food Science* 28: 233–241.

Drake, B.K. (1965). Food crushing sounds. Comparisons of subjective and objective data. *Journal of Food Science* 30: 556–559.

Drake, B.K. (1989). Sensory textural/rheological properties: a polyglot list. *Journal of Texture Studies* 20: 1–27.

Dreher, M.L., Dreher, C.J., Berry, J.W., and Fleming, S.E. (1984). Starch digestibility of foods: a nutritional perspective. *Critical Reviews in Food Science and Nutrition* 20: 47–71.

Duizer, L. (2001). A review of acoustic research for studying the sensory perception of crisp, crunchy and crackly textures. *Trends in Food Science and Technology* 12: 17–24.

Duizer, L.M., Campanella, O.H., and Barnes, G.R.G. (1998). Sensory, instrumental and acoustic characteristics of extruded snack food products. *Journal of Texture Studies* 29: 397–411.

Du Pont, M.S., Kirby, A.R., and Smith, A.C. (1992). Instrumental and sensory tests of texture of cooked frozen French fries. *International Journal of Food Science and Technology* 27: 285–295.

Fan, L., Zhang, M., and Mujumdar, A. (2005). Vacuum frying of carrot chips. *Drying Technology* 23: 645–656.

Feeney, R.D., Hawthorne, N.J., Prosise, R.L. et al. (1989). Potato based dough containing highly pectinated cellulosic fibers. United States Patent 4,876,102.

Franck, A. (2002). Technological functionality of inulin and oligofructose. *British Journal of Nutrition* 87 (Suppl. 2): S287–S291.

Garcia-Segovia, P., Igual, M., Noguerol, A.T., and Martinez-Monzo, J. (2020). Use of insects and pea powder as alternative protein and mineral sources in extruded snacks. *European Food Research and Technology* 246: 703–712.

Gates, J.C. (1987). *Basic Foods*, 3e. New York: Rinehart and Winston.

Glicksman, M. (1986). *Food Hydrocolloids*, vol. III. Boca Raton, FL: CRC Press.

Guest, S., Catmur, C., Lloyd, D., and Spence, C. (2002). Audio tactile interactions in roughness perception. *Experimental Brain Research* 146: 161–171.

Guraya, H.S. and Toledo, R.T. (1996). Microstructural characteristics and compression resistance as indices of sensory texture in a crunchy snack product. *Journal of Texture Studies* 27: 687–701.

Gutcho, M. (1973a). Potato chips. In: *Prepared Snack Foods*, 14–15, 35–36. Munich: Noyes Data Corporation.

Gutcho, M. (1973b). French fried potatoes. In: *Prepared Snack Foods*, 288. UK: Noyes Data Corporation.

Guy, R. (2001). Snack foods. In: *Extrusion Cooking, Technologies and Applications* (ed. R. Guy), 161–181. Cambridge, UK: Woodhead Publishing.

Harker, F.R., Maindonald, J., Murray, S.H. et al. (2002). Sensory interpretation of instrumental measurements. 1: Texture of apple fruit. *Postharvest Biology and Technology* 24: 225–239.

Huang, D.P. and Rooney, L.W. (2001). Starches for snack foods. In: *Snack Foods Processing* (ed. E.W. Lusas and L.W. Rooney), 115–116. Lancaster, PA: Technomic Publishing Co.

Huang, V.T., Panda, F.A., Rosenwald, D.R., and Chida, K. (2002). Starch hydrolysate and sugars for bakery products with enhanced crispness. PCT Application WO2002039820A2.

Igual, M., Garcıa-Segovia, P., and Martınez-Monzo, J. (2020). Effect of *Acheta domesticus* (house cricket) addition on protein content, colour, texture, and extrusion parameters of extruded products. *Journal of Food Engineering* 282: 110032.

Ioannides, Y., Howarth, M.S., Raithatha, C. et al. (2007). Texture analysis of red delicious fruit: towards multiple measurements on individual fruit. *Food Quality and Preference* 18: 825–833.

Jakubczyk, E. and Kaminska, A. (2007). Mechanical properties of porous agar gels. *Agricultural Engineering* 5: 195–203.

Jeon, I.J., Breene, W.M., and Munson, S.T. (1975). Texture of fresh-pack whole cucumber pickles: correlation of instrumental and sensory measurements. *Journal of Texture Studies* 5: 399–409.

Karam, L.B., Grossmann, M.V.E., and Silva, R.S.S. (2001). Oat flour/waxy corn starch blends for snacks production. *Ciencia e Technologia de Alimentos* 21: 158–163.

Katz, E.E. and Labuza, T.P. (1981). Effect of water activity on the sensory crispness and mechanical deformation of snack food products. *Journal of Food Science* 46: 403–409.

Klüver, E. and Meyer, M. (2015). Thermoplastic processing, rheology, and extrudate properties of wheat, soy and pea proteins. *Polymer Engineering and Science* 55: 1912–1919.

Kocchar, P.S. and Certz, C. (2004). New theoretical and practical aspects of the frying process. *European Journal of Lipid Science and Technology* 106: 722–726.

Kreger, J.W., Lee, Y., and Lee, S.-Y. (2012). Perceptual changes and drivers of liking in high protein extruded snacks. *Journal of Food Science* 77: S161–S169.

Krokida, M.K., Oreopoulou, V., Maroulis, Z.B., and Marinos-Kouris, D. (2001). Effect of osmotic dehydration pretreatment on quality of French fries. *Journal of Food Engineering* 49: 339–345.

Lachmann, A. (1969). Batter mixes. In: *Snacks and Fried Products*, 172–176. Munich: Noyes Data Corporation.

Langan, R.E. (1987). Food industry. In: *Modified Starches: Properties and Uses*, 3e (ed. O.B. Wurzburg), 199–212. Boca Raton, FL: CRC Press.

Lee, W.E. III, Deibel, A.E., Glembin, C.T., and Munday, E.G. (1998). Analysis of food crushing sounds during mastication: frequency-time studies. *Journal of Texture Studies* 19: 27–38.

Li, S.Q., Zhang, H.Q., Jin, Z.T., and Hsieh, F.-H. (2005). Textural modification of soya bean/corn extrudates as affected by moisture content, screw speed and soya bean concentration. *International Journal of Food Science and Technology* 40: 731–741.

Lucas, B.F., de Morais, M.G., Santos, T.D., and Vieira Costa, J.A. (2018). Spirulina for snack enrichment: nutritional, physical and sensory evaluations. *LWT – Journal of Food Science and Technology* 90: 270–276.

Luyten, H. and van Vliet, T. (2006). Acoustic emission, fracture behavior and morphology of dry crispy foods: a discussion article. *Journal of Texture Studies* 37: 221–240.

Mah, E. and Brannan, R.G. (2009). Reduction of oil absorption in deep-fried, battered, and breaded chicken patties using whey protein isolate as a post-breading dip: effect on flavor, color, and texture. *Journal of Food Science* 74: S9–S16.

Mallikarjunan, P., Chinnan, M.S., Balasubramaniam, V.M., and Phillips, R.D. (1997). Edible coating for deep-fat frying of starchy products. *LWT – Food Science and Technology* 30: 709–714.

Marengo, M., Akoto, H.F., Zanoletti, M. et al. (2016). Soybean-enriched snacks based on African rice. *Foods* 5: 38–48.

Martines, S., Villagran, M.D., Villagran, F.V. et al. (2002) Dough compositions used to prepare reduced and lowcalorie snacks. United States Patent 6,432,465.

Marzec, A. and Ziołkowski, T. (2007). Structure analysis of selected cereal products in the aspect of their acoustic properties. *Polish Journal of Food and Nutrition Sciences* 57: 89–93.

Meilgaard, M.C., Civille, G.V., and Carr, B.T. (2007). *Sensory Evaluation Techniques*, 4e. Boca Raton, FL: CRC Press.

Merriam-Webster's Collegiate Dictionary (2008). *Merriam-Webster's Collegiate Dictionary*, 11e. Springfield, MA: Merriam-Webster.

Michael, H.T., Charles, I.O., Audrey, E.T. et al. (2013). Critical evaluation of crispy and crunchy textures: a review. *International Journal of Food Properties* 16: 949–963.

Mohamed, A.A.A., Jowitt, R., and Brennan, J.G. (1982). Instrumental and sensory evaluation of crispness: I—In friable foods. *Journal of Food Engineering* 1: 55–75.

Moore, C.O., Tuschhoff, J.V., Hastings, C.W., and Schanefelt, R.V. (1984). Applications of starches in foods. In: *Starch: Chemistry and Technology*, 2e (ed. R.L. Whistler, J.N. BeMiller and E.F. Paschall), 575–591. Orlando, FL: Academic Press.

Mukjprasirt, A., Herald, T.J., Boyle, D.L., and Boyle, E.A.E. (2001). Physicochemical and microbiological properties of selected rice flour-based batters for fried chicken drumsticks. *Poultry Science* 80: 988–996.

Munes, Y. and Moreira, R. (2009). Effect of osmotic dehydration and vacuum frying parameters to produce high-quality mango chips. *Journal of Food Science* 74: 355–362.

Mustapha, N.A., Rahmat, F.F.B., Ibadullah, W.Z.W., and Hussin, A.S.M. (2015). Development of jackfruit crackers: effects of starch type and jackfruit level. *International Journal on Advanced Science, Engineering and Information Technology* 5: 330–333.

Nussinovitch, A., Corradini, M.G., Normand, M.D., and Peleg, M. (2001). Effect of starch, sucrose and their combinations on the mechanical and acoustic properties of freeze-dried alginate gels. *Food Research International* 34: 871–878.

Nussinovitch, A., Jaffe, N., and Gillilov, M. (2004). Fractal pore size distribution of freeze dried agar-texturized fruit surfaces. *Food Hydrocolloids* 18: 825–835.

Nussinovitch, A. and Hirashima, M. (2019). *More Cooking Innovations, Novel Hydrocolloids for Special Dishes*. Boca Raton, FL: CRC Press.

Onions, C.T. (1982). *The Oxford Dictionary of English Etymology*. Oxford: Oxford University Press.

Pandi, G., Rathnakumar, K., Velayutham, P. et al. (2019). Extruded fish snack from low valued fatty fish: an evaluation of nutritional and organoleptic characteristics. *Journal of Coastal Research* 86: 61–64.

Pardeshi, I.L. and Chattopadhyay, P.K. (2010). Hot air puffing kinetics for soy-fortified wheat-based ready-to-eat (RTE) snacks. *Food Bioprocess Technology* 3: 415–426.

Patel, J.R., Patel, A.A., and Singh, A.K. (2016). Production of a protein-rich extruded snack base using tapioca starch, sorghum flour and casein. *Journal of Food Science and Technology* 53: 71–87.

Peleg, M. (1983). The semantics of rheology and texture. *Food Technology* 37: 54–61.

Peleg, M. (1994). A mathematical model of crunchiness/crispness loss in breakfast cereals. *Journal of Texture Studies* 25: 403–410.

Peleg, M. and Normand, M.D. (1993). Determination of the fractal dimension of the irregular compressive stress-strain relationships of brittle crumbly particulates. *Particle System Characterization* 10: 301–307.

Phillips, G.O. and Williams, P.A. (2021). *Handbook of Hydrocolloids*, 3e. Kidlington, UK: Woodhead Publishing Series in Food Science, Technology and Nutrition.

Pons, M. and Fiszman, S.M. (1996). Instrumental texture profile analysis with particular reference to gelled systems. *Journal of Texture Studies* 27: 597–624.

Primo-Martín, C. (2012). Cross-linking of wheat starch improves the crispness of deep-fried battered food. *Food Hydrocolloids* 28: 53–58.

Primo-Martín, C., Sanz, T., Steringa, D.W. et al. (2010). Performance of cellulose derivatives in deep-fried battered snacks: oil barrier and crispy properties. *Food Hydrocolloids* 24: 702–708.

Primo-Martín, C., Sözer, N., Hamer, R.J., and van Vliet, T. (2009). Effect of water activity on fracture and acoustic characteristics of a crust model. *Journal of Food Engineering* 90: 277–284.

Primo-Martín, C. and van Deventer, H. (2011). Deep-fat fried battered snacks prepared using super heated steam (SHS): crispness and low oil content. *Food Research International* 44: 442–448.

Radley, J.A. (1976). The food industry. In: *Industrial Uses of Starch and Its Derivatives*, 51–116. London: Applied Science Publishers.

Rashid, S.N.A.A., Sukri, S.M., and Aziz, R. (2016). Determination of pectin content in preparation of banana peel chips. https://www.researchgate.net/publication/307138697 (accessed 3 March 2022).

Rassis, D.K., Saguy, I.S., and Nussinovitch, A. (2002). Collapse, shrinkage and structural changes in dried alginate gels containing fillers. *Food Hydrocolloids* 16: 139–151.

Riaz, M.N. (2000). Introduction to extruders and their applications. In: *Extruders in Food Applications* (ed. M.N. Riaz), 3. Boca Raton, FL: CRC Press.

Rimac-Brncic, S., Lelas, V., Rade, D., and Simundic, B. (2004). Decreasing of oil absorption in potato strips during deep fat frying. *Journal of Food Engineering* 64: 237–241.

Rohde, F., Normand, M.D., and Peleg, M. (1993). Characterization of the power spectrum of force-deformation relationships of crunchy foods. *Journal of Texture Studies* 24: 45–62.

Roudaut, G., Dacremont, C., Vallès-Pàmies, B. et al. (2002). Crispness: a critical review on sensory and material science approaches. *Trends in Food Science and Technology* 13: 217–227.

Saguy, I.S. and Pinthus, I.J. (1995). Oil uptake during deep-fat frying: factors and mechanisms. *Food Technology* 49: 142–145, 152.

Sajilata, M.G. and Singhal, R.S. (2005). Specialty starches for snack foods. *Carbohydrate Polymers* 59: 131–151.

Saklar, S., Ungan, S., and Katnas, S. (1999). Instrumental crispness and crunchiness of roasted hazelnuts and correlations with sensory assessment. *Journal of Food Science* 64: 1015–1019.

Saunders, F.R. and McLaughlin, R.L. (1980). Potato segment and process for preparing frozen French fried potatoes suitable for microwave preheating. United States Patent 4,219,575.

Sauveageot, F. and Blond, G. (1991). Effect of water activity on crispness of breakfast cereals. *Journal of Texture Studies* 22: 423–442.

Schaller-Povolny, L.A., Smith, D.E., and Labuza, T.P. (2000). Effect of water content and molecular weight on the moisture isotherms and glass transition properties of inulin. *International Journal of Food Properties* 3 (2): 173–192.

Seymour, S.K. and Hamann, D.D. (1988). Crispness and crunchiness of selected low moisture foods. *Journal of Texture Studies* 19: 79–95.

Seymour, S.K., Hamann Colas, B., and Sauvageot, F. (1991). Contribution of air- and bone-conduction to the creation of sounds perceived during sensory evaluation of foods. *Journal of Texture Studies* 22: 443–456.

Shah, F., Sharif, M.K., Butt, M.S., and Shahid, M. (2017). Development of protein, dietary fiber, and micronutrient enriched extruded corn snacks. *Journal of Texture Studies* 48: 221–230.

Sharifi, S., Majzoobi, M., and Farahnaky, A. (2021). Development of healthy extruded maize snacks; effects of soybean flour and feed moisture content. *International Journal of Food Science and Technology* 56: 3179–3187.

Singthong, J. and Thongkaew, C. (2009). Using hydrocolloids to decrease oil absorption in banana chips. *LWT – Food Science and Technology* 42: 1199–1203.

Sothornvit, R. (2011). Edible coating and post-frying centrifuge step effect on quality of vacuum-fried banana chips. *Journal of Food Engineering* 107: 319–325.

Spence, C. (2015). Eating with our ears: assessing the importance of the sounds of consumption on our perception and enjoyment of multisensory flavor experiences. *Flavour* 4, article no. 3.

Srisawas, W. and Jindal, V.K. (2003). Acoustic testing of snack food crispness using neural networks. *Journal of Texture Studies* 34: 401–420.

Suderman, D.R. and Cunningham, F.E. (1983). *Batter and Breading Technology*. Westport, CT: AVI Publishing.

Sundaram, J. and Durance, T.D. (2008). Water sorption and physical properties of locust bean gum-pectin-starch composite gel dried using different drying method. *Food Hydrocolloids* 22: 1352–1361.

Szczesniak, A.S. (1988). The meaning of textural characteristics – crispness. *Journal of Texture Studies* 19: 51–59.

Szczesniak, A.S. and Kahn, E.L. (1971). Consumer awareness and attitudes to food texture. I: Adults. *Journal of Texture Studies* 2: 280–295.

Szczesniak, A.S. and Kahn, E.L. (1985). Texture contrasts and combinations: a valued consumer attribute. *Journal of Texture Studies* 15: 285–301.

Tesch, R., Normand, M.D., and Peleg, M. (1995). On the apparent fractal dimension of sound bursts in acoustic signatures of two crunchy foods. *Journal of Texture Studies* 26: 685–694.

The Culinary Institute of America (2007). *Techniques of Healthy Cooking*. Professional Edition. Hoboken, NJ: Wiley.

Tunick, M.H., Onwulata, C.I., Thomas, A.E. et al. (2013). Critical evaluation of crispy and crunchy textures: a review. *International Journal of Food Properties* 16: 949–963.

van Vliet, T., Visser, J.E., and Luyten, H. (2007). On the mechanism by which oil uptake decreases crispy/crunchy behaviour of fried products. *Food Research International* 40: 1122–1128.

Varela, P., Aguilera, J.M., and Fiszman, S. (2008a). Quantification of fracture properties and microstructural features of roasted Marcona almonds by image analysis. *LWT – Food Science and Technology* 41: 10–17.

Varela, P., Chen, J., Fiszman, S., and Povey, M.J.W. (2006). Crispness assessment of roasted almonds by an integrated approach to texture description: texture, acoustics, sensory and structure. *Journal of Chemometrics* 20: 311–320.

Varela, P. and Fiszman, S.M. (2011). Hydrocolloids in fried foods. A review. *Food Hydrocolloids* 25: 1801–1812.

Varela, P., Salvador, A., and Fiszman, S. (2008b). On the assessment of fracture in brittle foods: the case of roasted almonds. *Food Research International* 41: 544–551.

Varela, P., Salvador, A., and Fiszman, S. (2008c). Methodological developments in the assessment of texture in solid foods. An integrated approach for the quantification of crispness/crunchiness. In: *Progress in Food Engineering Research and Development* (ed. J.M. Cantor), 17–60. New York: Nova Science Publishers.

Varela, P., Salvador, A., and Fiszman, S.M. (2008d). Methodological developments in crispness assessment: effects of cooking method on the crispness of crusted foods. *LWT – Food Science and Technology* 41: 1252–1259.

Veronica, A.O., Olusola, O.O., and Adebowale, E.A. (2006). Qualities of extruded puffed snacks from maize/soybean mixture. *Journal of Food Process Engineering* 29: 149–161.

Vickers, Z. (1977). What sounds good for lunch? *Cereal Foods World* 22: 246–247.

Vickers, Z.M. (1984a). Crackliness: relationships of auditory judgments to tactile judgments and instrumental acoustical measurements. *Journal of Texture Studies* 15: 49–58.

Vickers, Z.M. (1984b). Crispness and crunchiness—a difference in pitch? *Journal of Texture Studies* 15: 157–163.

Vickers, Z.M. (1985). The relationships of pitch, loudness and eating technique to judgments of the crispness and crunchiness of food sounds. *Journal of Texture Studies* 16: 85–95.

Vickers, Z.M. (1987). Sensory, acoustical, and force-deformation measurements of potato chip crispness. *Journal of Food Science* 52: 138–140.

Vickers, Z. and Bourne, M.C. (1976a). Crispness in foods—a review. *Journal of Food Science* 41: 1153–1157.

Vickers, Z. and Bourne, M.C. (1976b). A psychoacoustical theory of crispness. *Journal of Food Science* 41: 1158–1164.

Vickers, Z.M. and Christensen, C.M. (1980). Relationships between sensory crispness and other sensory and instrumental parameters. *Journal of Texture Studies* 11: 291–308.

Villagran, M.D.M.-S., Villagran, F.V., Lanner, D.A., and Hsieh, Y.-P.C. (2002). Dough compositions used to prepare reduced and low-calorie snacks. United States Patent 6,432,465.

Villagran, M.D.M.-S., Li, J., Yang, D.K., Chang, D.S.-J., and Evans, J.F. (2001). Fabricated potato chips. United States Patent 6,703,065B2.

Vincent, J.F.V. (1998). The quantification of crispness. *Journal of the Science of Food and Agriculture* 78: 162–168.

Vincent, J.F.V., Saunders, D.E.J., and Beyts, P. (2002). The use of critical stress intensity factor to quantify "hardness" and "crunchiness" objectively. *Journal of Texture Studies* 33: 149–159.

Voort, F.R. and Stanley, D.W. (1984). Improved utilization of dairy proteins: coextrusion of casein and wheat flour. *Journal of Dairy Science* 67: 749–758.

Wang, S.W. (1997). Starches and starch derivatives in expanded snacks. *Cereal Foods World* 42: 743–745.

Wang, X., Njehia, N.S., Katsuno, N., and Nishizu, T. (2020). An acoustic study on the texture of cellular brittle foods. *Reviews in Agricultural Science* 8: 170–185.

Williams, R. and Mittal, G.S. (1999). Low-fat fried foods with edible coatings: modeling and simulation. *Journal of Food Science* 64: 317–322.

Wurzburg, O.B. and Vogel, W.F. (1984). Modified food starch-safety and regulatory aspects. In: *Gums and Stabilizers for the Food Industry 2* (ed. G.O. Phillips, D.J. Wedlock and P.A. Williams), 405–415. Oxford: Pergamon Press.

Zampini, M. and Spence, C. (2010). Assessing the role of sound in the perception of food and drink. *Chemosensory Perception* 3: 57–67.

Zhu, Y., Zhang, M., and Wang, Y. (2015). Vacuum frying of peas: effect of coating and pre-drying. *Journal of Food Science and Technology* 52: 3105–3110.

6

Use of Hydrocolloids to Improve the Texture of Hard and Chewy Foods

Among the many textural properties of foods are hardness/softness, juiciness, and chewability, which may be variously favored by different cultures. In this chapter, we first define some textural terms and suggest their importance and similarities/differences among different cultures. Objective and subjective methods for estimating these properties will be very briefly mentioned and several representative food families and models will be discussed – including bread, dairy products, and fish products – for improvement in textural features and special applications. This chapter also includes recipes that demonstrate how hydrocolloids serve to create/improve textural and sensory effects.

6.1 Texture Definitions

6.1.1 Hardness

The word "hardness" is defined in the Cambridge Academic Content Dictionary as "the quality of being firm and solid, or not easy to bend, cut, or break" (https://dictionary.cambridge.org/dictionary/english/hardness). In Finnish and English, the terms categorizing hardness are different from those describing firmness (Lawless et al. 1997). In Japanese, the word *katai* relates to the English "rigid," "stiff," "hard," "firm," and "tough," providing an example of a single word that can be used for properties defined by quite dissimilar terms in other languages (Drake 1989). In Japanese, *katai* designates not only characteristics of materials but also emotional and situational states, among numerous other things. It refers to resistance to breaking, deforming, moving, or changing due to pressure from an external source (The Editorial Committee of Nihon Kokugo Daijiten 2001). To describe food texture, the word *katai* is written using three different Chinese characters, each with their corresponding nuances; these are not all that well-known, even amid food scholars (Hayakawa et al. 2012). Hardness is also defined as the perceived force necessary to break the sample into pieces by the first bite with the molars (Guraya and Toledo 1996). Albert Ferdinand Shore (1876–1936) was an American metallurgist who designed the quadrant durometer and its indenters (Figure 6.1) in 1915 to quantify the hardness of polymers and other elastomers (https://en.wikipedia.org/wiki/Albert_Ferdinand_Shore). It was neither the first hardness tester nor the first to be called a durometer, but today his name has come to refer to Shore hardness;

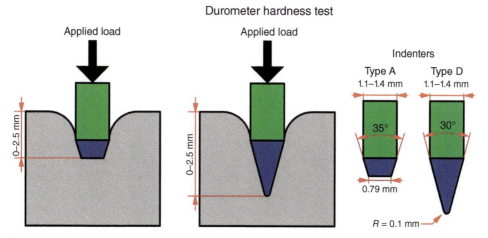

Figure 6.1 Illustration of Shore A and Shore D durometer indenters and how they are used to measure the hardness of a material (usually elastomers, rubbers, and polymers). *Source:* Adapted from http://www.substech.com/dokuwiki/lib/exe/fetch.php?cache=cache&media=durometer.png.

other apparatuses used other measures, which gave corresponding results, such as Rockwell hardness. The reader is referred to the excellent book Food Texture and Viscosity: Concept and Measurement (Bourne 2002) for information on the development and science of instruments used to quantify hardness as well as other measures of food texture. An assortment of models has been proposed to relate firmness or hardness to the physicochemical properties of foods (Guinard and Mazzucchelli 1996). Even if the perception of texture occurs throughout the course of the mastication process, the first bite and its force are used to evaluate hardness, which, consequently, depends on the extent of the applied force and the magnitude of the food's deformation during that first bite (Peyron et al. 1994).

6.1.2 Chewiness

Chewiness is the mouthfeel sensation of labored chewing due to sustained, elastic resistance from a food (https://en.wikipedia.org/wiki/Chewiness). Other definitions of the word "chewy" are "Tending to remain in the mouth without rapidly breaking up or dissolving. Requiring mastication" (Anonymous 1964) and "Possessing the textural property manifested by a low resistance to breakdown on mastication" (Jowitt 1974). Caramel, steak cooked rare, and chewing gum are regarded as typical chewy foods (Sasaki et al. 2012). Other foods for which this is an important part of the eating experience include springy cheeses (Chen et al. 1979) and apples (Li et al. 2015). Chewiness can be empirically estimated by chew counts and chew rate (Harrington and Pearson 1962).

6.1.3 Juiciness

Juiciness is the amount of liquid released from the food after, for example two bites with the molars (Yahia and Carrillo-López 2019). Juiciness is a highly desirable food feature. It has mutually hedonic and commercial characteristics (Szczesniak and Ilker 1988). Juicy foodstuffs are preferred to non-juicy products and therefore command a higher price. Word-association tests conducted in the United States (Szczesniak and Kleyn 1963; Szczesniak 1971) and Japan (Yoshikawa et al. 1970) presented juiciness as one of the three most often requested textural qualities (Szczesniak and Ilker 1988). A review of consumers' views on texture led to the following classification: juiciness is a well-liked and commonly mentioned texture parameter. Juiciness is highly appreciated as an integral part of a served food or dish (i.e. meat, fresh fruit) (Szczesniak and Kahn 1971; Szczesniak and Ilker 1988). Juiciness combines particularly well with firm, tender, and crisp foods to deliver a strong contrast (Szczesniak and Kahn 1984). According to the Kernerman English Multilingual Dictionary, juice is the liquid part of fruits or vegetables; the fluid contained in meat; or the fluid contained in the organs of the body. The adjective "juicy" refers to the amount of juice present (https://www.merriam-webster.com/dictionary/juicy). Juicy fruits characteristically contain 80–90% water; juicy (succulent) vegetables typically contain upwards of 90% water (Szczesniak and Ilker 1988). The minimal quantity of water required for juiciness is not precisely known and it seems to differ depending on the foodstuff and additional aspects (Szczesniak and Ilker 1988). Different plant materials can undergo different degrees of moisture loss before they lose their juiciness. In one study, the quantity of water lost before the plant material became non-juicy ranged from approximately 16 to nearly 61%. Cucumber was most sensitive, followed by mushrooms. Orange was the least sensitive to water loss (Szczesniak and Ilker 1988). Processed fruit are, in general, not juicy, because the tissues have lost their turgor due to heat or cold injury (Szczesniak and Ilker 1988). Taken together, the most important parameters of juiciness in plant foodstuffs are water content, cellular structure, turgor and cellular integrity, the strength of the cell wall, and the rheological character of the expressed juice (Szczesniak and Ilker 1988). An instrument for measuring the moisture content of a fresh or processed vegetable product (such as an ear of corn) was invented, termed succulometer (https://www.merriam-webster.com/dictionary/succulometer). The succulometer (Kramer and Smith 1946) measures the volume of juice expressed by pressing 100 g of tissue in a special cell.

6.2 Use of Hydrocolloids to Improve Bread Texture

The concept of interrupted baking has created a competitive substitute for fully baked loaves of bread, providing the option of enjoying fresh bread any time during the day (Rouille et al. 2000; Giannou et al. 2003). This practice is well known and there is a growing request for part-baked bread (Barcenas et al. 2004). Several studies have concentrated on adjusting the baking temperature and time for the pre-baking process (Unbehend and Neumann 2000; Fik and Surowka 2002). Other studies have explored the microbial quality of part-baked bread (Doulia et al. 2000) and the value of the bread once baking is completed (Fik and Surowka 2002). The interrupted baking process differs from the full baking process in baking time. The bread is baked until the crumb is made but the crust color is not yet established (Fik and Surowka 2002). Part-baked bread is stored frozen for extended shelf life. Nevertheless, freeze–thaw cycles have intense effects on bread properties (Barcenas et al. 2003), and bread improvers that might neutralize such effects would be beneficial (Barcenas et al. 2004). The use of hydrocolloids in bread making is of value in this regard (Mettler et al. 1992; Armero and Collar 1998; Rosell et al. 2001; Guarda et al. 2003). Baked goods are foods made from dough or batter and cooked by baking https://en.wikipedia.org/wiki/List_of_baked_goods. The most common baked item is bread but many other types of foods are baked as well (Figure 6.2). Hydrocolloids are included in bakery products to improve shelf life, because they retain moisture content and delay staling (Twillman and White 1988; Davidou et al. 1996; Collar et al. 1999; Rojas et al. 1999). In addition, hydrocolloids can stabilize food products during freeze–thaw cycles (Sanderson 1996; Gurkin 2002), and in general, help minimize the damaging effects of freezing and frozen storage on starch-based products (Ferrero et al. 1993; Friend et al. 1993; Liehr and Kulicke 1996). Hydroxypropyl methylcellulose (HPMC) and κ-carrageenan were found to improve conventional bread manufacturing (Rosell et al. 2001), and their effect on the quality and staling of part-baked bread after frozen storage and rebaking was also assessed to determine their usefulness in interrupted baking with frozen storage (Barcenas et al. 2004). The basic recipe in that study consisted of wheat flour, compressed yeast, salt and water, and 0.5% (w/w, flour basis) HPMC and κ-carrageenan were tested. Ingredients were mixed, the dough was allowed to rest for 10 minutes, and then it was divided, kneaded, and mechanically sheeted and rolled. It was proofed at 28 °C and 85% relative humidity for an up to threefold increase in dough volume (Barcenas et al. 2004). Part-baking was performed at 165 °C for seven minutes. Part-baked bread was cooled at room temperature and then placed in a freezer at −35 °C for rapid cooling; or kept at room temperature for 30 minutes, and therefore not stored, then rebaked at 195 °C for 14 minutes and finally cooled at room temperature for 60 minutes.

Figure 6.2 Examples of baked goods: American biscuit (left) from Bob Evans Restaurant (Pittsburgh, PA) and British biscuits (right) from a packet of Britannia "Bourbon" biscuits (India), showing the difference between the American English and British English meaning of the word "biscuit." *Source:* Lou Sander/Wikipedia Commons/Public Domain.

Frozen part-baked breads were packed in polypropylene bags and kept at -25 °C. At 7, 14, 28, and 42 days, the bread loaves were thawed and rebaked. For the aging studies, the finished baked bread was repacked and stored at 25 °C for 24 hours (Barcenas et al. 2004). In terms of fresh bread quality, HPMC increased the specific volume and moisture retention of the bread and decreased its water activity. The crumb hardness was checked using a texturometer. The presence of hydrocolloids decreased this parameter, and the softest crumb was achieved with HPMC. The same result was obtained when hydrocolloids, specifically HPMC, were added to conventional bread-making practice (Nishita et al. 1976; Davidou et al. 1996; Armero and Collar 1998; Collar et al. 1999; Martinez et al. 1999; Rojas et al. 1999; Rosell et al. 2001). HPMC is beneficial in improving the texture of bread dough because substitution of the cellulose hydroxyl groups with methoxyl and hydroxypropyl enhances water solubility and confers some affinity for the non-polar phase (Bell 1990). Therefore, in a multiphase system such as bread dough, this double property enables maintaining a uniform dough and achieving emulsion stability during bread making. In addition, HPMC forms interfacial films at the boundaries of the gas cells that confer stability against gas expansion as well as other changes in manufacturing conditions (Bell 1990). Furthermore, during baking, the temperature increases and the HPMC produces a gel (Haque et al. 1993; Sarkar and Walker 1995), which strengthens the expanding dough by protecting against volume loss. Since the produced gel does not remain after cooling, improved texture and softness are delivered without any negative effects on the palatability of the fresh product (Bell 1990). The crumb hardness of the control demonstrated a progressive intensification with time in frozen storage. In contrast, the hardness of the crumb comprising HPMC was not affected by the time in frozen storage (Barcenas et al. 2004) and in the case of the κ-carrageenan-supplemented sample, an increase in hardness was only detected for up to 14 days of frozen storage; no increase was found after that time. On the other hand, the firming effect of frozen storage was not intense (Barcenas et al. 2004). Taken together, the hardness of the crumb increased with time of storage in all samples, but the sample with added HPMC showed the lowest increase. In the control and in the presence of κ-carrageenan, an increase in crumb hardness and faster staling were found with time in frozen storage. Surprisingly, in the presence of HPMC, the hardening rate was independent of the frozen storage time (Barcenas et al. 2004). Thus, inclusion of HPMC reduced the hardness of the bread crumb and inhibited the effect of frozen storage on bread staling. The overall results demonstrated that κ-carrageenan is not beneficial for part-baked frozen bread (Barcenas et al. 2004).

Cereal β-glucans (Figure 6.3) have long been considered health-beneficial in the scientific literature (Ames and Rhymer 2008) and in the consumer's mind (Stevenson and Inglett 2009). The quality of gluten-containing and gluten-free bread dough with the addition of oat β-glucan was evaluated (Londono et al. 2015). Following the addition, both doughs showed enhanced stiffness and reduced extensibility. β-glucan's high molecular weight has a negative impact on dough extension. Nevertheless, the presence of <2% β-glucan in the flour resulted in greater gas retention in the dough. Another study discussed the addition of oat or barley β-glucan (1.3–3.9 g/100 g flour) to gluten-free rice flour dough. The highest β-glucan content had the strongest effect on dough rheological properties, conferring upper specific bread volume and lower bread crumb hardness (Perez-Quirce et al. 2017). Prebiotic oat β-glucan was incorporated into white bread to enhance its

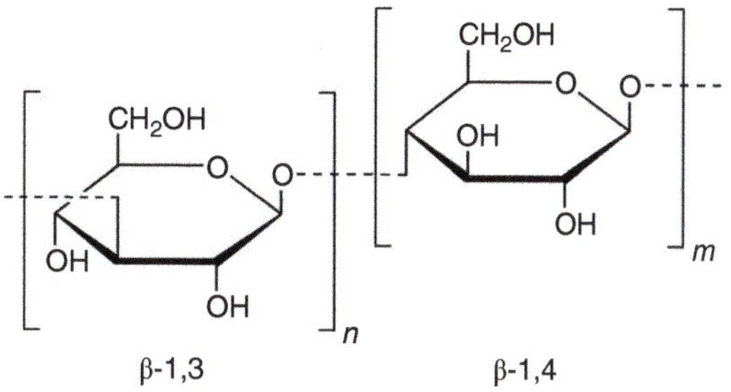

Figure 6.3 Oat β-glucan repeat structure. *Source:* Edgar 181/Wikimedia Commons/Public domain.

nutritional value, and to increase the dough's water absorption (Mohebbi et al. 2018). Dough stability and extensibility diminished with the addition of β-glucan, but the dough was softer. Oat β-glucan powder was added at 40–200 g/kg to pasta dough and dried using a convection oven or vacuum (Piwinska et al. 2016). Water uptake and swelling power were higher, whereas cooking loss was lower, for pasta dried under vacuum vs. in the convection oven, but all three variables increased with increasing β-glucan content. In addition, the pasta with added oat powder was darker, softer, and less sticky than durum wheat pasta with no β-glucan added (Piwinska et al. 2016).

6.3 Dairy Products

6.3.1 Dairy Foods

Dairy foods are basic components of the human diet (Yousefi and Jafari 2019). World milk production (roughly 81% cow milk, 15% buffalo milk, and 4% for goat, sheep, and camel milk combined) is expected to increase at 1.7% per year to 1020 Mt by 2030, more rapidly than most other key agricultural commodities. The projected growth in the number of milk-producing animals (1.1% per year) is higher than the projected average yield growth (0.7%), because herds are expected to grow faster in countries with lower yields, as are herds comprised of lower-yielding animals (i.e. goats and sheep) (https://www.fao.org/dairy-production-products/resources/publications/en/). It is expected that India and Pakistan, key milk producers, will contribute more than half of the growth in world milk production over the next 10 years, and will account for more than 30% of world production in 2030. Production in the European Union – the second largest global milk producer – is estimated to rise more gradually than the world average owing to policies on sustainable production and slower growth in domestic demand (https://www.fao.org/dairy-production-products/resources/publications/en/). Dairy foods are very important in the diet due to their beneficial effects in numerous illnesses, for example cardiovascular diseases, cancer, metabolic syndromes, and dental caries (Tanaka et al. 2010; Sonestedt et al. 2011; Louie et al. 2013). These advantages can be attributed to the occurrence of bioactive constituents in dairy foods, such as antioxidants, highly absorbable calcium, oligosaccharides, peptides, probiotic bacteria, vitamins, organic acids, and additional bioactive components (Dantas et al. 2016; Martins et al. 2017; Balthazar et al. 2018).

6.3.2 Cheeses

A unique feature of hydrocolloids in cheese production is their ability to replace or mimic fat by creating a texture resembling, or improving on, that of the natural cheese. For instance, supplementation of cheese with guar gum was found to reduce its hardness factor (Lashkari et al. 2014). Another study described the production of low-fat cheese containing 0.5% (w/w) konjac glucomannan. A 100% decrease in fat content was achieved with a hardness value analogous to that of control samples (i.e. whole-fat cheese without hydrocolloids) (da Silva et al. 2016).

Acid-coagulated cheese is one of the most popular types of soft cheese consumed in Egypt, especially in the countryside, due to its high protein, low fat, and low price. *Kariesh* cheese is manufactured from defatted cow or buffalo milk, or a mixture of the two. Excluding water, its composition is c. 16% protein, 4.0% sugar, and 0.1% fat (Abd-El-Salam et al. 1984; Kebary et al. 1997; Ahmed et al. 2005; Abou-Donia 2008). Numerous studies have aimed to improve the texture of this cheese, by adapting the traditional cheese-making technologies to increase moisture content, which improves texture (Ahmed et al. 2005). The effects of adding the hydrocolloids commercial pectin, citrus pectin or carboxymethylcellulose (CMC) at 0.2, 0.4, or 0.6% (w/w) on the chemical composition, yield, rheological and sensory characteristics of Egyptian kariesh cheese were studied (Korish and Abd Elhamid 2012). Hardness was influenced by the hydrocolloid additives. This rheological property decreased significantly (at the start of cold storage) when 0.2% commercial pectin was added to the cheese compared to the control. Under the same conditions, adhesiveness increased. The lowest values of hardness, springiness, and chewiness were obtained in the presence of 0.6% (w/w) commercial pectin. In general, texture analysis indicated that kariesh cheeses prepared with commercial pectin or CMC at 0.4% and 0.6%, respectively, had enhanced development of texture characteristics. The authors suggested that this is due to the increase in cheese moisture content, as a consequence of water adsorption by the hydrocolloid (Korish and Abd Elhamid 2012). Another study mentioned that the increase in moisture content weakens the casein micelle–pectin network, creating a less firm cheese (Beal and Mittal 2000; Kaya 2002). In addition, the texture of a high-moisture cheese was smoother than that obtained through the conventional production method (Mairfreni et al. 2002). Low-fat kashar cheese with fat replacers showed lower hardness values than the control (Koca and Metin 2004). Addition of 0.025% guar gum to low-fat Edam cheese (Figure 6.4) during its production improved its rheology and textural properties (Oliveira et al. 2010).

6.3.3 Functionality of Selected Hydrocolloids on Texture of Ice Cream

Ice cream is a colloidal system, comprised of air cells, ice crystals, and fat droplets dispersed in the serum phase (Marshall et al. 2003; Arana 2012). The functionality of hydrocolloids in ice cream relates to cryoprotection – interruption of the recrystallization phenomena, enhancement of ice cream mix viscosity, enhancement of texture and mouthfeel, and the product's shape retention (Marshall et al. 2003). The cryoprotective influence of hydrocolloids on ice cream can be described by three mechanisms. The first is control of ice crystal growth (Hagiwara and Hartel 1996; Miller-Livney and Hartel 1997); the second involves cryoprotection via hydrocolloids' ability to form cryogels as a result of temperature fluctuations during storage (Muhr and Blanshard 1986; Blond 1988; Patmore et al. 2003); and the third relates to the incompatibility of hydrocolloids with proteins, which induces phase separation that may contribute to the retardation of recrystallization (Regand and Goff 2002, 2003). κ-Carrageenan was included as a second stabilizing agent of ice cream at levels <0.05% to regulate phase separation initiated by incompatibility of other hydrocolloids with milk proteins (Bourriot et al. 1999; Langendorff et al. 2000; Thaiudom and Goff 2003). Consumers evaluate ice cream (Figure 6.5) in large part according to perceived texture and flavor (Soukoulis et al. 2008). There are numerous manuscripts dealing with the influence of hydrocolloids on texture perception and flavor release of dairy emulsions (Kilcast and Clegg 2002; Yanes et al. 2002; Akhtar et al. 2005, 2006; Cook et al. 2005). Hardness (in Newtons) of ice cream samples was determined as peak compression force

Figure 6.4 Edam cheese with crackers. *Source:* Jon Sullivan/Wikipedia Commons/Public Domain.

Figure 6.5 A single vanilla ice cream sandwich. *Source:* Renee Comet.

during penetration (Soukoulis et al. 2008). Prior to measuring hardness, samples were transferred from a −25 °C freezer to a −15 °C freezer and held there for 24 hours. A texture analyzer was used to define the textural attributes. Measurements were carried out with a 5-mm stainless-steel cylindrical probe attached to a load cell. The penetration depth at the geometrical center of the samples was 10 mm and the penetration speed was 2.0 mm/s (Soukoulis et al. 2008). Instrumental hardness is a measure of the extent of ice crystal size during storage, with increased size reflected by increased hardness. Furthermore, it may reveal the impact of hydrocolloids and other included ingredients, such as fats, sugars, and proteins, and processing conditions, for instance: ageing, freezing, and homogenization, used in manufacturing the frozen product (Guinard et al. 1997; Muse and Hartel 2004). The addition of the hydrocolloid sodium alginate to ice cream controlled ice recrystallization irrespective of storage time, thus improving the ice cream's texture. Nevertheless, cryoprotection by the stabilizers CMC, guar gum or xanthan gum was significantly influenced by storage time. Specifically, after four weeks of storage, no difference was observed among samples with xanthan gum, guar gum or CMC (Soukoulis et al. 2008). After eight weeks in storage, samples with xanthan or guar gum had comparable hardness, whereas CMC proved to be a poor cryoprotectant. After 16 weeks of storage, the stabilizing systems could be fully discriminated in terms of hardness in the following order: sodium alginate < xanthan gum < guar gum < CMC (Soukoulis et al. 2008). The addition of hydrocolloids expressly reinforced shear-thinning performance, mainly in the case of sodium alginate, xanthan gum, and κ-carrageenan, which was attributed to gelation phenomena. Sodium alginate achieved the best stabilizing effect, improving textural quality and acceptance of ice creams even after 16 weeks of storage, while the presence of κ-carrageenan was found to be a vital factor for cryoprotection. Xanthan gum was also evaluated as an operational stabilizing agent, demonstrating that gelling hydrocolloids may markedly alter ice cream shelf life (Soukoulis et al. 2008).

6.4 Fish Products

Application of food hydrocolloids in fish products offers novel prospects for the improvement in innovative foodstuffs founded upon surimi (Figure 6.6) or restructured product know-how. These products can satisfy the new low-salt requirement for healthy foods (Ramírez et al. 2011). Numerous additives, including microbial transglutaminase (MTG) and hydrocolloids, have been used to increase the mechanical properties of surimi gels (Ramírez et al. 2000a, 2000b). Fish products generally require minimal improvement in mechanical properties, which can be achieved simply by adding some hydrocolloids instead of MTG (Ramírez et al. 2011). Food hydrocolloids, i.e. proteins and polysaccharides, might have a significant role in the structure, stability and practical assets of numerous processed foods. Interactions between proteins and carbohydrates control useful properties in foods with proteins as their major ingredient, such as processed meat and fish products. Different hydrocolloids have been suggested to expand the functional and mechanical properties of surimi and restructured fish gels (Lee et al. 1992; Gomez-Guillén et al. 1997; Park 2000; Ramírez et al. 2002). Several hydrocolloids, for example carrageenan, konjac mannan and starch, are typically used to improve the mechanical properties of surimi gels (Gomez-Guillén et al. 1997;

Figure 6.6   *Surimi*. Left: Package of commercial *surimi*. Right: Commercial *surimi*.

Park 2000). Low methoxyl pectin enhanced the hardness of surimi gels and decreased their shear strain and cohesiveness, but had no observable effect on shear stress, springiness, or water-holding capacity. Hydrogen bonds between amidated low methoxyl (ALM) pectin and myofibrillar proteins resulted in improved mechanical properties – in particular hardness and breaking force – of fish gels when the ALM pectins were added at 10 g/kg (Ramírez et al. 2011). Wheat fiber with large particle size protected surimi from loss of gel strength and hardness during freezing (Sanchez-Alonso et al. 2006). The soluble fiber chicory root inulin has been used as an additive in restructured fish products, but is detrimental to their mechanical properties, specifically hardness. This undesirable effect could be circumvented by adding 20 g/kg carrageenan, but the latter was not well-matched with MTG (Cardoso et al. 2007a, 2007b, 2008a, 2008b, 2009). Pea fiber was added to minced fish and surimi at up to 40 g/kg to obtain restructured products with unaltered textural and valuable properties. It was well-matched with up to 20 g/kg carrageenan, tolerating increased hardness of restructured hake products. Pea fiber was compatible with MTG at 1 g/kg or more, and it enhanced the textural properties of surimi gels from Atlantic mackerel and chub mackerel (Cardoso et al. 2007a, 2007b, 2008b, 2009). Finally, addition of 10 g/kg chitosan to kamaboko gels from grass carp increased their hardness, chewiness, and many other textural attributes while decreasing peroxide contents and bacterial growth. A mixture of 300 and 10 kDa chitosan exhibited the strongest antibacterial activity (Mao and Wu 2007; Wu and Mao 2009). In the low-salt (10 g/kg NaCl) restructured product, higher values of hardness were obtained by adding MTG and dairy proteins than by adding only MTG, suggesting a positive interaction between meat proteins, dairy proteins, and MTG (Ramírez et al. 2006; Sun 2009).

6.5   Further Contributions of Hydrocolloids to Textural Improvement

Food scientists traditionally describe the textural properties of foods in terms of adhesiveness, chewiness, gumminess, hardness, fracturability, springiness, etc. (Sherman 1969; Tunick 2000). These textural properties are crucial in the first stage of oral processing, i.e. the first bite, and fundamentally comprise bulk rheology-controlled phenomena, i.e. the material characteristics of the food (Chen and Stokes 2012; Stokes et al. 2013; Sarkar et al. 2019).

Okra polysaccharide is a slimy and sticky food gum extracted from the pods of an extensively cultured vegetable, *Abelmoschus esculentus* L. Moench (Figure 6.7), comprised mostly of rhamnogalacturonan I with long galactan side branches. This non-gelling polysaccharide produced a synergistic gel with κ-carrageenan, with higher hardness, springiness, cohesiveness, gumminess, and thermal stability when compared to pure κ-carrageenan gel (Li et al. 2012; Chen et al. 2019).

Partial removal of the acetyl groups from *konjac glucomannan* significantly improved the hardness and springiness of mixed gels formed with κ-carrageenan. Decreasing the acetyl content can reduce the syneresis of mixed gels. By modifying the deacetylation degree of konjac, high-strength gels could be manufactured, offering a novel route for the creation of high-strength gelatin foods (Hu et al. 2019).

Figure 6.7 *Abelmoschus esculentus* (okra, gumbo, lady's finger). *Source:* Forest & Kim Starr/CC BY 4.0.

Arabinoxylans can potentially make up a large share of human dietary fiber intake because they are present in many different cereal crops. Arabinoxylans can be used in food preparations to improve the physicochemical properties of the product. Feruloylated arabinoxylans extracted from nixtamalized maize bran and incorporated into frankfurter sausage preparations enhanced sausage hardness, shear resistance, and water-holding ability. Furthermore, the antioxidant properties of the frankfurter sausages increased due to considerable enhancement of ferulic acid content (Herrera-Balandrano et al. 2019).

Food applications for *curdlan* are numerous. The compatibility of curdlan with animal meat proteins is superior to that of numerous other polysaccharides, and there are reports denoting the benefits of curdlan predominantly as a texture modifier in the meat system. In emulsified meatball from hogs, curdlan improved an extensive range of texture-related mechanical parameters: chewiness, gumminess, adhesiveness, etc., in contrast to the other hydrocolloids tested (Hsu and Chung 2000).

6.6 Other Miscellaneous Applications

6.6.1 Rice Starch Pastes

Starch is present in plants as insoluble granules. It is commercially extracted from various botanical sources, including maize, potato, rice, and wheat, to manufacture a comprehensive variety of marketable products (Tester and Karkalas 2002). The granules are mostly composed of amylose and amylopectin molecules (Hizukuri 1986; Cura et al. 1995). In general, (normal) starch comprises about 25% amylose and 75% amylopectin, and waxy genotypes have noticeably higher amounts of amylopectin, while high-amylose genotypes are rich in amylose (Suzuki et al. 1981; Hizukuri 1986). Texture profile analyses of freshly prepared paste samples (8% w/w) were performed at room temperature (c. 25 °C) using a texture analyzer. Weighed paste samples were poured into cylindrical containers and kept at room temperature for one hour prior to measurement. One compression cycle was performed at a constant crosshead velocity of 1 mm/s to a sample depth of 15 mm (50% gel height), followed by a return to the original position. The values for texture attributes were measured from the resulting force–time curve (Techawipharat et al. 2008).

Addition of cellulose derivatives have no effect on the textural properties of either normal or waxy rice starch pastes. When κ-carrageenan or ι-carrageenan were added quite the opposite was observed. They significantly increased the hardness and adhesiveness of the mixed pastes. The main reason for this might be that both gelling polysaccharides can produce network structures and enhance the textural properties of the pastes. Therefore, it could be concluded that the hardness and adhesiveness of the pastes mostly stemmed from the corresponding network structure, instead of from retrogradation of amylose, in agreement with a previous manuscript (Huang et al. 2007). Nevertheless, addition of λ-carrageenan to normal rice starch had no significant effect, perhaps because of λ-carrageenan's non-gelling property. Unexpectedly, addition of κ-carrageenan to waxy rice starch brought about a substantial reduction in the hardness and no noteworthy effect on the adhesiveness of the paste, whereas the other carrageenan displayed an effect that was comparable to that observed in the case of normal rice starch. This could be a result of an explicit interaction of amylose and κ-carrageenan, which enhanced the previously noted gelling properties of the latter (Tecante and Doublier 2002). With amylose being in effect absent from waxy rice starch, κ-carrageenan could not form a gel network structure and had no effect on the texture of the mixed paste. In contrast, ι-carrageenan shows synergistic interactions with all starches (Thomas 1997; Imeson 2000; Tischer et al. 2006), and it enhanced the textural properties of both normal and waxy rice pastes. λ-Carrageenan did not affect the textural properties of the waxy rice starch paste, again due to its non-gelling nature (Techawipharat et al. 2008).

6.6.2 Rice Starch–Polysaccharide and Other Mixed Gels

Mixed gels were prepared using rice starch (indica, japonica, or sticky rice starch), hydrocolloids (gellan, carrageenan, or konjac glucomannan), $CaCl_2$, and deionized water (Huang et al. 2007). The levels of starch, water, and $CaCl_2$ were kept constant. The effects of rice starch variety and concentrations of the different hydrocolloids on the texture of the mixed gels were studied by the profile analysis method. Addition of a non-suitable polysaccharide or unsuitable concentration led to weakening of the gel. Only a suitable polysaccharide at an appropriate concentration delivered a gel of first-rate quality (Huang et al. 2007). A low concentration (<0.2% w/w) of carrageenan and a high concentration (>0.3% w/w) of gellan markedly enhanced gel hardness and adhesiveness; the indica rice starch–polysaccharide gels retained the highest hardness, adhesiveness, and chewiness values at all time points. Inclusion of 0.2% (w/w) carrageen with the indica rice starch improved the texture of the rice gel foodstuff; when a higher adhesiveness-to-hardness ratio was needed, 0.3% (w/w) gellan could be added to indica rice starch. Konjac glucomannan did not increase the adhesiveness, chewiness, or hardness significantly for any of the rice starches. The three varieties of rice starch, polysaccharides, and their concentrations had no effect on the springiness of the mixed gels (Huang et al. 2007).

In the United States, agar is classified as generally recognized as safe (GRAS) by the Food and Drug Administration. It is permitted in foods up to a maximum level of 2.0% in confections and frostings, 1.2% in soft candy, 0.8% in baked goods, and 0.25% in all other food categories (https://www.ams.usda.gov/sites/default/files/media/Agar%20TR%202011.pdf). Mixed gels of agar with other biopolymers were also studied. In all cases, a decrease in gelation point with increasing added biopolymer was described. The rheological/mechanical properties were also affected by the addition of the second biopolymer. Upon addition of xanthan, the gel was less brittle, softer, and showed lower syneresis (Nordqvist and Vilgis 2011; Maurer et al. 2012). Phase separation (with the agar-rich phase forming the continuous one) seemed to take place in mixed gels with κ-carrageenan, and a large reduction of gel rigidity, and decreases in firmness, gel strength, and brittleness were observed (Norziah et al. 2006).

6.6.3 Hydrocolloid Effects on Pea Starch

The effects of adding yellow mustard mucilage (YMM) to buckwheat and pea starch on practical and rheological gel properties were studied (Liu et al. 2006). The addition resulted in a noticeable intensification in peak viscosity for both buckwheat and pea starches. Dynamic rheological measurements demonstrated that the storage modulus, loss modulus, and dynamic viscosity of buckwheat and pea starches were greater than before inclusion of YMM (Liu et al. 2006). The gel textures of both starches were largely modified by the presence of YMM, which increased hardness, chewiness, and adhesiveness, but reduced resilience. Adding a YMM–locust bean gum mixture (9 : 1) in the same way increased the viscosity of buckwheat and pea starches but decreased gel hardness (Liu et al. 2003). The solubility of buckwheat starch and the swelling abilities of both starches were somewhat decreased in the presence of YMM. Addition of YMM slowed the syneresis of buckwheat and pea starch gels (Liu et al. 2006).

In recent years, more and more hydrocolloids have been explored and applied for the modification of native starches. *Mesona chinensis* Benth gum (MCG) is considered a novel and economical starch-modifying hydrocolloid. The effects of xanthan gum, guar gum, and MCG on the pasting, rheology, texture, and thermal properties of pea starch were studied (Rong et al. 2022). Pasting results showed that similar to xanthan and guar gums, MCG significantly increased the pasting temperature, and retarded the starch's water-absorption and disintegration in a heated environment. The inhibitory effect of MCG on short-term retrogradation was more evident than that of the other gums when the effective concentration of starch was increased. The strain resistibility, elastic property, and hardness of pea starch gels were enhanced after the addition of xanthan gum, guar gum, and MCG, given that a more compact structure was formed by hydrogen bonding in the pea starch–hydrocolloid matrix. Furthermore, pea starch–MCG gels exhibited more prominent viscoelasticity, hardness, chewiness, and thermal stability than other pea starch–hydrocolloid gels, indicating that MCG effectively affected the gel properties of pea starch, and thus plays an important role in the field of starch modification (Rong et al. 2022).

6.7 Demonstrating the Use of Hydrocolloids in Creating/Controlling Food Hardness and Chewiness

There are several options for the use of hydrocolloids to create/improve the texture of different foods. One common one is to use a suitable hydrocolloid to create/change the texture of a fluidic, semi-solid or solid food. This is exemplified in creating the dried sweet dessert *Seiryu*. In Japan, it is termed *kohakuto* (琥珀糖), which literally means amber sugar (https://yunomi.life/blogs/recipes/wagashi-sweets-recipe-kohakutou). It is a traditional candy that has been made since the early Edo Period (about 400 years ago, 1603 and 1867). Economic growth, strict social order, nationalist foreign policies, a stable population, and perpetual peace characterized this period, along with popular enjoyment of arts and culture (https://en.wikipedia.org/wiki/Edo_period). The texture of this product is unique. It has a very hard shell on the outside and a softer jelly texture on the inside. Crystallized sugar creates the crunchy and brittle external shell. The unique texture on the inside is created by agar (Figure 6.8). In the recipe presented in Section 6.7.1, agar was deliberately chosen for the preparation of *Seiryu* for several reasons. As a gelling agent, agar forms firm,

Figure 6.8 (a) String-type agar. (b) Soaking string-type agar, (c) heating agar with stirring, (d) adding granulated sugar to the agar solution. (e) Sugar strings. (f) Pouring agar solution into a rectangular mold, (g) adding food coloring, (h) cooling at room temperature, (i) cutting into 1.5 cm rectangular parallelepiped shape, (j) cutting into random shapes, (k) putting on wax or parchment paper, (l) drying at room temperature.

hard, resilient, and transparent gels that are suited to a number of products (Davidson 1980). Moreover, pure agar gels are fairly stable. Those prepared with high-gel strength agar appear to be as stable as dry agar itself if sterilized and stored in airtight containers (Davidson 1980). The addition/inclusion of sugar and the drying creates a product with the desired hard texture. The product develops a crunchy outer crust after a few days but remains delightfully chewy and jelly-like inside. Flavoring and citric acid addition are possible options (https://sugargeekshow.com/recipe/kohakutou-crystal-gummy-candy/). This sweet dessert can be colored. Blue and transparent agar jelly represent coolness in hot summer and in fact, the meaning of the word *seiryu* is cool stream.

Another originally Japanese dessert known as *mizu Shingen mochi* (水信玄餅) is an evolution of the traditional Japanese dessert *Shingen mochi* (信玄餅). The dessert is prepared basically from water and a gelling hydrocolloid and is supposed to resemble a raindrop (https://kinseiken.co.jp/news/4231), although it is much bigger than a raindrop (Figure 6.9). It first became popular in Japan in 2014, and later gained international attention (https://en.wikipedia.org/wiki/Raindrop_cake). The Kinseiken Seika Company has the only stores that sell this dessert, and only on weekends from June to September. It is also sold in kits to be prepared at home (https://raindropcake.com/). The dish was introduced to the United States in New York City at the April 2016 Smorgasburg food fair

Figure 6.9 (a) Commercial carrageenan powder. (b) Pouring water into carrageenan powder and sugar mixture, (c) heating agar with stirring, (d) cooling in ice water bath with stirring, (e) pouring into an ice ball tray. (f) Carrageenan jelly. (g) Serve.

(Maitland 2018). In the recipe in Section 6.7.2, carrageenan powder (Cool Agar®) is used. The food item has a very soft, easily melted and slippery texture, and the appearance is transparent. In other words, for both recipes – *Seiryu* and *mizu Shingen mocha* – the type and concentration of the hydrocolloid have a profound influence on the hardness/softness of the product and its overall textural characteristics.

Many types of konjac noodle are sold in Japan (Figure 6.10), with many suggestions for serving them (https://konnyaku.co.jp/en/). There are also many varieties of "Miracle Noodle" konjac noodles (https://miraclenoodle.com/). Figure 6.10 illustrates three cases in which, to achieve improved texture, konjac noodles are enriched with starch from buckwheat, rice flour or glutinous rice powder. In many cases, konjac noodles are chewier and more elastic than wheat noodles. However, this also depends on the konjac content in the recipe; in a few cases, they are softer and not elastic. For example, Kahada noodle is made from Japanese konjac yam kneaded with brown rice flour. Although the noodle is thin, it has a firm/hard texture; it is gluten-free, non-GMO, and has only 16 kcal/100 g. As it is low in both calories and sugar, it serves as a suitable noodle type for healthy lifestyles at all ages (https://konnyaku.co.jp/en/). The konjac gum in konjac pasta and konjac buckwheat noodles improves their water-holding capacity (Nussinovitch and Hirashima 2014). It is

Figure 6.10 (a) Commercial dried buckwheat konjac noodles. (b) Serving example for boiled buckwheat konjac noodles. (c) Commercial konjac noodles with rice flour. (d) Serving example for konjac noodles with rice flour. (e) Commercial Vietnamese noodles, pho, made with konjac, soy pulp, and glutinous rice powder. (f) Serving example for pho with boiled chicken and vegetables.

Figure 6.11 (a) Commercial frozen meatballs with inulin. (b) Serving example.

well known that the addition of konjac glucomannan can improve the overall quality of noodles made from low-protein wheat flour by reinforcing the gluten network (Zhou et al. 2013). Additional studies have described interactions of konjac glucomannan with gluten proteins via noncovalent interactions (Wang et al. 2017; Li et al. 2019). Konjac glucomannan has also been described to affect the rheological and retrogradation properties of starch from various sources, which rely on structural compatibility and molecular interactions (Yoshimura et al. 1998; Funami et al. 2005; Khanna and Tester 2006).

Figure 6.11 shows frozen commercial meatballs with inulin, which are sold in Japan. Inulin gels can potentially replace fats and in general, replacing fat has a positive influence on the consistency of the modified products (Nussinovitch and Hirashima 2019). Nevertheless, there are controversial studies regarding the texture-modifying effects of inulin inclusion in meat and poultry products. It appears that issues such as type of inulin (powder/gel), inulin concentration, fat percentage, and the item for consumption govern the resultant texture (Yousefi et al. 2018). It is likely that in the future, there will be more applications for inulin. Studies have shown that incorporation of inulin into formulations of meat and poultry products reduces their energy content (Mendoza et al. 2001). In preparing reduced-fat cooked meat sausages, inulin was added as a dietary fiber at levels of 2.5, 5, and 7.5%, and a reduction of 33–37% and 25% in fat content and calories, respectively, was observed (Selgas et al. 2005). The type of inulin to be used as a fat replacer should form microcrystals when mixed with water or any liquid, should be only discretely perceptible, and should have a smooth, creamy mouthfeel. These microcrystals can act together to establish small aggregates which can entrap a large amount of water to provide a smooth and creamy texture (Kalyani Nair et al. 2010; Karimi et al. 2015).

6.7.1 Agar Jelly, *Seiryu*

(Serves 10)

5 g string-type agar (Figure 6.8a)
200 g (⅘ cup) water
360 g (1½ cups) granulated sugar
Food coloring (blue)

For preparation, see Figure 6.8.

1) Put string-type agar in a large amount of water for more than two hours (Figure 6.8b), then squeeze it dry.
2) Put squeezed agar (1) and 200 mL water in a pan, heat with stirring until agar dissolves in the water, and strain (Figure 6.8c) (*Hint 1*).
3) Add granulated sugar to the agar solution (2) (Figure 6.8d) and heat it until it dissolves.
4) Reduce heat, and continue to heat for four to five minutes until the agar and sugar solution becomes stringy (Figure 6.8e).
5) Pour half of agar solution (4) into a rectangular mold (Figure 6.8f); stir a little food coloring into the remaining agar solution (Figure 6.8g), and pour it into another rectangular mold.
6) Cool the molds (5) at room temperature for at least eight hours (Figure 6.8h) (*Hint 2*).
7) Remove the agar gels from their molds and cut them into 1.5 cm rectangular parallelepiped shapes (Figure 6.8i), then cut these into random pieces (Figure 6.8j).
8) Put them on wax or parchment paper (Figure 6.8k), and dry them for four to five days at room temperature (Figure 6.8l).

Preparation hints:

1) If agar is not completely dissolved in water, gelation will not occur.
2) Complete gelation must occur to become stable.

6.7.2 Low-Concentration Carrageenan Jelly, *mizu-Shingen mochi*

(Serves 6)

250 g (1 cup) water
6 g (1 tsp) carrageenan powder (Cool Agar®) (Figure 6.9a)
8 g (a little less than 1 Tbsp) white sugar
Brown sugar syrup, soybean flour

For preparation, see Figure 6.9.

1) Mix carrageenan powder and sugar in a bowl, and pour water into it gradually while stirring with a spatula (Figure 6.9b). Then bring to a boil over high heat, with stirring (Figure 6.9c), and continue to heat on low heat for two minutes.
2) Cool the mixture (1) in an ice water bath with stirring until it becomes thick (around 40 °C) (Figure 6.9d).
3) Pour the mixture (2) into an ice ball tray, and cool it until it sets (Figure 6.9e).
4) Remove carrageenan jelly from the ice ball tray (3) (Figure 6.9f), put it on the plate, and sprinkle with brown sugar syrup and soybean flour (Figure 6.9g).

References

Abd-El-Salam, M.H., El-Shibiny, S., and El-Alamy, H.A. (1984). Production of skim milk (*Kareish* cheese) from ultrafiltered reconstituted milk. *Egyptian Journal of Dairy Science* 12: 111–115.

Abou-Donia, S.A. (2008). Origin, history and manufacturing process of Egyptian dairy products: an overview. *Alexandria Journal of Food Science and Technology* 5: 51–62.

Ahmed, N.H., Elsoda, M., Hassan, A.N., and Frank, J. (2005). Improving the textural properties of an acid-coagulated (Karish) cheese using exopolysaccharide producing cultures. *LWT – Food Science and Technology* 38: 843–847.

Akhtar, M., Stenzel, J., Murray, B.S., and Dickinson, E. (2005). Factors affecting the perception of creaminess of oil-in-water emulsions. *Food Hydrocolloids* 19: 521–526.

Akhtar, M., Murray, B.S., and Dickinson, E. (2006). Perception of creaminess of model oil-in-water dairy emulsions: influence of the shear-thinning nature of a viscosity-controlling hydrocolloid. *Food Hydrocolloids* 20: 839–847.

Ames, N.P. and Rhymer, C.R. (2008). Issues surrounding health claims for barley. *The Journal of Nutrition* 138: 1237S–1243S.

Anonymous (1964). Sensory testing guide for panel evaluation of foods and beverages. Prepared by the Committee on Sensory Evaluation of the Institute of Food Technologists. *Food Technology* 18: 1135–1141.

Arana, I. (2012). *Physical Properties of Foods: Novel Measurement Techniques and Applications*, 385. Boca Raton, FL: CRC Press.

Armero, E. and Collar, C. (1998). Crumb firming kinetics of wheat breads with anti-staling additives. *Journal of Cereal Science* 28: 165–174.

Balthazar, C.F., Silva, H.L., Esmerino, E.A. et al. (2018). The addition of inulin and *Lactobacillus casei* 01 in sheep milk ice cream. *Food Chemistry* 246: 464–472.

Barcenas, M.E., Haros, M., Benedito, C., and Rosell, C.M. (2003). Effect of freezing and frozen storage on the staling of part-baked bread. *Food Research International* 36: 863–869.

Barcenas, M.E., Benedito, C., and Rosell, C.M. (2004). Use of hydrocolloids as bread improvers in interrupted baking process with frozen storage. *Food Hydrocolloids* 18: 769–774.

Beal, P. and Mittal, G.S. (2000). Vibration and compression responses of Cheddar cheese at different fat content and age. *Milchwissenschaft* 55: 139–142.

Bell, D.A. (1990). Methylcellulose as a structure enhancer in bread baking. *Cereal Foods World* 35: 1001–1006.

Blond, G. (1988). Velocity of linear crystallization of ice in macromolecular systems. *Cryobiology* 25: 61–66.

Bourne, M.C. (2002). *Food Texture and Viscosity: Concept and Measurement*. Amsterdam: Elsevier Science and Technology Books.

Bourriot, S., Garnier, C., and Doublier, J.-L. (1999). Micellar-caseine κ-carrageenan mixtures. I. Phase separation and ultrastructure. *Carbohydrate Polymers* 40: 145–157.

Cardoso, C., Mendes, R., and Nunes, M.L. (2007a). Dietary fibers' effect on the textural properties of fish heat-induced gels. *Journal of Aquatic Food Product Technology* 16: 19–30.

Cardoso, C., Mendes, R., and Nunes, M.L. (2007b). Effect of transglutaminase and carrageenan on restructured fish products containing dietary fibres. *International Journal of Food Science and Technology* 42: 1257–1264.

Cardoso, C., Mendes, R., and Nunes, M.L. (2008a). Development of a healthy low-fat fish sausage containing dietary fibre. *International Journal of Food Science and Technology* 43: 276–283.

Cardoso, C., Mendes, R., Pedro, S., and Nunes, M.L. (2008b). Quality changes during storage of fish sausages containing dietary fiber. *Journal of Aquatic Food Product Technology* 17: 73–95.

Cardoso, C., Mendes, R., Vaz-Pires, P., and Nunes, M.L. (2009). Effect of dietary fibre and MTG on the quality of mackerel surimi gels. *Journal of the Science of Food and Agriculture* 89: 1648–1658.

Chen, J. and Stokes, J.R. (2012). Rheology and tribology: two distinctive regimes of food texture sensation. *Trends in Food Science and Technology* 25: 4–12.

Chen, A.H., Larkin, J.W., Clark, C.J., and Irwin, W.E. (1979). Textural analysis of cheese. *Journal of Dairy Science* 62: 901–907.

Chen, J., Chen, W., Duan, F. et al. (2019). The synergistic gelation of okra polysaccharides with kappa-carrageenan and its influence on gel rheology, texture behavior and microstructures. *Food Hydrocolloids* 87: 425–435.

Collar, C., Andreu, P., Martinez, J.C., and Armero, E. (1999). Optimization of hydrocolloid addition to improve wheat bread dough functionality: a response surface methodology study. *Food Hydrocolloids* 13: 467–475.

Cook, D.J., Hollowood, T.A., Linforth, R.T.S., and Taylor, A.J. (2005). Correlating instrumental measurements of texture and flavor release with human perception. *International Journal of Food Science and Technology* 40: 631–641.

Cura, J.A., Jansson, P.-E., and Krisman, C.R. (1995). Amylose is not strictly linear. *Starch/Starke* 47: 207–209.

Dantas, A.B., Jesus, V.F., Silva, R. et al. (2016). Manufacture of probiotic Minas Frescal cheese with *Lactobacillus casei* Zhang. *Journal of Dairy Science* 99: 18–30.

Davidou, S., Le Meste, M., Debever, E., and Bekaert, D. (1996). A contribution to the study of staling of white bread: effect of water and hydrocolloid. *Food Hydrocolloids* 10: 375–383.

Davidson, R.L. (1980). *Handbook of Water-Soluble Gums and Resins*. New York, NY: McGraw-Hill.

Doulia, D., Katsinis, G., and Mougin, B. (2000). Prolongation of the microbial shelf life of wrapped part-baked baguettes. *International Journal of Food Properties* 3: 447–457.

Drake, B. (1989). Sensory textural/rheological properties – a polyglot list. *Journal of Texture Studies* 20: 1–27.

Ferrero, C., Martino, M.N., and Zaritzky, N.E. (1993). Stability of frozen starch pastes: effect of freezing, storage and xanthan gum addition. *Journal of Food Processing and Preservation* 17: 191–211.

Fik, M. and Surowka, K. (2002). Effect of prebaking and frozen storage on the sensory quality and instrumental texture of bread. *Journal of the Science of Food and Agriculture* 82: 1268–1275.

Friend, C.P., Waniska, R.D., and Rooney, L.W. (1993). Effects of hydrocolloids on processing and qualities of wheat tortillas. *Cereal Chemistry* 70: 252–256.

Funami, T., Kataoka, Y., Omoto, T. et al. (2005). Effects of non-ionic polysaccharides on the gelatinization and retrogradation behavior of wheat starch. *Food Hydrocolloids* 19: 1–13.

Giannou, V., Kessoglou, V., and Tzia, C. (2003). Quality and safety characteristics of bread made from frozen dough. *Trends in Food Science and Technology* 14: 99–108.

Gomez-Guillén, C., Borderias, A.J., and Montero, P. (1997). Thermal gelation properties of two different composition sardine (*Sardina pilchardus*) muscles with addition of non-muscle and hydrocolloids. *Food Chemistry* 58: 81–87.

Guarda, A., Rosell, C.M., Benedito, C., and Galotto, M.J. (2003). Different hydrocolloids as bread improvers and antistaling agents. *Food Hydrocolloids* 18: 241–247.

Guinard, J.-X. and Mazzucchelli, R. (1996). The sensory perception of texture and mouthfeel. *Trends in Food Science and Technology* 71: 213–219.

Guinard, J.X., Zoumas-Morse, C., Mori, L. et al. (1997). Sugar and fat effects on sensory properties of ice cream. *Journal of Food Science* 62: 1087–1094.

Guraya, H.S. and Toledo, R.T. (1996). Microscopical characteristics and compression resistance as indices of sensory texture in a crunchy snack product. *Journal of Texture Studies* 27: 687–701.

Gurkin, S. (2002). Hydrocolloids—ingredients that add flexibility to tortilla processing. *Cereal Foods World* 47: 41–43.

Hagiwara, T. and Hartel, R.W. (1996). Effect of sweetener, stabilizer and storage temperature on ice recrystallization in ice cream. *Journal of Dairy Science* 79: 735–744.

Haque, A., Richardson, R.K., Morris, E.R. et al. (1993). Thermogelation of methylcellulose. Part II: Effect of hydroxypropyl substituents. *Carbohydrate Polymers* 22: 175–186.

Harrington, G. and Pearson, A.M. (1962). Chew count as a measure of tenderness of pork loins with various degrees of marbling. *Journal of Food Science* 27: 106–110.

Hayakawa, F., Kazami, Y., Nishinari, K. et al. (2012). Classification of Japanese texture terms. *Journal of Texture Studies* 44: 140–159.

Herrera-Balandrano, D.D., Baez-Gonzalez, J.G., Carvajal-Millán, E. et al. (2019). Feruloylated arabinoxylans from nixtamalized maize bran byproduct: a functional ingredient in frankfurter sausages. *Molecules* 24: 2056. https://doi.org/10.3390/molecules24112056.

Hizukuri, S. (1986). Polymodal distribution of the chain lengths of amylopectins, and its significance. *Carbohydrate Research* 147: 342–347.

Hsu, S.Y. and Chung, H.Y. (2000). Interactions of konjac, agar, curdlan gum, κ-carrageenan and reheating treatment in emulsified meatballs. *Journal of Food Engineering* 44: 199–204.

Hu, Y., Tian, J., Zou, J. et al. (2019). Partial removal of acetyl groups in konjac glucomannan significantly improved the rheological properties and texture of konjac glucomannan and kappa-carrageenan blends. *International Journal of Biological Macromolecules* 123: 1165–1171.

Huang, M., Kennedy, J.F., Li, B. et al. (2007). Characters of rice starch gel modified by gellan, carrageenan, and glucomannan: a texture profile analysis study. *Carbohydrate Polymers* 69: 411–418.

Imeson, A.P. (2000). Carrageenan. In: *Handbook of Hydrocolloids* (ed. G.O. Phillips and P.A. Williams), 87–102. Cambridge, UK: Woodhead Publishing.

Jowitt, R. (1974). The terminology of food texture. *Journal of Texture Studies* 5: 351–358.

Kalyani Nair, K., Kharb, S., and Thompkinson, D.K. (2010). Inulin dietary fiber with functional and health attributes – a review. *Food Reviews International* 26: 189–203.

Karimi, R., Azizi, M.H., Ghasemlou, M., and Vaziri, M. (2015). Application of inulin in cheese as prebiotic, fat replacer and texturizer: a review. *Carbohydrate Polymers* 119: 85–100.

Kaya, S. (2002). Effect of salt on hardness and whiteness of Gaziantep cheese during short term brining. *Journal of Food Engineering* 52: 155–159.

Kebary, K.M.K., Salem, O.M., Hamed, A.I., and El-Sisi, A.S. (1997). Flavour enhancement of direct acidified Kareish cheese using attenuated lactic acid bacteria. *Food Research International* 30: 265–272.

Khanna, S. and Tester, R.F. (2006). Influence of purified konjac glucomannan on the gelatinisation and retrogradation properties of maize and potato starches. *Food Hydrocolloids* 20: 567–576.

Kilcast, D. and Clegg, S. (2002). Sensory perception of creaminess and its relationship with food structure. *Food Quality and Preference* 13: 609–623.

Koca, N. and Metin, M. (2004). Textural, melting and sensory properties of low-fat fresh kashar cheeses produced by using fat replacers. *International Dairy Journal* 14: 365–373.

Korish, M. and Abd Elhamid, A.M. (2012). Improving the textural properties of Egyptian *Kariesh* cheese by addition of hydrocolloids. *International Journal of Dairy Technology* 65: 237–242.

Kramer, A. and Smith, H.R. (1946). The succulometer, an instrument for measuring the maturity of raw and canned whole kernel corn. *Food Packer* 27: 56–60.

Langendorff, V., Cuvelier, G., Michon, C. et al. (2000). Effects of carrageenan type on the behavior of carrageenan/milk mixtures. *Food Hydrocolloids* 14: 273–280.

Lashkari, H., Madadlou, A., and Alizadeh, M. (2014). Chemical composition and rheology of low-fat Iranian white cheese incorporated with guar gum and gum Arabic as fat replacers. *Journal of Food Science and Technology* 51: 2584–2591.

Lawless, H., Vanne, M., and Tuorila, H. (1997). Categorization of English and Finnish texture terms among consumers and food professionals. *Journal of Texture Studies* 28: 687–708.

Lee, C.M., Wu, M.M., and Okada, M. (1992). Ingredient and formulation technology for surimi-based products. In: *Surimi Technology* (ed. T.C. Lanier and C.M. Lee), 273–302. New York: Marcel Dekker.

Li, X., Fang, Y., Al-Assaf, S. et al. (2012). Complexation of bovine serum albumin and sugar beet pectin: structural transitions and phase diagram. *Langmuir* 28: 10164–10176.

Li, G., Ren, Y., Ren, X., and Zhang, X. (2015). Non-destructive measurement of fracturability and chewiness of apple by FT-NIRS. *Journal of Food Science and Technology* 52: 258–266.

Li, J.X., Zhu, Y.P., Yadav, M.P., and Li, J.L. (2019). Effect of various hydrocolloids on the physical and fermentation properties of dough. *Food Chemistry* 271: 165–173.

Liehr, M. and Kulicke, W.M. (1996). Rheological examination of the influence of hydrocolloids on the freeze–thaw stability of starch gels. *Starch/Starke* 48: 52–57.

Liu, H., Eskin, N.A.M., and Cui, S.W. (2003). Interaction of wheat and rice starches with yellow mustard mucilage. *Food Hydrocolloids* 17: 863–869.

Liu, H., Eskin, N.A.M., and Cui, S.W. (2006). Effects of yellow mustard mucilage on functional and rheological properties of buckwheat and pea starches. *Food Chemistry* 95: 83–93.

Londono, D.M., Gilissen, L.J.W.J., Visser, R.G.F. et al. (2015). Understanding the role of oat β-glucan in oat-based dough systems. *Journal of Cereal Science* 62: 1–7.

Louie, J., Flood, V., Rangan, A. et al. (2013). Higher regular fat dairy consumption is associated with lower incidence of metabolic syndrome but not type 2 diabetes. *Nutrition, Metabolism, and Cardiovascular Diseases* 23: 816–821.

Mairfreni, M., Marino, M., Pittia, P., and Rondinini, G. (2002). Textural and sensorial characterization of Montasio cheese produced using proteolytic starters. *Milchwissenschaft* 57: 23–26.

Maitland, H. (2018). Everything you need to know about raindrop cakes. *British Vogue*. https://www.vogue.co.uk/article/everything-you-need-to-know-about-raindrop-cake (accessed 18 April 2022).

Mao, L.C. and Wu, T. (2007). Gelling properties and lipid oxidation of kamaboko gels from grass carp (*Ctenopharyngodon idellus*) influenced by chitosan. *Journal of Food Engineering* 82: 128–134.

Marshall, R.T., Goff, H.D., and Hartel, R.W. (2003). *Ice Cream*. New York: Aspen Publishers.

Martinez, J.C., Andreu, P., and Collar, C. (1999). Storage of wheat breads with hydrocolloids, enzymes and surfactants: anti-staling effects. *Leatherhead Food RA Food Industry Journal* 2: 133–149.

Martins, N., Oliveira, M.B.P.P., and Ferreira, I.C.F.R. (2017). Development of functional dairy foods. In: *Bioactive Molecules in Food* (ed. J.M. Mérillon and K.G. Ramawat), 1–19. Cham, Switzerland: Springer International Publishing.

Maurer, S., Junghans, A., and Vilgis, T.A. (2012). Impact of xanthan gum, sucrose and fructose on the viscoelastic properties of agarose hydrogels. *Food Hydrocolloids* 29: 298–307.

Mendoza, E., García, M.L., Casas, C., and Selgas, M.D. (2001). Inulin as fat substitute in low fat, dry fermented sausages. *Meat Science* 57: 387–393.

Mettler, E., Seibel, W., Bruemmer, J.M., and Pfeilsticker, K. (1992). Experimentelle studien der emulgator- und hydrokolloidwirkung zur optimierung der funktionellen eigenschaften von weizenbroten. V. Einfluss der emulgatoren und hydrokolloide auf die funktionelleneigenschaften von weizenbroten. *Getreide Mehl und Brot* 46: 43–47.

Miller-Livney, T. and Hartel, R.W. (1997). Ice recrystallization in ice cream: interactions between sweeteners and stabilizers. *Journal of Dairy Science* 80: 447–456.

Mohebbi, Z., Homayouni, A., Azizi, M.H., and Hosseini, S.J. (2018). Effects of beta-glucan and resistant starch on wheat dough and prebiotic bread properties. *Journal of Food Science and Technology* 55: 101–110.

Muhr, A.H. and Blanshard, J.M. (1986). Effect of polysaccharide stabilizers on the rate of growth of ice. *Journal of Food Technology* 21: 683–710.

Muse, M.R. and Hartel, R.W. (2004). Ice cream structural elements that affect melting rate and hardness. *Journal of Dairy Science* 87: 1–10.

Nishita, K.D., Roberts, R.L., and Bean, M.M. (1976). Development of a yeast-leavened rice bread formula. *Cereal Chemistry* 53: 626–635.

Nordqvist, D. and Vilgis, T.A. (2011). Rheological study of the gelation process of agarose based solutions. *Food Biophysics* 6: 450–460.

Norziah, M.H., Foo, S.L., and Karim, A.A. (2006). Rheological studies on mixtures of agar (*Gracilaria changii*) and κ-carrageenan. *Food Hydrocolloids* 20: 204–217.

Nussinovitch, A. and Hirashima, M. (2014). *Cooking Innovations, Using Hydrocolloids for Thickening, Gelling and Emulsification*, 209–210. Boca Raton, FL: CRC Press.

Nussinovitch, A. and Hirashima, M. (2019). *More Cooking Innovations Novel Hydrocolloids for Special Dishes*, 120–121. Boca Raton, FL: CRC Press.

Oliveira, N.M., Dourado, F.Q., Peres, A.M. et al. (2010). Effect of guar gum on the physicochemical, thermal, rheological and textural properties of green Edam cheese. *Food and Bioprocess Technology* 4: 1414–1421.

Park, J.W. (2000). Ingredient technology and formulation development. In: *Surimi and Surimi Seafood* (ed. J.W. Park), 343–391. New York: Marcel Dekker.

Patmore, J.V., Goff, H.D., and Fernandes, S. (2003). Cryogelation of galactomannans on ice cream model systems. *Food Hydrocolloids* 17: 161–169.

Perez-Quirce, S., Lazaridou, A., Billiaderis, C.G., and Ronda, F. (2017). Effect of β-glucan molecular weight on rice flour dough rheology, quality parameters of breads and in vitro starch digestibility. *LWT – Food Science and Technology* 82: 446–453.

Peyron, M.A., Mioche, L., and Culioli, J. (1994). Bite force and sample deformation during hardness assessment of viscoelastic models of foods. *Journal of Texture Studies* 25: 59–76.

Piwinska, M., Wyrwisz, J., Kurek, M., and Wierzbicka, A. (2016). Effect of oat β-glucan fiber powder and vacuum-drying on cooking quality and physical properties of pasta. *CyTA – Journal of Food* 14: 101–108.

Ramírez, J.A., Rodríguez-Sosa, R., Morales, O.G., and Vázquez, M. (2000a). Surimi gels from striped mullet (*Mugil cephalus*) employing microbial transglutaminase. *Food Chemistry* 70: 443–449.

Ramírez, J.A., Santos, I.A., Morales, O.G. et al. (2000b). Application of microbial transglutaminase to improve mechanical properties of surimi from silver carp. *Ciencia y Tecnología Alimentaria* 3: 21–28.

Ramírez, J.A., Barrera, M., Morales, O.G., and Vázquez, M. (2002). Effect of xanthan gum and locust bean gums on the gelling properties of myofibrillar protein. *Food Hydrocolloids* 16: 11–16.

Ramírez, J.A., Del Ángel, A., Velázquez, G., and Vázquez, M. (2006). Production of low-salt restructured fish products from Mexican flounder (*Cyclopsetta chittendeni*) using microbial transglutaminase or whey protein concentrate as binders. *European Food Research and Technology* 223: 341–345.

Ramírez, J.A., Uresti, R.M., Velazquez, G., and Vázquez, M. (2011). Food hydrocolloids as additives to improve the mechanical and functional properties of fish products: a review. *Food Hydrocolloids* 25: 1842–1852.

Regand, A. and Goff, H.D. (2002). Effects of biopolymers on structure and ice recrystallization in dynamically-frozen ice cream model systems. *Journal of Dairy Science* 85: 2722–2732.

Regand, A. and Goff, H.D. (2003). Structure and ice recrystallization in frozen stabilized ice cream model systems. *Food Hydrocolloids* 17: 95–102.

Rojas, J.A., Rosell, C.M., and Benedito de Barber, C. (1999). Pasting properties of different wheat flour-hydrocolloid systems. *Food Hydrocolloids* 13: 27–33.

Rong, L., Shen, M., Wen, H. et al. (2022). Effects of xanthan, guar and *Mesona chinensis* Benth gums on the pasting, rheological, texture properties and microstructure of pea starch gels. *Food Hydrocolloids* 125: https://doi.org/10.1016/j.foodhyd.2021.107391.

Rosell, C.M., Rojas, J.A., and Benedito de Barber, C. (2001). Influence of hydrocolloids on dough rheology and bread quality. *Food Hydrocolloids* 15: 75–81.

Rouille, J., Le Bail, A., and Courcoux, P. (2000). Influence of formulation and mixing conditions on breadmaking qualities of French frozen dough. *Journal of Food Engineering* 43: 197–203.

Sanchez-Alonso, I., Haji-Maleki, R., and Borderias, A.J. (2006). Effect of wheat fibre in frozen stored fish muscular gels. *European Food Research and Technology* 223: 571–576.

Sanderson, G.R. (1996). Gums and their use in food systems. *Food Technology* 50: 81–84.

Sarkar, N. and Walker, L.C. (1995). Hydration–dehydration properties of methylcellulose and hydroxypropylmethylcellulose. *Carbohydrate Polymers* 27: 177–185.

Sarkar, A., Andablo-Reyes, E., Bryant, M. et al. (2019). Lubrication of soft oral surfaces. *Current Opinion in Colloid & Interface Science* 39: 61–75.

Sasaki, K., Motoyama, M., and Narita, T. (2012). Increased intramuscular fat improves both 'chewiness' and 'hardness' as defined in ISO5492:1992 of beef Longissimus muscle of Holstein × Japanese black F1 steers. *Animal Science Journal* 83: 338–343.

Selgas, M.D., Cáceres, E., and García, M.L. (2005). Long-chain soluble dietary fiber as functional ingredient in cooked meat sausages. *Food Science and Technology International* 11: 41–47.

Sherman, P. (1969). A texture profile of foodstuffs based upon well-defined rheological properties. *Journal of Food Science* 34: 458–462.

da Silva, D.F., de Souza Ferreira, S.B., Bruschi, M.L. et al. (2016). Effect of commercial konjac glucomannan and konjac flours on textural, rheological and microstructural properties of low fat processed cheese. *Food Hydrocolloids* 60: 308–316.

Sonestedt, E., Wirfält, E., Wallström, P. et al. (2011). Dairy products and its association with incidence of cardiovascular disease: the Malmö diet and cancer cohort. *European Journal of Epidemiology* 26: 609–618.

Soukoulis, C., Chandrinos, I., and Tzia, C. (2008). Study of the functionality of selected hydrocolloids and their blends with κ-carrageenan on storage quality of vanilla ice cream. *LWT – Food Science and Technology* 41: 1816–1827.

Stevenson, D.G. and Inglett, G.E. (2009). Cereal β-glucans. In: *Handbook of Hydrocolloids*, 2e (ed. G.O. Phillips and P.A. Williams), 615–652. Cambridge, UK: Woodhead Publishing.

Stokes, J.R., Boehm, M.W., and Baier, S.K. (2013). Oral processing, texture and mouthfeel: from rheology to tribology and beyond. *Current Opinion in Colloid and Interface Science* 18: 349–359.

Sun, X.D. (2009). Utilization of restructuring technology in the production of meat products: a review. *CyTA – Journal of Food* 7: 153–162.

Suzuki, A., Hizukuri, S., and Takeda, Y. (1981). Physicochemical studies of kuzu starch. *Cereal Chemistry* 58: 286–290.

Szczesniak, A.S. (1971). Consumer awareness of texture and of other food attributes II. *Journal of Texture Studies* 2: 196–206.

Szczesniak, A.S. and Ilker, R. (1988). The meaning of textural characteristics – juiciness in plant foodstuffs. *The Journal of Texture Studies* 19: 61–78.

Szczesniak, A.S. and Kahn, E.L. (1971). Consumer awareness of and attitudes to food texture. I. Adults. *Journal of Texture Studies* 2: 280–295.

Szczesniak, A.S. and Kahn, E.L. (1984). Texture contrasts and combinations. A valued consumer attribute. *Journal of Texture Studies* 15: 285–301.

Szczesniak, A.S. and Kleyn, D.H. (1963). Consumer awareness of texture and other attributes. *Food Technology* 17: 74–77.

Tanaka, K., Miyake, Y., and Sasaki, S. (2010). Intake of dairy products and the prevalence of dental caries in young children. *Journal of Dentistry* 38: 579–583.

Tecante, A. and Doublier, J.L. (2002). Rheological investigation of the interaction between amylose and κ-carrageenan. *Carbohydrate Polymers* 49: 177–183.

Techawipharat, J., Suphantharika, M., and BeMiller, J.N. (2008). Effects of cellulose derivatives and carrageenans on the pasting, paste, and gel properties of rice starches. *Carbohydrate Polymers* 73: 417–426.

Tester, R.F. and Karkalas, J. (2002). Starch. In: *Biopolymers. Polysaccharides II: Polysaccharides from Eukaryotes*, vol. 6 (ed. E.J. Vandamme, S. De Baets and A. Steinbuchel), 381–438. Weinheim: Wiley.

Thaiudom, S. and Goff, H.D. (2003). Effect of κ-carrageenan on milk protein polysaccharide mixtures. *International Dairy Journal* 13: 763–771.

The Editorial Committee of Nihon Kokugo Daijiten (2001). Katai. In: *Nihon Kokugo Daijiten*, 2e, vol. 3, 688–689. Tokyo, Japan: Shogakukan (in Japanese).

Thomas, W.R. (1997). Carrageenan. In: *Thickening and Gelling Agents for Food*, 2e (ed. A. Imeson), 45–59. Gaithersburg, MD: Aspen Publishers.

Tischer, P.C.S.F., Noseda, M.D., Freitas, R.A. et al. (2006). Effects of iota-carrageenan on the rheological properties of starches. *Carbohydrate Polymers* 65: 49–57.

Tunick, M.H. (2000). Rheology of dairy foods that gel, stretch, and fracture. *Journal of Dairy Science* 83: 1892–1898.

Twillman, T.J. and White, P.J. (1988). Influence of monoglycerides on the textural shelf life and dough rheology of corn tortillas. *Cereal Chemistry* 65: 253–257.

Unbehend, G. and Neumann, H. (2000). Production of prebaked, deep frozen wheat rolls to finish at home. *Getreide, Mehl und Brot* 54: 110–120.

Wang, Y., Chen, Y., Zhou, Y. et al. (2017). Effects of konjac glucomannan on heat-induced changes of wheat gluten structure. *Food Chemistry* 229: 409–416.

Wu, T. and Mao, L.C. (2009). Application of chitosan to maintain the quality of kamaboko gels made from grass carp (*Ctenopharyngodon idellus*) during storage. *Journal of Food Processing and Preservation* 33: 218–230.

Yahia, E.M. and Carrillo-López, A. (2019). *Postharvest Physiology and Biochemistry of Fruits and Vegetables*. Cambridge, UK: Woodhead Publishing.

Yanes, M., Duran, L., and Costell, E. (2002). Effect of hydrocolloid type and concentration on flow behaviour and sensory properties of milk beverages model systems. *Food Hydrocolloids* 16: 605–611.

Yoshikawa, S., Nishimaru, S., Tashiro, T., and Yoshida, M. (1970). Collection and classification of words for description of food texture. I. Collection of words. *Journal of Texture Studies* 1: 437–442.

Yoshimura, M., Takaya, T., and Nishinari, K. (1998). Rheological studies on mixtures of corn starch and konjac-glucomannan. *Carbohydrate Polymers* 35: 71–79.

Yousefi, M. and Jafari, S.M. (2019). Recent advances in application of different hydrocolloids in dairy products to improve their techno-functional properties. *Trends in Food Science and Technology* 88: 468–483.

Yousefi, M., Khorshidian, N., and Hosseini, H. (2018). An overview of the functionality of inulin in meat and poultry products. *Nutrition and Food Science* 48: 819–835.

Zhou, Y., Cao, H., Hou, M. et al. (2013). Effect of konjac glucomannan on physical and sensory properties of noodles made from low-protein wheat flour. *Food Research International* 51: 879–885.

7

Use of Hydrocolloids to Control the Texture of Multilayered Food Products

7.1 Introduction

A simple technique for achieving different textures and tastes in the same bite is to make a food product consisting of different layers. This chapter starts with a description of a handful of such multilayered hydrocolloid products. It explains layer adhesion, and how some of the mechanical properties of the layered array can be estimated from the properties of the individual layers. Different adhesion methods for multilayered hydrocolloid gels might lead to innovative products in the food industry and to the development of novel foods and dishes. Multilayered confectionery compositions made up of hydrocolloids and formed using extrusion or co-extrusion processes with no less than two different confectionery compositions which differ visually or sensorially are designated. This chapter considers multilayered confectionery products, cream-filled products, multilayered gel products, and multilayered films, coatings, particles, and liposomes. Problems related to multilayered and colored products are described. Recipes for prepared multilayered gelatin jelly and beer jelly, and photographs of multilayered Japanese sweets are also included to demonstrate how simple, and sometimes complicated, it is to construct such products, as well as the usefulness and beauty of these foods.

7.2 Multilayered Hydrocolloid-Based Foodstuffs

7.2.1 Confectionery Products

Confectionery is a term with indistinct boundaries. Nonetheless, in the main it refers to a delicacy that is sweet and usually eaten with the hands, and that keeps for some time. The word confection is related to the Medieval Latin *confecta* and the English word "comfit," with meanings related to the preparation of a mixture and its preservation in sugar. Most confectionery substances are manufactured with large amounts of sugar (Davidson 2014). Middle Eastern confectionery customarily makes use of cereal ingredients. For example, Turkish delight is thickened with corn flour and many halva formulas have a base of semolina (i.e. the coarse, purified wheat middling [intermediate milling stage] of durum wheat).

Use of Hydrocolloids to Control Food Appearance, Flavor, Texture, and Nutrition, First Edition.
Amos Nussinovitch and Madoka Hirashima.
© 2023 John Wiley & Sons Ltd. Published 2023 by John Wiley & Sons Ltd.

In numerous regions where sugar-based sweets are prepared, there is no division between sugar confectionery and pastries or other sweet dishes (Davidson 2014).

Confectionery products with multiple layers and coatings are common, such as Mentos candies which have a hard-shelled sugar exterior and a chewy center (Figure 7.1). These candies are composed of sugar, wheat glucose syrup, hydrogenated coconut oil, natural flavors, sucrose esters of fatty acids, carnauba wax, beeswax, rice starch, gum arabic, and gellan gum. Mentos Chewy Mints are available in many refreshing flavors. Mentos Mints offer the advantage of a mint with an enjoyable chew (https://web.archive.org/web/20140928001346/http://us.mentos.com/mentos-mint). Another such candy, Golia Active Plus, has a liquid filling and a hard amorphous candy layer surrounding it. Nevertheless, the chewy candies on the market that have a liquid filling, particularly those with a coating, do not exhibit consistent shapes or dimensional homogeneity. European Patent 1845799 revealed a technique for making a confectionery product by coextruding a chewy material, which was coated with the candy material. The jacketed material was then formed into discrete pieces which were coated with a hard shell (Overly and Doerr 2007).

Preparation of multilayered confectioneries manufactured from sugar-free compositions poses exceptional challenges in terms of processing and stability, in addition to consumer acceptance (Campomanes Marin et al. 2016). Such novel confectionery compositions should deliver the preferred advantage of an immediate burst of flavor together with a long-term flavor profile. In addition, innovative multilayered confectionery and chewing gum compositions should deliver the anticipated benefits of new textural features, for instance matching or mismatched textures (Campomanes Marin et al. 2016). The texturizing agent can be selected from gelatin, albumin, modified starch, cellulose, or their combinations. The gel strength of the gelatin can be greater than or equal to about 130–250 Bloom. The Bloom value, determined by a Bloom gelometer, is proportional to the average molecular mass and serves as an indication of the strength of a gel formed from a solution of known gelatin concentration. Other hydrocolloids might include gelling and non-gelling gums (Campomanes Marin et al. 2016). Numerous publications include descriptions of systems and methods to form multilayered confectionery, such as a method and apparatus, that includes at least two sets of forming drums

Figure 7.1 Mentos Chewy Cinnamon Mints. *Source:* Evan-Amos/Wikimedia Commons/Public Domain.

to form and laminate confectionery sheets (Jani and Miladinov 2014). Hydrocolloids and modified natural gums are also used in such products as mouth moisteners (Jani and Miladinov 2014).

As stated, several patents describe methods for the preparation of a multilayered confectionery product. Such products can include a confectionery paste center, surrounded by an intermediate layer of chewy candy material, enveloped in a layer of amorphous candy material. The multilayered product can be further coated with a hard coating (De Jong et al. 2019). In this latter case, the chewy candy material included 4.7% gelatin on a wet basis. The syrup for the hard coating contained 2.0% gum arabic (De Jong et al. 2019). It is important to note that the recognized techniques for manufacturing a paste center enclosed by a chewy coating and hard shell (i.e. multilayered structure) are the result of much working through technical problems. For example, a fondant material, which has a semisolid consistency, mixed with the chewy material at the exit of the coextrusion nozzle, resulting in loss of the multilayered structure (De Jong et al. 2019). This problem was overcome by adding a specific nozzle in the center of a cone of the amorphous candy material to permit extrusion of the chewy material as a hollow rope with clearly differentiated concentric layers. Extruding the chewy material ahead of the paste material allowed for the frictional energy of the system, consisting of the chewy material plus the amorphous candy, to dissipate before the extrusion of the paste material. In this way, the chewy material was not forced to mix with the paste material (De Jong et al. 2019). Yet another patent consisting of methods and an apparatus for forming layered confections (Miladinov et al. 2017) described a dough-like confectionery material that contained solid particles, a liquid, and a diffusion controller. The dough-like confectionery material was an effective replacement for panned coatings, and could be applied to an edible substrate, such as candy or chewing gum, to form a layered confection (Miladinov et al. 2017).

7.2.2 Cream-Filled Multilayered Food Products

The construction of an edible item made up of dissimilar layers provides a consumable with different textures and tastes in the same bite. Today, a diversity of multilayered consumables is available, e.g. crispy wafers (i.e. a crisp, often sweet, very thin, flat, and dry biscuit) that have a sweetened plant fat-based chocolate or vanilla flavor filling, and multilayered, sweetened, agar–agar-based confections for children (Nussinovitch 2017). The cream filling can serve as an adhesive material for the different layers of the food product. Examples include sandwich cookies, which occupy a noteworthy place in the world market for biscuits, where a soft cream filling is sandwiched between either cookies or wafer sheets (Shamsudin 2009; Nussinovitch 2017). Well-known trademarks of such products are *Oreo* (Figure 7.2), which is manufactured by Nabisco, and *Tim Tam* (Figure 7.3) created by Arnott's Biscuits (Yuan-Kwan 2008; Shamsudin 2009). Filling creams used in baked goods are grouped into heavy and light creams, with or without water. The choice of processing equipment is governed by the cream classification. Typical products using heavy creams include biscuit and sandwich creams; characteristic products with light filling creams are chocolate coverings, dough improvers mixed with fat nougat paste and wafer creams, and fillings for chocolate bars and biscuits (Gerstenberg Schröder 2009). The functionality and value of the cream vary depending on the application but overall bakery creams have distinct properties of adhesion,

Figure 7.2 Regular Oreo cookies. *Source:* Evan-Amos/Wikipedia Commons/Public Domain.

Figure 7.3 *Tim Tams* – Two flavors for the South East Asian market, Choco-chocolate and Choco-cappuccino, sold in Indonesia. *Source:* Kcdtsg/Wikipedia Commons/Public Domain.

firmness, sweetness, and texture. The wafer application necessitates a somewhat softer cream with the ability to hold a number of wafer layers. The industrial process is comparable to that with heavy creams, as the manufacturing lines should be able to handle an abrasive product under adequate pressure (Gerstenberg Schröder 2009).

7.2.3 Gelled Multilayered Food Products

The layers in agar–agar confections are of comparable consistency, but the flavors and hues may vary. In the Far East, where unrelated gel textures are better accepted than in the West, a sweetened curdlan-based multilayered gel is manufactured (Ikeda et al. 1976). Although the layers are all produced from the same hydrocolloid (curdlan), two dissimilar layer types are formulated from its powder by heating the suspension to different temperatures. Multilayered hydrocolloid-based foodstuffs are important for foods of the future. Cautiously

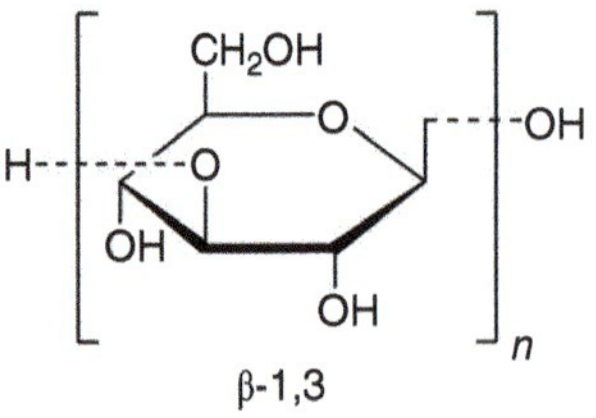

Figure 7.4 Chemical formula of curdlan depicted as Haworth projection. *Source:* Edgar 181/ Wikimedia Commons/ Public domain.

worked-out models describing the mechanical properties of multilayered gels, foods, and other items manufactured for consumption have been proposed. One example is a model predicting the stress–strain relationship of layered polymeric sponges made from polyurethane (Swyngedau et al. 1991b). Another model (Nussinovitch et al. 1991) was developed to estimate the compressive deformability of double-layered curdlan gels. This latter model was used to study the compressive deformabilities of gels whose layers were glued together by one of three adhesion techniques (Ben-Zion and Nussinovitch 1995).

Agar–agar can be used in food items that are rich in sugar and carbohydrates (such as confections) at concentrations of ~0.3–1.8%. Agar gel is used to assemble multilayered, sweetened, ready-to-eat products (Nussinovitch 1997). As already noted, the compressive force vs. deformation relationships of double-layered curdlan gels (prepared with 2.5 and 3.5% gels) were computed from those of the individual components. The calculation was based on three assumptions: that the normal force (i.e. the component of the contact force exerted on an object, perpendicular to its surface) in the layers is the same; that their deformations are additive; and that the consequences of lateral (direction or position relative to the shape of an object spanning the width of a body) stresses and viscoelasticity can be neglected. A mathematical model whose constants were determined from the behavior of the individual layers was used. The calculated relationships were compatible with experimental relationships over a substantial range of strains. Failure of the arrays, however, preceded that of the layers, resulting in their separation, a phenomenon that was unrelated to the model but limited the range of strains for which it was applicable (Nussinovitch et al. 1991).

Curdlan is valuable in formulating novel varieties of jelly products (Figure 7.4). The polymer can be used at a final concentration of 0.4–6.0% in foods. The character of curdlan gels lies between the elasticity of gelatin and the brittleness of agar gels (Kimura et al. 1973). The gel is stable between pH 3.0 and 9.5 and it can absorb sugars rapidly and at high concentrations from a syrup (i.e. a thick, viscous liquid containing a large amount of dissolved sugar in water but showing little tendency to deposit crystals), and thus it can be used to formulate sweet jellies (Harada 1979; Nussinovitch and Hirashima 2014b). The gelling properties of curdlan indicate that it can replace, to some extent or completely, polymers such as agar, gelatin, or carrageenan. Curdlan is insoluble in water; nevertheless, its aqueous suspension can form two types of heat-induced gel, depending on the temperature. A "low-set gel" is formed when the aqueous suspension is heated to between 55 and 60 °C and after that cooled to below 40 °C. This gel is thermoreversible, and junction zones are formed by hydrogen bonds, analogous to agar–agar or gelatin. The other gel type – "high-set gel" – is formed by heating the aqueous suspension to over 80 °C. It is thermoirreversible, and the junction zones in the gel are formed by hydrogen bonds and hydrophobic interactions (Nussinovitch and Hirashima 2014b). The high-set gel is stable under high-temperature treatments, for instance retorting, and remains tasteless, odorless, and colorless. The incipient temperature for thermoirreversible gelation decreases with increasing curdlan concentration. Furthermore, curdlan forms thermoirreversible gels in food

Figure 7.5 Red and white kamaboko made by Kibun Foods Inc. *Source:* Kinori/Wikimedia Commons/Public Domain.

systems, even when processed at lower than 80 °C, due to increases in its components (Nussinovitch and Hirashima 2014b). A canned, multilayered jelly can be formulated containing both high-set and low-set gels. This product has been extensively evaluated as a new gel type (Harada 1979).

Additional products can be manufactured using this double-gel-set procedure. Kamaboko (Figure 7.5), a gelled seafood product made from frozen surimi, has distinct textural properties. Classification of those properties, by means of an integrated approach to rheological studies, was achieved through an instrumental texture-profile analysis and assessment of consequential stress–strain relationships. The material had near-ideal area expansion, even at a compression of 60%, while retaining its highly elastic texture. The product did not yield to up to 80% compression. Hardness of the kamaboko at 80% compression was characterized by a local maximum at 37.5 °C which might have been related to the processing temperature of the initial surimi gel used in the double-gel-set procedure. Assessment of stress–strain associations confirmed the incompressible nature of the gel and demonstrated comparatively minor variations between Young's and deformability moduli. The elastic limit of the kamaboko increased markedly with a temperature increase from 25 to 50 °C (Konstance 1991).

7.2.4 Multilayered Films

Biodegradable, ecofriendly, and safe films for food packaging have been developed at various research institutes (Etxabide et al. 2017). More than a few proteins are known for their film-forming abilities and outstanding barrier capacity against diffusion of carbon dioxide and oxygen. For example, with the aid of suitable plasticizers, proteins such as collagen, kafirin, gelatin, soy isolates, and zein readily produce continuous solid matrices upon drying from their solutions (Alves et al. 2017; Liu et al. 2017). Numerous approaches have been recommended to increase the moisture resistance of gelatin films: chemical, enzymatic cross-linking, and physical methods (Martucci et al. 2012); compositing with supplementary moisture-resistant polymers (Rhim et al. 2006; Moreno et al. 2017); and constructing multilayered films (Martucci and Ruseckaite 2009, 2010a, 2010b). The production of multilayered films by combining the advantages of gelatin and kafirin film matrices in one sheet is an encouraging way to avoid their particular deficiencies (Wang et al. 2018). With the kafirin film as the outer layer, the gelatin film as the internal layer, and a hybrid kafirin/gelatin film acting as the intermediate transition layer, the kafirin–kafirin/gelatin–gelatin multilayered film is expected to perform as a one-way water barrier (Wang et al. 2018). Thus, when the film is applied with the kafirin layer facing the outside and gelatin layer facing the inner packaged item, one obtains resistance to environmental moisture and retention of moisture in the packaged product (Wang et al. 2018).

7.2.5 Nano-Multilayer Coatings

Nanoscale materials could conceivably advance the performance of edible films and coatings, presenting novel properties, specifically the option of joining different layers with distinct functionalities, such as antimicrobial and antioxidant properties, with enhanced gas-barrier features (Medeiros et al. 2012a). Polysaccharide-based coatings can control the internal atmosphere of fruit and slow their senescence (Nisperos-Carriedo 1994), consequently extending their shelf life; they do this by providing a partial barrier to carbon dioxide, oxygen and moisture, and preventing volatile losses (Olivas and Barbosa-Cánovas 2005). Surface modifications of polypropylene films using chitosan and chitosan/pectin multilayers were reported by Elsabee et al. (2008). They studied tomato packing in bags of transparent polypropylene film coated with 12 alternating layers of chitosan and pectin, as a new concept of active packaging for fruit preservation (Elsabee et al. 2008). A nano-multilayer coating consisting of five nanolayers of the hydrocolloids pectin and chitosan was also produced (Medeiros et al. 2012a). This coating was characterized in terms of carbon dioxide, oxygen, and water vapor permeabilities. Other reports studied a multilayer film of alginate and chitosan. Contact angles of a similar magnitude to those observed for polyethylene terephthalate (PET) and aminolized PET (Xu et al. 2008). Medeiros et al.'s (2012a) nano-multilayer system was applied on whole "Tommy Atkins" mangoes and the layers' adsorption was confirmed by changes in the contact angle of the coated fruit skin. Following 45 days of storage, uncoated mangoes presented a greater mass loss, and higher total soluble solids and lower titratable acidity than coated mangoes. Uncoated mangoes had a damaged/wrinkly exterior, presenting symptoms of microbial damage, as well as slightly brownish flesh color, in comparison to the coated mangoes. As a result, the shelf-life of the coated mangoes was extended (Medeiros et al. 2012a).

Minimally processed fruit are one of the most important growing sectors in food retail marketing (Robles-Sánchez et al. 2013). An alginate–chitosan nano-multilayer coating was applied to "Rocha" pears (Medeiros et al. 2012b) and fresh-cut mangoes (Souza et al. 2015). The nano-multilayer coating of mango was obtained by electrostatic layer-by-layer self-assembly. Coated and uncoated fresh-cut mangoes were stored under refrigeration (8 °C) for 14 days. At the end of the storage period, lower values of mass loss, pH, malondialdehyde content, browning rate, soluble solids, and microorganism proliferation, and higher titratable acidity were observed in the coated mangoes (Souza et al. 2015). However, the nano-multilayer coating did not improve the retention of vitamin C during storage of fresh-cut mangoes. Results suggested that the chitosan–alginate nano-multilayer edible coating extends the shelf life of fresh-cut mangoes for up to eight days (Souza et al. 2015).

7.2.6 Multilayered Liposomes and Capsules

Multilayered liposomes can be manufactured by using an oppositely charged compound to create an additional layer. This method is described as layer-by-layer deposition (Decher 2003; McClements et al. 2009). This technique, among others, enables the encapsulation of useful compounds (Ogawa et al. 2003a). For example, high-pressure homogenization was used to prepare soy lecithin liposomes with incorporated grape seed extract (GSE) (Gibis et al. 2014). The GSE and the uncoated and coated liposomes were mixed with native and heat-treated protein solutions of bovine serum albumin (BSA), whey protein isolate (WPI) and sodium caseinate (Na-caseinate) to investigate their interactions. Chitosan-coated liposomes show reduced precipitation. The chitosan-coated liposomes containing GSE could be used as a supplement in beverages or food, conferring the functional properties of GSE and milk proteins (Gibis et al. 2014).

Microencapsulation of a probiotic and prebiotic in alginate-chitosan capsules improves survival in simulated gastro-intestinal conditions (Chávarri et al. 2010). The approach involved the specific addition of each of the encapsulated strains to the food preparation, making it difficult to guarantee the consistency of the mixture and the absolute quantity of each administered strain upon consumption of the product. These complications needed to be overcome (Marañón García et al. 2016), and European Patent # 3 205 216 A1 provided a solution, in the form of a microcapsule including: (i) a core consisting of at least one probiotic, prebiotic or mixture thereof embedded in a polymer selected from the hydrocolloids alginate and pectin; (ii) an inner layer surrounding the core containing a probiotic, prebiotic or mixture thereof embedded in an alginate or pectin polymer; (iii) an outer layer surrounding the inner layer containing hydrophobic material with a melting point of 35–90 °C, where the probiotic(s) or prebiotic(s) included in the core is different from the probiotic(s) or prebiotic(s) contained in the inner layer (Marañón García et al. 2016).

7.2.7 Multilayered Particles

Many techniques can be used to manufacture particles. Among them, the mild method of ionic gelation has been used to produce gel particles via the interaction of multivalent ions such as Ca^{2+} with COO^- groups present in hydrocolloids (Bjapai and Tankhiwale 2006). Such particles/beads can encapsulate cells, enzymes, probiotic microorganisms, and hydrophilic and hydrophobic compounds (De Vos et al. 2006), and can serve for food applications (Willaert and Baron 1996). Such microcapsules have been used in the dietary and agricultural industries (Jafari et al. 2008). Microencapsulation protects active compounds from adverse conditions, such as light, oxygen, and pH (Shahidi and Han 1993). Various methods have been attempted to encapsulate probiotic microorganisms (Truelstrup et al. 2002; Anal and Sing 2007; Burey et al. 2008).

During ionic gelation, not all alginate carboxyl groups act together with the crosslinking agent, thus producing a matrix with an additional negative charge that allows interaction of the gel surface with oppositely charged polyelectrolytes. The result is a reduction in gel particle porosity and a concomitant increase in their barrier and protective properties (De Vos et al. 2007). This technique, based on the alternation of oppositely charged polyelectrolytes, is known as layer-by-layer assembly (Decher 1997).

To increase the protection of immobilized probiotic microorganisms such as *Lactobacillus plantarum*, the embedding microparticles (produced by ionic gelation) were coated with whey protein (Gbassi et al. 2009, 2011). Another approach described carriers consisting of whey protein coated by polysaccharide, in several alternating layers (Rosenberg and Lee 2004; Hebrard et al. 2009, 2013; Doherty et al. 2011, 2012). Alginate is an anionic, and therefore negatively charged hydrocolloid when its pH is above its pK_a (pH 3.65–3.80) (Annan et al. 2008). Consequently, whey proteins, comprising predominantly mixtures of the globular proteins β-lactoglobulin (82% w/w, 18.5 kDa) and α-lactoglobulin (15% w/w, 14.5 kDa), can interact when the pH is adjusted below pH 4.4–5.2 (Damodaran 2008).

Hebrard et al. (2013) and Gbassi et al. (2009) used particles covered with only one layer of protein or alginate. Nogueira et al. (2017) produced alginate particles that were covered with two layers of whey protein interspersed with a layer of alginate. Their assumption was that these three layers would provide additional resistance for these multilayered particles against gastric/intestinal fluids. As already mentioned, the multilayered gelled particles were produced by the layer-by-layer method using whey proteins and alginate solution (Nogueira et al. 2017). The protein adsorbing on the particles reached 64.9% of the dry weight of the multilayered particle. This produced a reduction in the moisture content of the particles and an increase in their average size with the successive adsorption of new layers. Multilayered particles with a greater amount of adsorbed protein remained stability over a wide pH range (4.0–8.0), with a substantial reduction in size and increased protein solubility in the medium when the pH value was 2.0. Increasing the ionic strength of the medium led to a significant increase in the loss of particle protein (Nogueira et al. 2017). Another report discussed crosslinking gelatin microspheres with the non-cytotoxic genipin and then coating with alginate crosslinked by Ca^{2+} from external or internal sources to improve the survival of the probiotic *Bifidobacterium adolescentis* 15703T during exposure to adverse environmental conditions (Annan et al. 2008).

7.3 Methods to Estimate Properties of Multilayered Products

7.3.1 Assessment of Stiffness and Compressive Deformability of Multilayered Texturized Fruit and Gels

It is possible to predict the deformability modulus (stiffness) of a layered gel array made up of a number of gelling agents and supplementary components. The prediction, based on the deformability moduli and heights of the separate layers, is fairly precise, and the approach is suitable for calculating the stiffness of multilayered gels, texturized fruit and additional similar edible products (Ben-Zion and Nussinovitch 1996). E_{D_A}, E_{D_B}, and $E_{D_{AB}}$ represent the deformability moduli of layers A, B and the entire array, respectively, in a joined gel comprised of two layers. H_{0_A}, H_{0_B}, and $H_{0_{AB}}$ are the heights of layers A, B, and their sum, respectively, and the stress is expressed as: $\sigma = E_D\varepsilon$. These moduli can be determined using Eqs. (7.1) and (7.2). Equation (7.1) is valid when the engineering strain is being contemplated, whereas Eq. (7.2) is appropriate for calculating Hencky's strain (Nussinovitch 2017).

$$E_{D_{AB}} = \frac{1}{\left(\dfrac{H_{0_A}}{H_{0_{AB}} E_{D_A}} + \dfrac{H_{0_B}}{H_{0_{AB}} E_{D_B}} \right)} \tag{7.1}$$

$$E_{D_{AB}} = \frac{\sigma}{\ln\left[\dfrac{H_{0_A}}{H_{0_{AB}}} e^{\frac{\sigma}{E_{D_A}}} + \dfrac{H_{0_B}}{H_{0_{AB}}} e^{\frac{\sigma}{E_{D_B}}} \right]} \tag{7.2}$$

This experimental, carefully worked out model effectively predicted the deformability modulus of a multilayered gel made up of agar and one of four galactomannans in three arrangements with non-identical layer thicknesses, and of a four-layered gel array of texturized fruit. The model was constructed on the postulation that the uniaxial stress in the layers is identical and that their deformations are additive (Nussinovitch 2017). No noteworthy differences were observed between the experimental and model-calculated deformability moduli. Both Eqs. (7.1) and (7.2), which are simply expanded forms of the Takayanagi isostress blending law (Takayanagi et al. 1963), gave comparable estimates of predicted E_D values. The model offers a tool for approximating multilayered gel stiffness when there is no protrusion during compression, and may be suitable for other food systems that perform similarly (Ben-Zion and Nussinovitch 1996). Another study dealt with the compressive force–deformation relationships of multilayered hydrocolloid gels comprised of dissimilar combinations of agar, xanthan, carrageenan and konjak mannan and four galactomannans, and texturized fruit based on banana, apple, kiwi and strawberry pulp and agar–locust bean gum combinations, adhered using three different gluing techniques. These relationships were computed from those of the separate layers (Ben-Zion and Nussinovitch 1997). The gluing techniques consisted of (i) pouring hot hydrocolloid solution on a gelled layer; (ii) using melted agar as a glue between already gelled layers; or (iii) pouring the simultaneously prepared pregelled (gum solution before setting) hydrocolloid solutions together (Figure 7.6). The assumptions of identical normal force in the layers and additive deformations were made. The effects of lateral stress were considered negligible. The calculation was performed using a mathematical model previously developed for double-layered curdlan gels (Nussinovitch et al. 1991). The model constants were determined from the behavior of the individual layers. Good agreement was found between experimental and fitted results over a substantial range of strains. Thus, the model's applicability to a specific gel system was validated, suggesting a very convenient tool for analyzing and predicting the compressive behavior of any number of arrays with different layer combinations (Figure 7.7) (Nussinovitch et al. 1991; Ben-Zion and Nussinovitch 1997; Nussinovitch 2017).

Figure 7.6 Three different techniques to "glue" the layers of a multilayered gel. (a) Hot presetting solution of hydrocolloid or hydrocolloid mixture is poured onto already-gelled layers with the same or different compositions at room temperature; this usually (but not always) produces multilayered gels. (b) The surface of one layer is thoroughly smeared (using a fine brush) with a 2% agar solution at 95 °C, then the two layers are pressed together. (c) The two solutions are prepared simultaneously and poured together. After a short time, the layers separate, while the solution is still hot. After gelation, a two-layered gel system is easily observed. *Source:* Adapted from Ben-Zion and Nussinovitch (1997).

7.3.2 Calculating the Stress–Strain Relationships of a Layered Array of Cellular Solids

The word cellular solid originates from the word "cell," derived from the Latin "cella" – a small enclosed space. Unique structures, which can be observed in many natural instances, e.g. wood, cork, sponge and coral, are cellular solids (Gibson and Ashby 1988). Man-made natural or synthetic cellular products include items from disposable coffee cups to crash padding in aircraft cockpits. Foamed polymers, metals, ceramics and glass are used for insulation, cushioning and absorption of impact kinetic energy (Nussinovitch 1997b, 2003). Cellular solids contain different ordered and disordered structures; of major importance is the distinction between open-cell (interconnected) and closed-cell (a cell sealed off from its

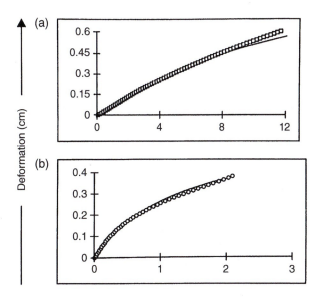

Figure 7.7 Deformation vs. force relationships of (a) four-layered texturized fruit and (b) a four-layered gel. Both multilayered gels were constructed by pouring hot hydrocolloid solutions on already gelled layers. Symbols denote experimental data and solid lines represent model predictions. *Source:* Reproduced from Ben-Zion and Nussinovitch (1997)/with permission of Elsevier.

neighbors by membrane-like faces) cellular solids (Gibson and Ashby 1988; Jeronomidis 1988). Tailor-made cellular solids are and will continue to be utilized as affordable carriers for numerous food supplements and biotechnological operations (Nussinovitch 2003, 2017).

Different cellular materials can be arranged in layers of similar or different thicknesses. When a flat layered array of different cellular materials, each having a dissimilar (or identical) thickness, is compressed uniaxially (i.e. on only one axis), its cross-sectional area, like that of its individual layers, can be assumed to be practically unchanged (Peleg 1997). Therefore, the stress along the array can be considered identical in all of the layers, while the total deformation may be denoted as the sum of the deformations of each layer (Figure 7.8). Expressed mathematically:

$$\sigma_{total} = \sigma_i \tag{7.3}$$

where σ_{total} is the array's stress and σ_i is the stress in an individual layer, i, and

$$\varepsilon_{total} = \left(\frac{1}{H_{0_{total}}}\right) \Sigma H_{0_i} \varepsilon_i(\sigma) \tag{7.4}$$

where ε_{total} is the array's strain, H_{0_i} is the individual layer's thickness, and ε_i is its strain as a function of the stress. The array's initial overall thickness is the sum of that of the individual layers:

$$H_{0_{total}} = \Sigma H_{0_i} \tag{7.5}$$

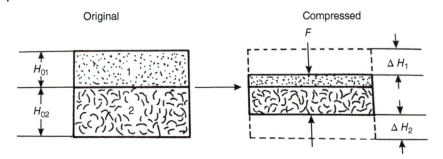

Figure 7.8 Geometry of a uniaxially compressed double-layered array. Source: Reproduced from Swyngedau and Peleg (1992)/with permission of American Association of Cereal Chemists; permission conveyed through Copyright Clearance Center, Inc.

The use of equations as deformability models for cellular solids is especially convenient for expressing strain as an explicit algebraic function of stress $\varepsilon(\sigma)$ (Peleg 1997). Inserting the terms $\varepsilon_i(\sigma)$ and the corresponding H_{0_i} into Eq. (7.5) allows calculating the stress–strain relationships of any layered array of sponges, as long as the assumption of practically unchanged cross-sectional area remains valid (Swyngedau et al. 1991a, 1991b; Swyngedau and Peleg 1992):

$$\sigma(\varepsilon) = C_1' \varepsilon^{n_1'} + C_2' \varepsilon^{n_2'} \quad (n_1 \langle 1, n_2 \rangle 1) \tag{7.6}$$

The problem with deformability models is that a determination of their constants by non-linear regression requires accurate guesses for the initial values (Peleg 1997). This difficulty can be eliminated if Eq. (7.6) or a polynomial model is used instead. Arriving at sufficiently close initial values for Eq. (7.6) is fairly easy, because we know that $n_1 < 1$ and $n_2 > 1$.

Constants of a polynomial model, such as

$$\sigma(\varepsilon) = C_1 \varepsilon + C_2 \varepsilon^2 + C_3 \varepsilon^3 + C_4 \varepsilon^4 \tag{7.7}$$

can be determined by a generalized linear regression computer program with no need for initial guesswork (Peleg 1997).

Once all of the $\sigma_i(\varepsilon)$ values are expressed in terms of any model, and the H_{0_i} values are known, the stress–strain relationship of the array can be calculated numerically using standard equation-solving software. All that is needed is to find the root of each σ_i for any desired level of stress or series of stress levels:

$$\varepsilon_i(\sigma) = \text{root}\left[\sigma - \sigma_i(\varepsilon) = 0, \varepsilon\right] \tag{7.8}$$

and insert the solution into Eq. (7.6) to generate the array's stress–strain relationships (Peleg 1993). Once the program is in place, one can change the values of H_{0_i} to produce the stress–strain relationship of any desired array:

$$\varepsilon_{\text{total}} = \left(\frac{1}{H_{0_{\text{total}}}}\right) \Sigma H_{0_i} \text{root}\left[\sigma - \sigma_i(\varepsilon) = 0, \varepsilon\right] \tag{7.9}$$

Although the calculation is performed by finding the strain that corresponds to a particular stress, creating the more conventional relationship or plot of σ vs. ε_{total} is a trivial task (Peleg 1997). Computer programs can be used with a variety of mixed-model combinations, irrespective of the mathematical structure of the model. All of these methods were developed assuming no practical change in the cross-sectional area of the systems. Another implementation involves closed-cell cellular solids exhibiting small expansions under large compressive strains (Peleg 1993).

7.3.3 Other Techniques to Assess Multilayered Products

The software and techniques used to evaluate multilayered products are founded in numerous branches of science. Software for the optical categorization of thin films by means of spectrophotometric and/or ellipsometric measurements has been developed. The program facilitates examining an extensive range of multilayered structures, taking into account the composition, microstructure and thickness of their individual layers (Leinfellner et al. 2000). X-ray diffraction is a powerful tool for studying residual stress states in microsystems, because it enables the non-destructive testing of materials and components, and the measurement of very small volumes. Owing to the small thicknesses of the individual layers in multilayered microsystems, in numerous cases, the X-rays penetrate deeply into the samples, and diffraction patterns reflect the layers of different materials. Novel detection systems, for example, area detectors, might be used effectively, pending their evaluation (Kampfe 2000).

Multilayered food products based on hydrocolloids are still rather rare. Nevertheless, the apparent increasing trend in the availability of these food products cannot be ignored. Consequently, methods from diverse, more technical areas need to be adopted to improve on and analyze existing and future products. A good example might be a proposed analytical technique for studying propagation of transient elastic waves in multilayered media (Lee and Ma 1998). A dynamic displacement response can be obtained by setting up experimental conditions that simulate planar stress conditions for a layered half-space. To date, experimental and theoretical solutions have gone hand in hand and the methods can also be applied to a three-dimensional (3D) space (Lee and Ma 1998). Another method deals with application of the Rayleigh–Ritz method, a variation on a method for non-equilibrium statistical dynamics, to analyze multilayered plates with residual stresses for membrane and bending deformation. The benefit of this method is its straightforwardness and ease of application to simple sample shapes (Lee and Kim 1999). Other methods that might be appropriate for such complex products are ultra micro hardness, adhesion, and residual-stress analyses, particularly in systems where one layer is very thin; i.e. it behaves like a film coating a surface (Tavares et al. 1999).

7.4 Current Systems and Methods to Prepare Multilayered Products

7.4.1 Extrusion and Coextrusion

Numerous food industries, e.g. those producing cereals, puff pastry, sugar wafer sticks, and snacks, include multilayered products. A number of processes have been developed to deal with the complexities of their layering (Vardi and Tslaf 2013). Extrusion has provided means of manufacturing novel foodstuffs and has modernized many of the conventional snack-manufacturing processes (Simitchiev et al. 2010). Extrusion equipment offers various basic design benefits that reduce cost, energy, and time, while concomitantly increasing the degree of adaptability and flexibility (Sevatson and Huber 2000).

The term coextrusion was first used in the 1950s, referring to a method for manufacturing polymer materials with different structures. It introduced the use of dies to spread and thin the component masses and then join them at the end of the process. In the food industry, the coextrusion method became widespread in 1984 (Frame 1994), when the M&M/Mars company used it to create the products "Combos" and "Corn Quistos" (https://www.dailymotion.com/video/x926s4). Both of these products were crispy cereal-based tubes stuffed with spicy soft cream.

Coextrusion is an industrial process that is widely used to form multilayered sheets or films that are suitable for a variety of products, ranging from food packaging materials to reflective polarizers (Han 1973, 1981; Yu and Han 1973). The applications of multilayer coextrusion are vast; studies have dealt with many of its aspects, ranging from equipment design (Mitsoulis 1988), fluid mechanics and analysis of multilayer flow to interfacial defects, and optimization of processing conditions (Schrenk et al. 1978; Wilson and Khomami 1993). Coextrusion enables the concurrent molding of products that include two or more components. It is performed through an extruder (single or twin screw), pumps, and a molding device. In tubular-type instruments, the outer pipe is shaped by the extrusion of wheat, corn or rice meal, or some combination of these (Huber and Rokey 1990). Maize, rice, and wheat are the main cereal sources of starch, with the world production of cereals in 2019 reaching 2800 million tons (Food and Agriculture Organization 2021). Other major sources of starch are tubers (e.g. potatoes and yam), roots (cassava and taro), and legume seeds (e.g. pea).

A filled confectionery, a cooked multilayered food product, was described. A food base product was subjected to a cooking extrusion process through a multilayer-shaped extrusion die to provide the cooked multilayered food product (Vardi and Tslaf 2013). This product could be produced in a starfish shape with three arms, similar to the cereal layers surrounding the cereal tube. The dry mix of the food base contained 14.5% starch. The list of ingredients included corn grits, calcium carbonate, soy flour, salt, crystalline sucrose, powdered soy lecithin and green pea fiber. The dry mix was constantly fed into a twin-screw cooking extruder and in parallel, water was added to it. Sweet cream was fed into the extruder tube during the coextrusion process to produce the interior filling for the product. The dry mixture and water were mixed, cooked, and shaped by the extruder, and no expansion of the filled confectionery was observed at the extruder die exit. The product was then transported for about 5 m on a conveyor belt to the cutting device. The cut product went through a drying oven to bring it to a final moisture content that would enable a shelf-stable product. Packaging was the last step of the production process (Vardi and Tslaf 2013).

Extrusion can thus be used to manufacture multilayered products. A recent patent described processes to form food products that include one or more extruded components (e.g. vegetable, fruit, dairy, meat, flavoring, spice, coloring, particles, or combinations thereof), and one or more extruded collagen layers essentially encasing the extruded component(s) (Carlson et al. 2015). Another representative model food product included an extruded component, and a carrier coextruded with the first extruded component; the carrier contained a matrix that adhered to the first extruded component and an additive suspended within it (Carlson et al. 2015). Coextrusion is the simultaneous extrusion of two or more different but compatible materials through the same die. Multilayered confectionery compositions were formed using a coextrusion process with at least two different confectionery compositions that had at least one visual or sensory distinction (Aldridge et al. 2014). The processing parameters for treating the various confectionery compositions prior to cutting and wrapping the pieces were altered to keep the average piece size within a predetermined tolerance range. The tolerance of the resulting pieces of multilayered confectionery was thus retained, such that they could withstand rigid packaging (Aldridge et al. 2014).

One method of manufacturing multilayered, laminated, lipid-based sweet confections included tempering the lipid-based formulation for each layer, and depositing these layers separately onto a moving conveyor belt that passes through a cooling tunnel between each layer's placement station (Miller and Miller 2005). To provide for the production of more than two layers, additional stations were used further along the conveyor belt. A set of "fingers" at each station striated the passing layers from above, governing the width of each formed ribbon. The width of the fingers in each successive set was never greater than the width of the fingers in the preceding set. A mixer was arranged with a pump to inject discrete predetermined quantities of a syrup additive which had a color component and sometimes, a flavor component as well, into the center of a stream of tempered formulation for at least one of the layers as it was being fed into the mixer, so that upon its exit from the mixer, the formulation had at least one color that differed from that of the contiguous layer of the confection being formed (Miller and Miller 2005).

7.4.2 Injection Molding

Injection molding is a manufacturing process that produces pieces by injecting molten material into a mold. Injection molding can be performed with various materials, including metals, glass, elastomers, confections, and most commonly, thermoplastic and thermosetting polymers (Collyer et al. 1990). A process for forming multilayered pet treats or animal chews was described (Axelrod and Gajria 2012), in which a moveable mold portion is aligned with the first of a number of stationary mold portions fed a first composition by a first injection molding unit to form a first layer in the cavity space formed by the aligned mold portions. The moveable mold portion and the first layer may be indexed to align with a second of several stationary mold portions fed a second composition by a second injection molding unit, thereby forming a second layer over the first one in the cavity space formed by the aligned mold portion. The compositions fed by the first and second injection molding units may both be edible. Moreover, multicomponent pet treats can be formed from two materials which may differ in their physical, optical, sensual, nutritional or compositional properties. One of the materials might specifically include a textured vegetable protein (Axelrod and Gajria 2012). Coextrusion was also valuable in manufacturing pet chews (Axelrod and Gajria 2014). For such products, multiple compositions were coextruded to form a layered construction. Separate extruders were used to tailor processing conditions to heat- and/or shear-sensitive compositions, such as nutritional additives, and the respective extrudates were combined such that the compositions with the relatively higher nutritional values made up the outer layer (Axelrod and Gajria 2014).

7.4.3 3D-Printing and Layered Products

Chapter 8 of this book is fully devoted to textural control of 3D-printed foods by hydrocolloids. The technique of 3D printing has been quite useful to food processing, and is widely appreciated by food scientists (Guo et al. 2019). It has been suggested that fused deposition modeling (FDM) printed tablets with multilayered structures can be an effective means for dual-drug release. These studies, which applied FDM to pharmaceutical engineering, established the distinct advantages of 3D printing in the production of personalized, controlled-release products (Goyanes et al. 2015). Accordingly, a 3D-printed multilayer-structured food can offer surprising and varied taste experiences for consumers (Guo et al. 2019). Food 3D printing is largely considered to be one of the top current advances; in 2018, its potential for accuracy, customization, and modernization of the structures and textures of various foods was demonstrated (Le Tohic et al. 2018; Feng et al. 2019). The very first 3D printing used in the food trade was a paste extrusion of a mixture that consisted of sugar, starch, yeast, corn syrup, and cake frosting for the "cake mix" (Yang et al. 2001; Lille et al. 2018). The process of 3D food printing consists of three steps: 3D model building, object printing, and post-treatment. Controlling factors such as layer thickness, printing height, nozzle diameter, printing rate, nozzle-movement rate, and temperature are quite important for the success of the resultant printed food products (Hao et al. 2010).

Depending on the characteristics and the number of materials involved in food printing, parameters that can be assigned and changed to optimize print settings include: size, shape, extrusion rate, cooking rate, and layer height (Hertafeld et al. 2019). The fruitful use of fluids in a layered food manufacturing application can be credited to their non-Newtonian properties. For Bingham plastic fluids, a minimum shear stress, known as the "yield stress," must be exceeded before flow begins. Once the material is extruded and the stress decreases, it exhibits solid-like properties (Hertafeld et al. 2019). Hydrocolloids are often used to achieve a tendency to resist flow (Cohen et al. 2009). The main issue in using hydrocolloids is the stability of the final product. This can be controlled by proper selection of the hydrocolloid. It is vital to avoid degradation of the polymer, which will result in viscosity reduction and thus product deterioration (Nussinovitch and Hirashima 2014a).

7.4.4 Multilayered Emulsions

Oil-in-water (O/W) emulsions can be manufactured by homogenizing lipids with an aqueous emulsifier solution (Güzey and McClements 2006b). A multilayered emulsion is made of emulsifier-coated oil droplets covered by one or more supplementary layers of charged hydrocolloids (Araiza-Calahorra et al. 2018). The layer-by-layer electrostatic deposition technique is used to coat O/W emulsion droplets with charged biopolymers, to increase the resistance of emulsions to environmental stresses (Gu et al. 2004). The so-called "multilayered" emulsions can be formed by alternatingly adding oppositely charged proteins or hydrocolloids at suitable concentrations to the droplet interface (Aoki et al. 2005). Ionic strength and pH influence the degree of ionization of the charged functional groups on the respective biopolymers and intrinsically, the assembly of multilayered emulsions. The charge of the biopolymers affects the electrostatic interactions that are accountable for the attractive forces that result in the deposition of biopolymers onto the droplet surfaces (Schmitt et al. 1998). Nevertheless, alterations in environmental surroundings after deposition of the biopolymer layers may lead to desorption of multilayers and destabilization of multilayered emulsions (Guzey and McClements 2006a). The layer-by-layer process can be repeated many times by means of a combination of anionic and cationic hydrocolloids. As an example, multilayered emulsions have been fabricated from caseinate and κ-carrageenan (Perrechil and Cunha 2013), as well as gelatin, whey protein, and sugar beet pectin (Zeeb et al. 2012). Double-layer emulsions offered better stability than conventional single-layer emulsions with respect to droplet aggregation, lipid oxidation, and thermal processing (Ogawa et al. 2003b; Klinkesorn et al. 2005).

Multiple enzymes can potentially be used to stabilize the deposited biopolymer membranes (Gübitz and Paulo 2003), for instance, the crosslinking of protein residues such as whey protein isolates, caseinates and gelatin via γ-carboxyl (glutamine)-ε-amino (lysine) isopeptide bond formation by the enzyme transglutaminase (Hernàndez-Balada et al. 2009). Catalyzed crosslinking of β-casein micelles by transglutaminase eliminated the micelles' ability to dissociate on cooling and disruption, and extensive polymerization of β-lactoglobulin enabled the materialization of a weak gel (Tanimoto and Kinsella 2002; O'Connell and de Kruif 2003; Smiddy et al. 2006). Another class of enzymes, the laccases, joined the group of oxidoreductases with a multinuclear copper-containing active site (Riva 2006). They oxidized tyrosine and tyrosine residues in proteins (Mattinen et al. 2005, 2008) and induced the establishment of oligomers and polymers of whey protein isolates (Færgemand et al. 1998). Laccase also induced crosslinking of sugar beet pectin in multilayered emulsions and has been suggested as a useful tool to covalently crosslink biopolymers in multilayered emulsions (Zeeb et al. 2012). It helped in the enzymatic crosslinking of adsorbed sugar beet pectin in multilayered emulsions, improving emulsion stability over a wide range of pH values. Laccase-treated emulsions could tolerate alkaline surroundings for a certain period. Laccase did not affect fish gelatin-stabilized emulsions (Zeeb et al. 2012). Emulsions with enhanced stability features can be designed by cautiously governing the characteristics of the interfacial coatings (Zeeb et al. 2012).

7.5 Further Matters Related to Multilayered Products

7.5.1 Natural Food-Grade Emulsifiers and Interfacial Layers

Emulsifiers are surface-active substances that facilitate emulsion formation and promote emulsion stability. Proteins, polysaccharides, phospholipids, and surfactants are used in the food industry for this purpose (Kralova and Sjoblom 2009). Many proteins are surface-active due to the mixture of hydrophilic and hydrophobic amino acids along their polypeptide chains (Lam and Nickerson 2013). Proteins can therefore adsorb to oil–water interfaces and coat the formed oil droplets throughout homogenization. Proteins prevent oil-droplet aggregation through their negatively and positively charged amino acids, creating electrostatic repulsion (McClements 2004; Dickinson 2010). Aggregation is inhibited by steric repulsion that forms thick interfacial layers, or by attached carbohydrate moieties (Wooster and Augustin 2006). Phospholipids are natural amphiphilic molecules found in the cell membranes of animals, plants, and microbes (Erickson 2008). Their surface activity stems from their hydrophobic fatty acid tail groups and hydrophilic head groups (Klang and

Valenta 2011). Because of this surface activity, phospholipids are sometimes poor emulsifiers due to the formation of interfacial layers that are susceptible to coalescence (McClements 2004). Proteins are comparatively small molecules (~10–50 kDa) that quickly adsorb to droplet surfaces and produce thin, electrically charged interfacial layers (Bouyer et al. 2012). The flexible proteins casein and gelatin rapidly submit to changes, i.e. the hydrophilic groups protrude into water and the hydrophobic groups protrude into oil. The rigid globular proteins of egg, soy, pea or whey may partially unfold following adsorption and form cohesive viscoelastic layers (Ozturk and McClements 2016). Polysaccharides are large molecules (~100–1000 kDa) that slowly adsorb to droplet surfaces and produce thick hydrophilic interfacial layers (Bouyer et al. 2012). Due to their large size, polysaccharides have greater surface loads than proteins, thus more emulsifier is required to cover the droplet surfaces (Ozturk et al. 2014). Saponins are customarily more effective at forming small droplets at low concentrations than biopolymers (Yang et al. 2013; Ozturk et al. 2014). Their relatively low molecular weight (~1.67 kDa) makes them ideal for quickly adsorbing to droplet surfaces to form thin interfacial layers (Ozturk and McClements 2016). Phospholipids seem to behave somewhere between saponins and polysaccharides, regardless of their somewhat low molecular weight (~0.760 kDa). This may be why they tend to form large supramolecular structures – such as bilayers or vesicles – in solution instead of existing as discrete molecules or small micelles (Ozturk and McClements 2016).

7.5.2 Multilayer Adsorption

Intermolecular electrostatic interactions may be involved in emulsion preparation (Gashua et al. 2016). An understanding of the mechanisms underlying the alterations in emulsification properties displayed by gums from *Acacia senegal* and *Acacia seyal* trees was achieved by using polystyrene latex particles as a model system to define the adsorbed layer thickness (Gashua et al. 2016). For the *A. senegal* gum sample, the adsorbed layer thickness increased over time, reaching 61 nm after 14 days (Gashua et al. 2016). A previous study also reported that *A. senegal* gum alone self-associates in solution through electrostatic interactions between its protein and carboxylate moieties (Gashua et al. 2015), and another study described multilayer adsorption for sugar beet pectin through a similar mechanism (Siew et al. 2008). For the *A. seyal* gum sample, the adsorbed layer thickness was only ~3 nm and did not increase with time (Gashua et al. 2016). Transmission electron microscopy revealed the presence of a distinct, thick adsorbed layer for the *A. senegal* gum and the presence of a much thinner, more diffuse layer for the *A. seyal* gum. Emulsification studies showed that *A. senegal* gum is more effective at stabilizing limonene O/W emulsions than *A. seyal* gum, because markedly more *A. senegal* gum adsorbed at the oil–water interface compared to the *A. seyal* gum exudate.

7.5.3 Gelled Double-Layered Emulsions

Multilayer systems can present diverse practical properties in line with the number of layers deposited, the type of biopolymers, the arrangement of the biopolymer layers, and the solution properties utilized throughout deposition. Emulsion-based delivery systems have been extensively studied for the inclusion of lipophilic compounds in beverages and food

(Đorđević et al. 2015). In general, the functional properties of the multilayer coating can be controlled by changing the number of layers deposited, the type of biopolymers, the sequence of biopolymer layers, and the solution properties used during deposition (Li et al. 2010; Đorđević et al. 2015). Double-layered emulsions were atomized in a calcium chloride solution (Perrechil et al. 2011) and gelled to form microgels for the protection of flaxseed oil against oxidation (Đorđević et al. 2015). The emulsification process resulted in decreased content of the secondary oxidation products of flaxseed oil as compared to non-emulsified oil (Đorđević et al. 2015). The gelation process substantially enhanced oxidative stability with a noteworthy decline in both primary and secondary oxidation products throughout storage. These lipid-based microgels could potentially be used as delivery systems with developed oxidative stability (Đorđević et al. 2015).

7.6 Complications Related to Multilayered and Colored Products

A multilayered pudding has at least two layers. Each pudding layer is manufactured from conventional constituents, including non-fat milk, water, a sweetener, an emulsified fat and/or oil, a thickener – usually a starch thickener, and at least one emulsifier/stabilizer (Table 7.1). Supplementary ingredients consist of non-fat milk solids, pieces of fruit, salt, colorants, and flavorings (Hashisaka et al. 2004). Such multilayered products suffer from color migration, leading to the invention of a technique to minimize this problem (Hashisaka et al. 2004). Color migration can be defined as the unwelcomed movement of a colorant in a product formulation into an adjacent layer of the same product (Hashisaka et al. 2004). Such products are formed from a layer containing a first colorant and an adjacent layer containing a second, different colorant. Example multilayered products include, among others, pudding products, yogurt products, and non-milk gel-based dessert products. The color of a pudding or other gel-based dessert is one of the main merits for the

Table 7.1 Typical pudding formulation.

Ingredient	Preferred range (% weight)
Non-fat milk[a]	35–45
Water	10–15
Sweetener	0.5–25
Emulsified fat or oil[b]	0.5–10
Thickener (usually starch)	3–8
Salt	0.75–1.25
Emulsifier/stabilizer	0.05–1.50
Colorants	0.02–1.25
Flavorings	0.10–1.50

[a] Not all gel desserts contain milk products; some are "non-dairy."
[b] Certain ingredients may not be present in certain types of pudding, such as the intentional lack of fat or oil in "fat-free" pudding.

buyer and anticipated consumer. Typically, when a manufacturer produces pudding or some other dessert product that has more than one flavor layer, these layers are colored differently to render the product more noticeable and commercially tempting. For instance, manufacturers have tried to produce pudding products with a chocolate-flavored base that is dark brown in color next to a topping with a different flavor and color, for example tan, caramel, yellow or white (Sethi et al. 2002). When the bottom layer of a multilayered pudding is a dark-brown cocoa-flavored layer and the top layer is a lighter caramel color, the dark-brown cocoa powder pigment can migrate from the bottom layer into the upper layer, thereby darkening the caramel layer (Sethi et al. 2002). As a result, a modified cocoa powder was needed that does not migrate between differently colored layers of a multilayered, gel-based dessert product, along with preparation methods that are simple and inexpensive. In one patented process, the modified cocoa powder was prepared by extracting one part by weight of non-modified cocoa powder starting material with at least one part by weight of water, at a temperature and time period that extracted a substantial share of the water-soluble solids. The resultant supernatant containing the water-soluble solids was then separated from the water-extracted cocoa powder by centrifugation, membrane separation, or their combination. The water-extracted cocoa powder was, overall, beneficial for formulating multilayered, gel-based dessert products, such as puddings, yogurts, and non-milk gel-based dessert products with two gel-based layers, at least one of which contained about 1–10% (w/w) of the water-extracted cocoa powder. The layer comprising this powder characteristically had a total solids content of about 1–5% (w/w) protein (Sethi et al. 2002). Another patent described a process for preparing a multilayered food composition that includes at least one heat-treated layer (i.e. mousse, cream, jelly, and/or sauce) and at least one brittle, thin fat-containing layer, e.g. chocolate, so that during consumption, one experiences a cracking upon penetrating the fat layer with a spoon and then again, a novel mouth-feel due to the fat-containing composition layer (Eder et al. 2001).

There are several approaches to decreasing color migration. The first is to choose oil-soluble colorants such as beta-carotene, annatto, paprika oleoresin or lycopene. Beta-carotene (Figure 7.9) is a strongly colored red–orange pigment that is abundant in plants and fruit. It is an organic compound and chemically classified as a hydrocarbon, specifically a terpenoid (isoprenoid), reflecting its derivation from isoprene units. Annatto (Figure 7.10) is derived from the seeds of the achiote tree (Figure 7.11) which grows in tropical and subtropical regions worldwide. The seeds are sourced to produce a carotenoid-based yellow to orange food coloring and characteristic flavor (Scotter et al. 1998; Levy and Rivadeneira 2000). Paprika oleoresin (Figure 7.12) and lycopene – a bright-red carotene and carotenoid pigment and phytochemical, respectively – are found in tomatoes and other red fruit and vegetables, such as red carrot, watermelon, and papaya. These colorants can

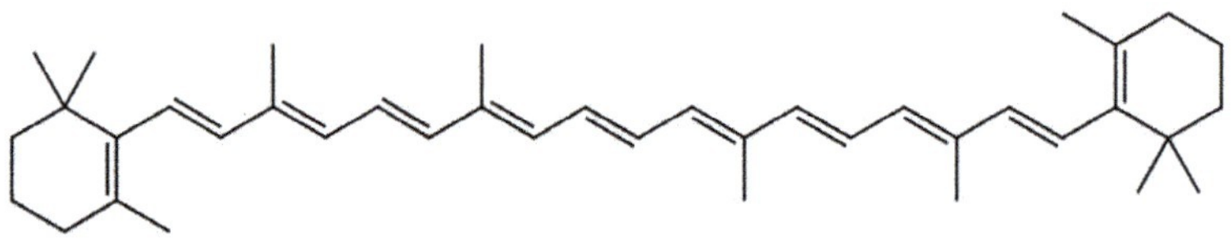

Figure 7.9 Structure of beta-carotene. *Source:* NEUROtiker/Wikimedia Commons/Public domain.

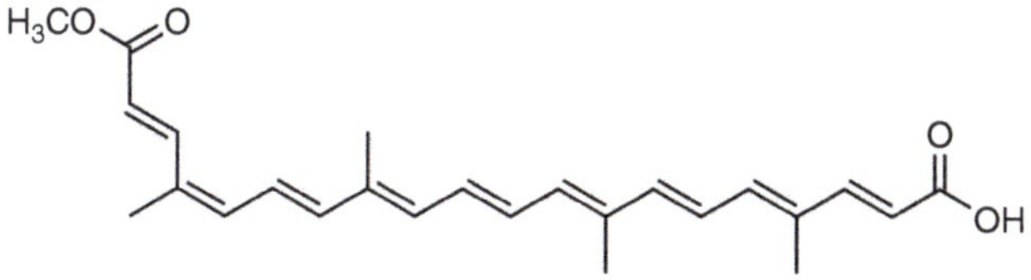

Figure 7.10 Bixin, the major apocarotenoid of annatto. *Source:* Edgar181/Wikimedia Commons/Public Domain.

Figure 7.11 *Bixa orellana* (annatto, achiote, lipstick tree, alaea laau, kumauna). Annatto fruit at Maui County Fair, Kahului, Maui, 3 October 2009. Image # 091003-7747. *Source:* Forest Starr & Kim Starr/CC BY 4.0.

be used either alone or in combination. Another approach to reducing color migration relates to charged colorants and/or their molecular weights. For example, the layer that contains the natural oil-soluble colorant also encloses a negatively charged caramel colorant with an average molecular mass of between about 200 000 and 650 000 Da, present at ~0.02–0.08% by weight (Hashisaka et al. 2004).

The construction of multilayered, multicolored gelled food products in which adjacent layers vary in color, and there is no color migration between the layers, has been described (Soedjak and Spradlin 1995). In this invention, each gelled component that contains a water-soluble colorant also contains an agent that complexes with the colorant to produce a water-soluble complex which, owing to its size, does not migrate either within the gel matrix or into adjacent matrices (Soedjak and Spradlin 1995). The formed complexes are believed to be held together by association of hydrophobic regions within the colorant and the complexing agent, by charge–charge interactions between attracting charges on the colorant and the complexing agent, or by a combination of these forces. For commercial purposes, the complexes must remain stable over time and not precipitate or unfavorably affect the hue or intensity of the colorant. This stability must be preserved under the conditions of use which, in the case of fruit-flavored gels, will include a pH of about 3.2–4.5 under either ambient conditions or refrigeration over a period of several months (Soedjak and

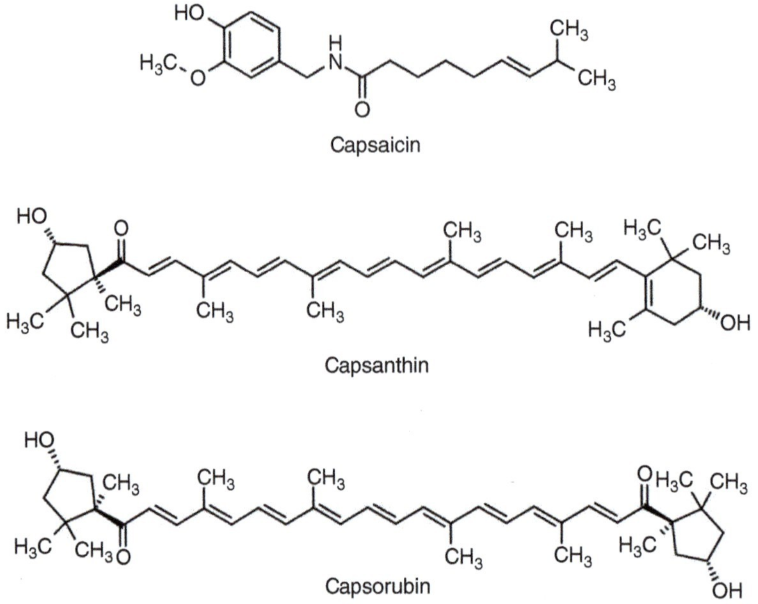

7 Use of Hydrocolloids to Control the Texture of Multilayered Food Products

Figure 7.12 Chemical structure of paprika oleoresin. *Source:* Ronhjones/Wikimedia Commons/ Public domain.

Spradlin 1995). It should be noted that the migration of a colorant within a gel depends on the pore size of the gel, which in turn is inversely proportional to the concentration of the gelling agent (Soedjak and Spradlin 1995).

Conventional pizzas fall into one of three categories: single-layer thin, single-layer thick or Sicilian, and single-layer deep dish. Thus, what pizzas all have in common is a single layer of pastry and a single layer of ingredients. More than 30 years ago, a double-layered pizza-type product was patented (Giordano and Giordano 1981). The preparation method was as follows: at least one layer of ingredients, which included tomato sauce, was applied on a prebaked pastry layer, followed by a layer of cheese – a blend of mozzarella and provolone, then a second layer of rolled out, unbaked pastry was stretched out over the entire surface of the cheese layer; the two pastry layers, with at least one layer of ingredients and the layer of cheese, were then partially baked, cooled, and a layer of ingredients was applied to the upper surface of the second pastry layer; the entire assembly was then baked until done (Giordano and Giordano 1981). There are a number of different products that call for two or more pastry layers. Therefore, a method and apparatus for continuously manufacturing multilayered dough materials consisting of dough or alternating layers of dough and some other material, such as in pies and Danish pastries, was invented (Hayashi et al. 1981). This invention made it possible to continuously and automatically manufacture multilayered dough materials containing 30 layers or more of dough, or dough and other arranged materials, without damaging the dough. This invention further enabled the manufacture of multilayered dough materials comprised of alternating or overlapping layers of dough and fat or oil, such as butter, without the usual problems of butter being squeezed out or dough adhering to machine parts during the stretching operation (Hayashi et al. 1981).

7.7 Future Potential Biotechnological Uses of Multilayered Gels

Biotechnological applications of natural and synthetic multilayered gels are not used much in the food industry, or in advancing future foods. However, some illustrations of their potential input to this industry and to non-food applications are presented here. For example, the encapsulation of reagents in a multilayered hydrogel formulation, and how their release relies on the layered structure, have been described for non-food applications (Moriyama et al. 1999). Additional food-oriented uses might be related to fibers with single or double gel layers for the production of alcohol. Globally, alcoholic beverages constitute over half of the gross value of fermented products. Worldwide alcohol consumption is on the rise, and the quantity of manufactured alcoholic beverages is vast. Three basic categories of alcoholic beverage are defined in line with their production: beers, wines and ciders, and distilled products (Sutherland et al. 1986). In a study of alcohol fermentation, immobilization of yeast cells in double-layered gel fibers led to decreased production of α-acetolactate (a diacetyl precursor) compared to immobilization in single-layered gel beads. The decrease in this diacetyl precursor might have been a consequence of the more anaerobic conditions in the gel fibers (Shindo et al. 1993). Double-layered gels can be used in immobilization processes, and these are described, and their properties defined elsewhere (Bucke 1983; Mattiasson 1983; Nussinovitch 2003). To sum up, there are numerous techniques for immobilizing cells: adsorption to neutral or charged supports, containment through entrapment by natural or synthetic polymers, covalent coupling, and flocculation (Nussinovitch 2003). In general, immobilized cell preparations must have the following properties: high biocatalytic activity, long-term stability of the biocatalyst, the possibility of regenerating the biocatalyst, low loss of activity during immobilization, low leakage of cells, non-compressible particles, high resistance to abrasion, resistance to microbial degradation, low diffusional limitations, spherical shape, high surface area, appropriate density for the reactor type, technique simplicity, inexpensive support materials, and non-toxicity (Tampion and Tampion 1987; Nussinovitch 1997b, 2003).

An additional study outlined the advantages of using bacteria [(Citr+) *Lactococcus lactis* subsp. *lactis* 3022] immobilized in a double-layered gel of calcium-alginate fibers for the production of diacetyl (an important carbonyl compound formed during fermentation; the flavor attributed to this compound has been described as "buttery," "honey- or toffee-like," or "butterscotch"), vs. immobilization in single-layer fine calcium-alginate fibers. Lactic acid bacteria are antagonistic toward numerous microorganisms because of their competition for nutrients and their manufacture of compounds such as acetic acid, diacetyl, hydrogen peroxide, and lactic acid. Moreover, citric acid fermentation in conjunction with lactic acid fermentation yields diacetyl and supplementary aroma and flavor compounds in dairy products (Doores 2002). In the presence of catalase, these bacteria produced a threefold higher yield of diacetyl. A stronger increase in diacetyl production was detected when a double-layered gel of calcium-alginate fibers was used to hold lactic acid bacteria (outer layer) and homogenized bovine liver as the growth substrate (inner layer) (Ochi et al. 1991).

Double-layered calcium-alginate gel fibers are beneficial in their ability to prevent cell leakage during the course of production. This was observed in an immobilized preparation of *Wasabia japonica* utilized for chitinase manufacturing. Chitinase production by the immobilized cells was superior to that by freely suspended cells under the same conditions. Aeration with pure oxygen brought about increased chitinase production – in the area of fivefold that of the freely suspended cells (Tanaka et al. 1996a). To maintain chitinase production at over 80% of its maximal rate, its chitinase concentration in the broth had to be lower than 2 U/mL. One particular manufacturing arrangement consisted of double-layered gel fibers coupled to a chitin column which maintained chitinase production for 40 days (Tanaka et al. 1996b). Additional biotechnological processes include encapsulation of adsorbents, cells, enzymes, magnetic materials, and proteins for food and non-food applications (Chang 1993; Bader et al. 1995; Grant et al. 2000). Cells can be cryopreserved to maintain their viability and for other purposes. Most of the evidence on such procedures can be culled from the microbiological or medical literature (Koebe et al. 1999).

7.8 Demonstrating the Use of Hydrocolloids to Prepare Multilayered Products/Recipes

The use of proteins/hydrocolloids in the preparation of multilayered products is demonstrated in this section by three examples. In the first recipe, a multilayered jelly (Figure 7.13) based on gelatin is introduced. In general, the uses of gelatin can be classified according to its function (Jones 1977): adhesiveness, emulsification, gelation, stabilization, and sedimentation (for fining) (Jones 1977). Gelatin was chosen for this recipe due to its thermoreversible gel formation, its "melt-in-the-mouth" texture and the gel's elasticity (Karim and Bhat 2008; Green et al. 2013). The first property also makes gelatin a key ingredient in a comprehensive variety of confectionery products and desserts (Schrieber and Gareis 2007; Nussinovitch and Hirashima 2014a). The oldest application of gelatin may be for gluing

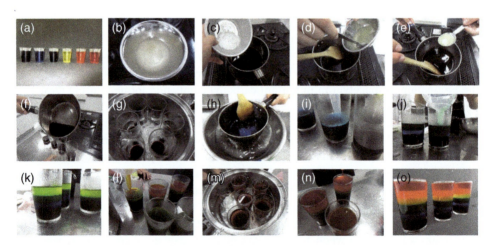

Figure 7.13 (a) Six types of colored water. (b) Swollen gelatin. (c) Adding white sugar to the purple-colored water, (d) adding swollen gelatin, (e) adding lemon juice, (f) pouring gelatin solution into a glass, (g) cooling in an ice-water bath, (h) cooling blue gelatin solution, (i) pouring blue gelatin solution on purple jelly, (j) pouring green gelatin solution on blue jelly, (k) pouring yellow gelatin solution on green jelly, (l) pouring orange gelatin solution on yellow jelly, (m) cooling gelatin jelly, (n) pouring red gelatin solution on orange jelly. (o) Serve.

purposes. A lukewarm gelatin solution needs to be used for successful adhesion. It should, however, be noted that the gelatin must not gel prior to contact between the surfaces that are to be joined. Examples of this use of gelatin can be found in the pharmaceutical and confectionery industries (Baziwane and He 2003). In the recipe presented here, the gelatin solution also serves as an adhesive agent (Nussinovitch 2017). In addition, the lukewarm gelatin solution has wetting ability which preserves contact with the already solidified differently colored bottom gelatin layer and the subsequent interactions when the two are introduced together (Nussinovitch 2017).

In the second recipe (Figure 7.14), the melting and setting abilities of gelatin are used, as well as its excellent foam-stabilizing ability (Baziwane and He 2003) and the adhesion between the two layers, i.e. the gelled and foamed ones, which characterize this unique product. It is important to note that different gelatins have different foam-stabilizing abilities and thus, the gelatin used for this purpose has to be carefully selected (Schrieber and Gareis 2007).

There are many kinds of Japanese sweets shaped to look like plants or fish. These are demonstrated by the commercial products shown in Figure 7.15. Some of them, such as those shown in Figure 7.15h and i, involve multiple layers (not necessarily parallel or similar), as well as objects embedded in the continuous mass. The other sweets

Figure 7.14 (a) Swollen gelatin. (b) Adding the swollen gelatin to warm apple juice, (c) cooling the gelatin solution, (d) gelatin solution in glasses, (e) mixing the gelatin solution with a hand mixer. (f) Whipped gelatin solution. (g) Serve.

Figure 7.15 Various kinds of multilayered sweets. (a) Japanese maple in autumn. (b) Under the sea in summer. (c–g) Fish and leaves in a pond. (h, i) Fly Me to the Moon. *Source:* Toyota Tsusho Corporation.

(Figure 7.15a–g) include shaped objects embedded in a jellified mass which, when cut perpendicularly, will present partial layers above and below the cut object. Figure 7.3h and i shows Nagatoya's latest line of wagashi (yokan), named "Fly Me to the Moon" (https://nagatoya.net/?mode=f7). These products are extraordinary. What looks like a standard block of yokan from the outside comprises complex designed scenes in each cube. The yokan is made with champagne gelatin (from seaweed, not actual gelatin) and sweet bean paste. The images on each cube portray a bird flying toward the moon with slight variations so that together, they create a "flip-book" type of effect. The sequence aims to convey the old-style wagashi technique to future generations. The design that appears each time you cut the yokan block changes, from a bird flapping toward a semicircular moon which then changes to a full moon. The bird, which has stopped at the crescent moon, progressively flaps toward the full moon, and the scenery slowly fades to a night sky. The taste of each yokan piece changes depending on where you cut it. In addition, there are some slices at each end that do not yet depict the moon or the bird. The dynamic darkening blue sky and the changing mountains are reminiscent of a jazz melody or a bird song (https://nagatoya.net/?mode=f7).

7.8.1 Multilayered Gelatin Jelly

(Serves 5–6) (Figure 7.13)

900 (150×6) mL water
150 (25×6) g white sugar
27 (4.5×6) g gelatin powder
135 (22.5×6) mL water
30–90 (5–15×6) lemon juice (*Hint 1*)
Coloring agent (red, yellow, green, blue) (*Hint 2*)

For preparation, see Figure 7.13

1) Make red, orange, yellow, green, blue, and purple (six rainbow colors) water (Figure 7.13a): add each coloring agent to 150 mL (⅗ cup) water to make red, yellow, green, and blue water. Mix red and blue coloring agent to make purple water, and mix red and yellow coloring agent to make orange water.
2) Make six gelatin dispersions (*Hint 3*): each 4.5 g (1½ tsp) gelatin to each 22.5 mL (1½ Tbsp) water (Figure 7.13b).
3) Boil purple colored water (1) and 25 g (a little under 3 Tbsp) white sugar (Figure 7.13c), remove from the heat, add gelatin dispersion (2) (Figure 7.13d) and dissolve.
4) Add 5–15 mL (⅓–1 Tbsp) lemon juice (Figure 7.13e).
5) Pour the gelatin solution (4) into a glass (Figure 7.13f) and cool it in the refrigerator or in an ice-water bath until set (Figure 7.13g).
6) Repeat preparation steps 2–4 using blue coloring water.
7) Cool the blue gelatin solution to room temperature in a water or ice-water bath (Figure 7.13h).
8) Pour the blue gelatin solution (7) into the glass in which the purple jelly has set (Figure 7.13i) (*Hint 4*), and cool it in the refrigerator or in an ice-water bath until set.
9) Repeat preparation steps 2, 3, 4, 7, and 8 using the other colors of water (Figure 7.13j–n) and serve (Figure 7.13o). The order of purple, blue, green, yellow, orange, and red is recommended.

Preparation hints:

1) You can use any citrus juice, such as lime juice.
2) For a layered jelly with five or six colors, four coloring agents are needed.
3) If the gelatin dissolves easily in water, this process can be omitted.
4) Use a funnel for easy pouring.

7.8.2 Beer-Like Jelly

(Serves 4) (Figure 7.14)

800 g (3⅕ cups) apple juice
15 g (1⅔ Tbsp) gelatin powder
100 g (⅖ cup) water
21 g (3 Tbsp) white sugar
pH 4.11

For preparation, see Figure 7.14

1) Put gelatin and water in a bowl, and soak for at least 20 minutes (Figure 7.14a).
2) Heat apple juice in a pan to 70 °C and remove from the heat.
3) Add white sugar and the swollen gelatin (1) to the heated apple juice (2) (Figure 7.14b), and mix until dissolved.
4) Cool the gelatin solution (3) to less than 30 °C in an ice-water bath until it thickens (Figure 7.14c).
5) Pour 180 g of the thickened the gelatin solution (4) into each of four glasses (Figure 7.14d), and cool to less than 10 °C in the refrigerator or in an ice-water bath until set.
6) Whip the remaining gelatin solution (4) with a hand mixer (Figure 7.14e) until it becomes white and bubbly (Figure 7.14f, Hint 1).
7) Garnish bubbled gelatin solution (6) on the set gelatin jelly, and cool to serve (Figure 7.14g).

Preparation hint:

1) Mix well until it looks like beer foam.

References

Aldridge, A., Degady, M., Elejalde, C.C. et al. (2014). Co-extruded layered candy and gum apparatus and methods. US Patent 08920856.

Alves, M.M., Gonçalves, M.P., and Rocha, C.M.R. (2017). Effect of ferulic acid on the performance of soy protein isolate-based edible coatings applied to fresh-cut apples. *Lebensmittel-Wissenschaft & Technologie* 80: 409–415.

Anal, A.K. and Sing, H. (2007). Recent advances in microencapsulation of probiotics for industrial applications and targeted delivery. *Trends in Food Science and Technology* 18: 240–251.

Annan, N.T., Borza, A.D., and Hansen, L.T. (2008). Encapsulation in alginate-coated gelatin microspheres improves survival of the probiotic *Bifidobacterium adolescentis* 15703T during exposure to simulated gastro-intestinal conditions. *Food Research International* 41: 184–193.

Aoki, T., Decker, E.A., and McClements, D.J. (2005). Influence of environmental stresses on stability of O/W emulsions containing droplets stabilized by multilayered membranes produced by a layer-by-layer electrostatic deposition technique. *Food Hydrocolloids* 19: 209–220.

Araiza-Calahorra, A., Akhtar, M., and Sarkar, A. (2018). Recent advances in emulsion-based delivery approaches for curcumin: from encapsulation to bioaccessibility. *Trends in Food Science & Technology* 71: 155–169.

Axelrod, G.S. and Gajria, A. (2012). Processes for forming multi-layered pet treats. US Patent 08124156.

Axelrod, G.S. and Gajria, A. (2014). Multi-layer extrusion. US Patent 08771775.

Bader, A., Knop, E., Boker, K. et al. (1995). A novel bioreactor design for in-vitro reconstruction of in-vivo liver characteristics. *Artificial Organs* 19: 368–374.

Baziwane, D. and He, Q. (2003). Gelatin: the paramount food additive. *Food Reviews International* 19: 423–435.

Ben-Zion, O. and Nussinovitch, A. (1995). Calculating the compressive deformabilities of multilayered gels and texturized fruits glued together by three different adhesion techniques. *The Eighth International Conference and Industrial Exhibition on Gums and Stabilizers for the Food Industry*, Cartrefle College, Wrexham, Clwyd, UK (1–14 July 1995). The North East Wales Institute.

Ben-Zion, O. and Nussinovitch, A. (1996). Predicting the deformability modulus of multi-layered texturized fruits and gels. *LWT – Food Science and Technology* 29: 129–134.

Ben-Zion, O. and Nussinovitch, A. (1997). A prediction of the compressive deformabilities of multilayered gels and texturized fruit, glued together by three different adhesion techniques. *Food Hydrocolloids* 11: 253–260.

Bjapai, S.K. and Tankhiwale, R. (2006). Investigation of water uptake behavior and stability of calcium alginate/chitosan bi-polymeric beads: part-1. *Reactive and Functional Polymers* 66: 645–658.

Bouyer, E., Mekhloufi, G., Rosilio, V. et al. (2012). Proteins, polysaccharides, and their complexes used as stabilizers for emulsions: alternatives to synthetic surfactants in the pharmaceutical field? *International Journal of Pharmaceutics* 436: 359–378.

Bucke, C. (1983). Immobilized cells. *Philosophical Transactions of the Royal Society of London, Series B* 300: 369–389.

Burey, P., Bhandari, B.R., Howes, T., and Gidley, M.J. (2008). Hydrocolloid gel particles: formation, characterization, and application. *Critical Reviews in Food Science and Nutrition* 48: 361–377.

Campomanes Marin, J.P., Davis, B., Levenson, D. et al. (2016). Sugar free confectionary; methods of making same and use in preparing multilayered confectionary. US Patent 9,247,761 B2.

Carlson, J.S., Pinkevich, Y.Y., Schmitt, B.K. et al. (2015). System and method for forming a multi-layer extruded food product. US Patent 08945643.

Chang, T.M.S. (1993). Living cells and microorganisms immobilized by microencapsulation inside artificial cells. In: *Fundamentals of Animal Cell Encapsulation and Immobilization* (ed. M.F.A. Goosen), 184. Boca Raton, FL: CRC Press.

Chávarri, M., Marañón, I., Ares, R. et al. (2010). Microencapsulation of a probiotic and prebiotic in alginate-chitosan capsules improves survival in simulated gastro-intestinal conditions. *International Journal of Food Microbiology* 142: 185–189.

Cohen, D.L., Lipton, J.I., Cutler, M. et al. (2009). Hydrocolloid printing: a novel platform for customized food production. http://citeseerx.ist.psu.edu/viewdoc/download?doi=10.1.1.477.665&rep=rep1&type=pdf (accessed 19 October 2021).

Collyer, A.A., Bennett, M.J., Mayhew, R.H. et al. (1990). Injection moulding of confectionery materials. In: *Third European Rheology Conference and Golden Jubilee Meeting of the British Society of Rheology* (ed. D.R. Oliver), 113–116. Dordrecht: Springer.

Damodaran, S. (2008). Amino acids, peptides, and proteins. In: *Food Chemistry* (ed. O. Fennema), 245–371. New York: Marcel Dekker.

Davidson, A. (2014). *The Oxford Companion to Food*, 213. Oxford: Oxford University Press.

De Jong, P.H., Hendrikx, H.J.C., and Vleugels, T.C.J. (2019). Method for preparing a multilayer confectionary product. US Patent Application 0328001 A1.

De Vos, P., Faas, M.M., Strand, B., and Calafiore, R. (2006). Alginate-based microcapsules for immunoisolation of pancreatic islets. *Biomaterials* 27: 5603–5617.

De Vos, P., De Haan, B.J., Kamps, J.A. et al. (2007). Zetapotentials of alginate-PLL capsules: a predictive measure for biocompatibility? *Journal of Biomedical Materials Research Part A* 80: 813–819.

Decher, G. (1997). Fuzzy nano-assemblies: toward layered polymeric multicomposites. *Science* 277: 1232–1237.

Decher, G. (2003). Layer-by-layer assembly (putting molecules to work). In: *Multilayer Thin Films: Sequential Assembly of Nanocomposite Materials* (ed. G. Decher and J.B. Schlenoff), 1–22. Weinheim, Germany: Wiley-VCH.

Dickinson, E. (2010). Flocculation of protein-stabilized oil-in-water emulsions. *Colloids and Surfaces B – Biointerfaces* 81: 130–140.

Doherty, S.B., Gee, V.L., Ross, R.P. et al. (2011). Development and characterization of whey protein micro-beads as potential matrices for probiotic protection. *Food Hydrocolloids* 25: 1604–1617.

Doherty, S.B., Auty, M.A., Stanton, C. et al. (2012). Application of whey protein micro-bead coatings for enhanced strength and probiotic protection during fruit juice storage and gastric incubation. *Journal of Microencapsulation* 29: 713–728.

Doores, S. (2002). pH control agents and acidulants. In: *Food Additives* (ed. A.L. Branen, P.M. Davidson, S. Salminen and J. Thorngate), Chapt. 21, 621–660. New York and Basel: Marcel Dekker.

Đorđević, V., Balanč, B., Belščak-Cvitanović, A. et al. (2015). Trends in encapsulation technologies for delivery of food bioactive compounds. *Food Engineering Reviews* 7: 452–490.

Eder, H.P., Elhaus, B., and Liebenspacher, F. (2001). Preparation and packaging of a multi-layered heat-treated dessert composition. US Patent 6,203,831 B1.

Elsabee, M.Z., Abdou, E.S., Nagy, K.S.A., and Eweis, M. (2008). Surface modification of polypropylene films by chitosan and chitosan/pectin multilayer. *Carbohydrate Polymers* 71: 187–195.

Erickson, M.C. (2008). Chemistry and function of phospholipids. In: *Food Lipids* (ed. C.C. Akoh), 39–62. Boca Raton, FL: CRC Press.

Etxabide, A., Uranga, J., Guerrero, P., and de la Caba, K. (2017). Development of active gelatin films by means of valorisation of food processing waste: a review. *Food Hydrocolloids* 68: 192–198.

Færgemand, M., Otte, J., and Qvist, K.B. (1998). Cross-linking of whey proteins by enzymatic oxidation. *Journal of Agricultural and Food Chemistry* 46: 1326–1333.

Feng, C., Zhang, M., and Bhandari, B. (2019). Materials properties of printable edible inks and printing parameters optimization during 3D printing: a review. *Critical Reviews in Food Science and Nutrition* 59: 3074–3081.

Food and Agriculture Organization (FAO) (2021). World food situation. http://www.fao.org/worldfoodsituation/csdb/en/ (accessed October 2021).

Frame, N.D. (1994). *The Technology of Extrusion Cooking*, 127. Dordrecht, London: Springer Science and Business Media.

Gashua, I.B., Williams, P.A., Yadav, M.P., and Baldwin, T.C. (2015). Characterization and molecular association of Nigerian and Sudanese Acacia gum exudates. *Food Hydrocolloids* 51: 405–413.

Gashua, I.B., Williams, P.A., and Baldwin, T.C. (2016). Molecular characteristics, association and interfacial properties of gum arabic harvested from both *Acacia senegal* and *Acacia seyal*. *Food Hydrocolloids* 61: 514–522.

Gbassi, G.K., Vandamme, T., Ennahar, S., and Marchioni, E. (2009). Microencapsulation of *Lactobacillus plantarum* spp in an alginate matrix coated with whey proteins. *International Journal of Food Microbiology* 129: 103–105.

Gbassi, G.K., Vandamme, T., Yolou, F., and Marchioni, E. (2011). in vitro effects of pH, bile salts and enzymes on the release and viability of encapsulated *Lactobacillus plantarum* strains in a gastrointestinal tract model. *International Dairy Journal* 21: 97–102.

Gerstenberg Schröder (2009). Continuous on-site processing of bakery filling creams. https://www.spxflow.com/assets/pdf/GS_continuous_on-site_processing_of_bakery_filling_creams_GB.pdf (accessed March 2015).

Gibis, M., Thellmann, K., Thongkaew, C., and Weiss, J. (2014). Interaction of polyphenols and multilayered liposomal-encapsulated grape seed extract with native and heat-treated proteins. *Food Hydrocolloids* 41: 119–131.

Gibson, L.J. and Ashby, M.F. (1988). *Cellular Solids Structure and Properties*. Oxford: Pergamon Press.

Giordano, G. and Giordano, F. (1981). Method of making a multi-layer pizza type product. US Patent 4,283,431.

Goyanes, A., Wang, J., Buanz, A. et al. (2015). 3D printing of medicines: engineering novel oral devices with unique design and drug release characteristics. *Molecular Pharmaceutics* 12: 4077–4084.

Grant, M.H., Anderson, K., McKay, G. et al. (2000). Manipulation of the phenotype of immortalised rat hepatocytes by different culture configurations and by dimethyl sulphoxide. *Human & Experimental Toxicology* 19: 309–317.

Green, A.J., Littlejohn, K.A., Hooley, P., and Cox, P.W. (2013). Formation and stability of food foams and aerated emulsions: hydrophobins as novel functional ingredients. *Current Opinion in Colloid and Interface Science* 18: 292–301.

Gu, Y.S., Decker, E.A., and McClements, D.J. (2004). Influence of pH and ι-carrageenan concentration on physicochemical properties and stability of β-lactoglobulin-stabilized oil-in-water emulsions. *Journal of Agricultural and Food Chemistry* 52: 3626–3632.

Gübitz, G.M. and Paulo, A.C. (2003). New substrates for reliable enzymes: enzymatic modification of polymers. *Current Opinion in Biotechnology* 14: 577–582.

Guo, C., Zhang, M., and Bhandari, B. (2019). Model building and slicing in food 3D printing processes: a review. *Comprehensive Reviews in Food Science and Food Safety* 18: 1052–1069.

Guzey, D. and McClements, D.J. (2006a). Formation, stability and properties of multilayer emulsions for application in the food industry. *Advances in Colloid and Interface Science* 128–130: 227–248.

Güzey, D. and McClements, D.J. (2006b). Influence of environmental stresses on O/W emulsions stabilized by β-lactoglobulin–pectin and β-lactoglobulin–pectin–chitosan membranes produced by the electrostatic layer-by-layer deposition technique. *Food Biophysics* 1: 30–40.

Han, C.D. (1973). A study of bicomponent coextrusion of molten polymers. *Journal of Applied Polymer Science* 17: 1289–1303.

Han, C.D. (1981). *Multiphase Flow in Polymer Processing*, 1e. New York: Academic Press.

Hao, L., Mellor, S., Seaman, O. et al. (2010). Material characterization and process development for chocolate additive layer manufacturing. *Virtual and Physical Prototyping* 5: 57–64.

Harada, T. (1979). Curdlan: a gel-forming β-1,3 glucan. In: *Polysaccharides in Food* (ed. J.M.V. Blanshard and J.R. Mitchell), 283–300. London: Butterworths.

Hashisaka, A.E.A., Sethi, V., Lammert, A., and Mikula, M. (2004). Method for reducing color migration in multi-layered and colored gel-based dessert products and the products so produced. US Patent 6808728 B2.

Hayashi, T., Kageyama, M., and Morikawa, M. (1981). Apparatus for continuously manufacturing multi-layered dough materials. US Patent 4,266,920.

Hebrard, G., Hoffart, V., Cardot, J.-M. et al. (2009). Investigation of coated whey protein/alginate beads as sustained release dosage form in simulated gastrointestinal environment. *Drug Development and Industrial Pharmacy* 35: 1103–1112.

Hebrard, G., Hoffart, V., Cardot, J.-M. et al. (2013). Development and characterization of coated-microparticles based on whey protein/alginate using the encapsulator device. *Drug Development and Industrial Pharmacy* 39: 128–137.

Hernàndez-Balada, E., Taylor, M.M., Phillips, J.G. et al. (2009). Properties of biopolymers produced by transglutaminase treatment of whey protein isolate and gelatin. *Bioresource Technology* 100: 3638–3643.

Hertafeld, E., Zhang, C., Jin, Z. et al. (2019). Multi-material three-dimensional food printing with simultaneous infrared cooking. *3D Printing and Additive Manufacturing* 6: 13–19.

Huber, G.R. and Rokey, G.J. (1990). Extruded snacks. In: *Snack Food* (ed. R.G. Booth), 107–138. Boston: Springer.

Ikeda, T., Moritaka, S., Sugiura, S., and Umeki, T. (1976). Method for preparing jelly foods. US Patent 3,969,536.

Jafari, S.M., Assadpoor, E., He, Y., and Bhandari, B. (2008). Encapsulation efficiency of food flavors and oils during spray drying. *Drying Technology* 26: 816–835.

Jani, B. and Miladinov, V.D. (2014). System and method of forming multilayer confectionery. US Patent Application 2014/0004224 A1.

Jeronomidis, G. (1988). Structure and properties of liquid and solid foams. In: *Food Structure* (ed. J.M.V. Blanshard and J.R. Mitchell), 59–75. London: Butterworth.

Jones, N.R. (1977). Uses of gelatin in edible products. In: *The Science and Technology of Gelatin* (ed. A.G. Ward and A. Courts), 366–392. London: Academic Press.

Kampfe, B. (2000). Investigation of residual stresses in microsystems using X-ray diffraction. *Materials, Science and Engineering A – Structural Material, Properties, Microstructure and Processing* 288: 119–125.

Karim, A. and Bhat, R. (2008). Gelatin alternatives for the food industry: recent developments, challenges and prospects. *Trends in Food Science and Technology* 19: 644–656.

Kimura, H., Moritaka, S., and Misaki, M. (1973). Polyscaccharide 13140: a new thermo-gelable polysaccharide. *Journal of Food Science* 38: 668–670.

Klang, V. and Valenta, C. (2011). Lecithin-based nanoemulsions. *Journal of Drug Delivery Science and Technology* 21: 55–76.

Klinkesorn, U., Sophanodora, P., Chinachoti, P. et al. (2005). Encapsulation of emulsified tuna oil in two-layered interfacial membranes prepared using electrostatic layer-by-layer deposition. *Food Hydrocolloids* 19: 1044–1053.

Koebe, H.G., Muhling, B., Deglmann, C.J., and Schildberg, F.W. (1999). Cryopreserved porcine hepatocyte cultures. *Chemico-Biological Interactions* 121: 99–115.

Konstance, R.P. (1991). Axial compression properties of kamaboko. *Journal of Food Science* 56: 1287–1291.

Kralova, I. and Sjoblom, J. (2009). Surfactants used in food industry: a review. *Journal of Dispersion Science and Technology* 30: 1363–1383.

Lam, R.S.H. and Nickerson, M.T. (2013). Food proteins: a review on their emulsifying properties using a structure–function approach. *Food Chemistry* 141: 975–984.

Le Tohic, C., O'Sullivan, J.J., Drapala, K.P. et al. (2018). Effect of 3D printing on the structure and textural properties of processed cheese. *Journal of Food Engineering* 220: 56–64.

Lee, B.C. and Kim, E.S. (1999). A simple and efficient method of analyzing mechanical behaviors of multi-layered orthotropic plates in rectangular shape. *Journal of Micromechanics and Microengineering* 9: 385–393.

Lee, G.S. and Ma, C.C. (1998). Transient elastic waves propagating in a multi-layered medium subjected to in-plane dynamic loadings. *Proceedings of the Royal Society of London, Series A: Mathematical, Physical and Engineering Sciences* 456: 1355–1374.

Leinfellner, N., Ferre-Borrull, J., and Bosch, S. (2000). A software for optical characterization of thin films for microelectronic applications. *Microelectronics Reliability* 40: 873–875.

Levy, L.W. and Rivadeneira, D.M. (2000). Annatto. In: *Natural Food Colorants, Science and Technology. IFT Basic Symposium Series* (ed. G.B. Lauro and F.J. Francis), Chapt. 6, 117–152. Boca Raton, FL: CRC Press.

Li, Y., Hu, M., Xiao, H. et al. (2010). Controlling the functional performance of emulsion-based delivery systems using multi-component biopolymer coatings. *European Journal of Pharmaceutics and Biopharmaceutics* 76: 38–47.

Lille, M., Nurmela, A., Nordlund, E. et al. (2018). Applicability of protein and fiber-rich food materials in extrusion-based 3D printing. *Journal of Food Engineering* 220: 20–27.

Liu, F., Avena-Bustillos, R.J., Chiou, B.-S. et al. (2017). Controlled-release of tea polyphenol from gelatin films incorporated with different ratios of free/nanoencapsulated tea polyphenols into fatty food simulants. *Food Hydrocolloids* 62: 212–221.

Marañón García, I., San Vicente Laurent, L., Hidalgo Lemus, N., and Chávarri Hueda, M.B. (2016). Multilayer probiotic microcapsules. European Patent Application EP 3 205 216 A1

Martucci, J.F. and Ruseckaite, R.A. (2009). Biodegradation of three-layer laminate films based on gelatin under indoor soil conditions. *Polymer Degradation and Stability* 94: 1307–1313.

Martucci, J.F. and Ruseckaite, R.A. (2010a). Biodegradable three-layer film derived from bovine gelatin. *Journal of Food Engineering* 99: 377–383.

Martucci, J.F. and Ruseckaite, R.A. (2010b). Three-layer sheets based on gelatin and poly(lactic acid), part 1: preparation and properties. *Journal of Applied Polymer Science* 118: 3102–3110.

Martucci, J.F., Accareddu, A.E.M., and Ruseckaite, R.A. (2012). Preparation and characterization of plasticized gelatin films cross-linked with low concentrations of glutaraldehyde. *Journal of Materials Science* 47: 3282–3292.

Mattiasson, B. (1983). *Immobilized Cells and Organelles*, vol. 1 and 2. Boca Raton, FL: CRC Press.

Mattinen, M.L., Kruus, K., Buchert, J. et al. (2005). Laccase-catalyzed polymerization of tyrosine-containing peptides. *FEBS Journal* 272: 3640–3650.

Mattinen, M.L., Hellman, M., Steffensen, C.L. et al. (2008). Laccase and tyrosinase catalysed polymerization of proteins and peptides. *Journal of Biotechnology* 136S: S318.

McClements, D.J. (2004). Protein-stabilized emulsions. *Current Opinion in Colloid & Interface Science* 9: 305–313.

McClements, D.J., Decker, E.A., Park, Y., and Weiss, J. (2009). Structural design principles for delivery of bioactive components in nutraceuticals and functional foods. *Critical Reviews in Food Science and Nutrition* 49: 577–606.

Medeiros, B.G.S., Pinheiro, A.C., Carneiro-da-Cunha, M.G., and Vicente, A.A. (2012a). Development and characterization of a nanomultilayer coating of pectin and chitosan – evaluation of its gas barrier properties and application on 'Tommy Atkins' mangoes. *Journal of Food Engineering* 110: 457–464.

Medeiros, B.G.S., Pinheiro, A.C., Teixeira, J.A. et al. (2012b). Polysaccharide/protein nanomultilayer coatings: construction, characterization and evaluation of their effect on 'Rocha' pear (*Pyrus communis* L.) shelf-life. *Food Bioprocess Technology* 5: 2435–2445.

Miladinov, V.D., Elejalde, C.C., Kiefer, J. et al. (2017). Multi-region chewing gum confectionery composition, article, method and apparatus. US Patent Application 2017/0258107 A1.

Miller, V. and Miller, R. (2005). Apparatus for production of striated, laminated lipid-based confections. US Patent 06935769.

Mitsoulis, E. (1988). Multilayer sheet coextrusion: analysis and design. *Advances in Polymer Technology* 8: 225–242.

Moreno, O., Gil, A., Atares, L., and Chiralt, A. (2017). Active starch-gelatin films for shelf-life extension of marinated salmon. *Lebensmittel-Wissenschaft & Technologie* 84: 189–195.

Moriyama, K., Ooya, T., and Yui, N. (1999). Pulsatile peptide release from multi-layered hydrogel formulations consisting of poly(ethyleneglycol)-grafted and ungrafted dextrans. *Journal of Biomaterials Science, Polymer Edition* 10: 1251–1264.

Nisperos-Carriedo, M.O. (1994). Edible coatings and films based on polysaccharides. In: *Edible Coatings and Films to Improve Food Quality* (ed. J.M. Krochta, E.A. Baldwin and M.O. Nisperos-Carriedo), 305–336. Basel, Switzerland: Technomic Publishing Co.

Nogueira, G.F., Prata, A.S., and Grosso, C.R.F. (2017). Alginate and whey protein based-multilayered particles: production, characterisation and resistance to pH, ionic strength and artificial gastric/intestinal fluid. *Journal of Microencapsulation* 34: 151–161.

Nussinovitch, A. (1997). *Hydrocolloid Applications: Gum Technology in the Food and Other Industries*. London: Blackie Academic & Professional.

Nussinovitch, A. (2003). *Water-Soluble Polymer Applications in Foods*. Oxford, UK: Blackwell Science.

Nussinovitch, A. (2017). *Adhesion in Foods, Fundamental Principles and Applications*. West Sussex, UK: Wiley Blackwell.

Nussinovitch, A. and Hirashima, M. (2014a). *Cooking Innovations Using Hydrocolloids for Thickening, Gelling and Emulsification*. Boca Raton, FL: CRC Press.

Nussinovitch, A. and Hirashima, M. (ed.) (2014b). Curdlan. In: *Cooking Innovations: Using Hydrocolloids for Thickening, Gelling, and Emulsification*, Chapt. 6, 101–116. Boca Raton, FL: CRC Press.

Nussinovitch, A., Lee, S.J., Kaletunc, G., and Peleg, M. (1991). Model for calculating the compressive deformability of double-layered curdlan gels. *Biotechnology Progress* 7: 272–274.

O'Connell, J.E. and de Kruif, C.G. (2003). β-Casein micelles; cross-linking with transglutaminase. *Colloids and Surfaces A: Physicochemical and Engineering Aspects* 216: 75–81.

Ochi, H., Takahashi, M., Kaneko, T. et al. (1991). Diacetyl production by co-immobilized citrate-positive *Lactococcus lactis* subsp *lactis* 3022 and homogenized bovine liver in alginate fibers with double gel layers. *Biotechnology Letters* 13: 505–510.

Ogawa, S., Decker, E.A., and McClements, D.J. (2003a). Production and characterization of O/W emulsions containing cationic droplets stabilized by lecithin-chitosan membranes. *Journal of Agricultural and Food Chemistry* 51: 2806–2812.

Ogawa, S., Decker, E.A., and McClements, D.J. (2003b). Influence of environmental conditions on the stability of oil in water emulsions containing droplets stabilized by lecithin–chitosan membranes. *Journal of Agricultural and Food Chemistry* 51: 5522–5527.

Olivas, G.I. and Barbosa-Cánovas, G.V. (2005). Edible coatings for fresh-cut fruits. *Critical Reviews in Food Science and Nutrition* 45: 657–670.

Overly, H.J., III and Doerr, C.M. (2007). Coated confectionary product. European Patent 1845799.

Ozturk, B. and McClements, D.J. (2016). Progress in natural emulsifiers for utilization in food emulsions. *Current Opinion in Food Science* 7: 1–6.

Ozturk, B., Argin, S., Ozilgen, M., and McClements, D.J. (2014). Formation and stabilization of nanoemulsion-based vitamin E delivery systems using natural surfactants: quillaja saponin and lecithin. *Journal of Food Engineering* 142: 57–63.

Peleg, M. (1993). Calculation of the compressive stress-strain relationships of layered arrays of cellular solids using equation-solving computer software. *Journal of Cellular Plastics* 29: 285–293.

Peleg, M. (1997). Review: mechanical properties of dry cellular solid foods. *Food Science and Technology International* 3: 227–240.

Perrechil, F.A. and Cunha, R.L. (2013). Stabilization of multilayered emulsions by sodium caseinate and kappa-carrageenan. *Food Hydrocolloids* 30: 606–613.

Perrechil, F.A., Sato, A.C.K., and Cunha, R.L. (2011). κ-Carrageenan–sodium caseinate microgel production by atomization: critical analysis of the experimental procedure. *Journal of Food Engineering* 104: 123–133.

Rhim, J.-W., Mohanty, K.A., Singh, S.P., and Ng, P.K.W. (2006). Preparation and properties of biodegradable multilayer films based on soy protein isolate and poly(lactide). *Industry and Engineering Chemistry Research* 45: 3059–3066.

Riva, S. (2006). Laccases: blue enzymes for green chemistry. *Trends in Biotechnology* 24: 219–226.

Robles-Sánchez, R.M., Rojas-Graü, M.A., Odriozola-Serrano, I. et al. (2013). Influence of alginate-based edible coating as carrier of anti-browning agents on bioactive compounds and antioxidant activity in fresh-cut Kent mangoes. *LWT – Food Science and Technology* 50: 240–246.

Rosenberg, M. and Lee, S.J. (2004). Water-insoluble, whey protein-based microspheres prepared by an all-aqueous process. *Journal of Food Science* 69: FEP50–FEP58.

Schmitt, C., Sanchez, C., Desobry-Banon, S., and Hardy, J. (1998). Structure and techno functional properties of protein-polysaccharide complexes: a review. *Critical Reviews in Food Science and Nutrition* 38: 689–753.

Schrenk, W.J., Bradley, N.L., Alfrey, J.R., and Maack, H. (1978). Interfacial flow instability in multilayer coextrusion. *Polymer Engineering and Science* 18: 620–623.

Schrieber, R. and Gareis, H. (2007). *Gelatin Handbook—Theory and Industrial Practice*. Weinheim, Germany: Wiley-VCH Verlag GmbH & Co.

Scotter, M.J., Wilson, L.A., Appleton, G.P., and Castle, L. (1998). Analysis of Annatto (*Bixa orellana*) food coloring formulations. 1. Determination of coloring components and colored thermal degradation products by high-performance liquid chromatography with photodiode array detection. *Journal of Agricultural and Food Chemistry* 46: 1031–1038.

Sethi, V., Lammert, A., Mikula, M., and Sandu, C. (2002). Cocoa powder for use in multi-layered gel-based dessert products and method for making same. US Patent 6,488,975 B1.

Sevatson, E. and Huber, G.R. (2000). Extruders in the food industry. In: *Extruders in Food Applications*, 1e (ed. M.N. Riaz), 167–204. Boca Raton, FL: CRC Press.

Shahidi, F. and Han, X.Q. (1993). Encapsulation of food Ingredients. *Critical Reviews in Food Science and Nutrition* 33: 501–547.

Shamsudin, S.Y. (2009). Non-lauric fats for cream filling. Malaysian Palm Oil Board Information Series, ISSN 1511-7871. www.palmoilworld.org/PDFs/FOOD/7_TT-434_Food.pdf (accessed March 2015).

Shindo, S., Sahara, H., Koshino, S., and Tanaka, H. (1993). Control of diacetyl precursor [alpha-acetolactate] formation during alcohol fermentation with yeast-cells immobilized in alginate fibers with double gel layers. *Journal of Fermentation and Bioengineering* 76: 199–202.

Siew, C.K., Williams, P.A., Cui, S.W., and Wang, Q. (2008). Characterization of the surface-active components of sugar beet pectin and the hydrodynamic thickness of the adsorbed pectin layer. *Journal of Agricultural and Food Chemistry* 56: 8111–8120.

Simitchiev, A., Nenov, V., and Lambrev, A. (2010). Coextrusion of food products – essence and technical characteristics. *Food Science, Engineering and Technologies* LVII (2): 617–623.

Smiddy, M.A., Martin, J.E.G.H., Kelly, A.L. et al. (2006). Stability of casein micelles cross-linked by transglutaminase. *Journal of Dairy Science* 89: 1906–1914.

Soedjak, H.S. and Spradlin, J.E. (1995). Ready-to-eat, multi-component, multi-colored gels. US Patent 5,417,990.

Souza, M.P., Vaz, A.F.M., Cerqueira, M.A. et al. (2015). Effect of an edible nanomultilayer coating by electrostatic self-assembly on the shelf life of fresh-cut mangoes. *Food Bioprocess Technology* 8: 647–654.

Sutherland, J.P., Varnam, A.H., and Evans, M.G. (1986). Foreign bodies and infestations. In: *A Colour Atlas of Food Quality Control*, 243. Weert, the Netherlands: A Wolfe Science Book.

Swyngedau, S. and Peleg, M. (1992). Characterization and prediction of the stress-strain relationship of layered arrays of spongy baked goods. *Cereal Chemistry* 69: 217–221.

Swyngedau, S., Nussinovitch, A., and Peleg, M. (1991a). Models for the compressibility of layered polymeric sponges. *Polymer Engineering and Science* 31: 140–144.

Swyngedau, S., Nussinovitch, A., Roy, I. et al. (1991b). Comparison of four models for the compressibility of breads and plastic foams. *Journal of Food Science* 56: 756–759.

Takayanagi, M., Harima, H., and Iwata, Y. (1963). Viscoelastic behaviour of polymer blends and its comparison with model experiments. *Memoirs – Faculty of Engineering, Kyushu University* 23: 1–13.

Tampion, J. and Tampion, M.D. (1987). *Immobilized Cells: Principles and Applications*. Cambridge, UK: Cambridge University Press.

Tanaka, H., Kaneko, Y., Aoyagi, H. et al. (1996a). Efficient production of chitinase by immobilized *Wasabia japonica* cells in double-layered gel fibers. *Journal of Fermentation and Bioengineering* 81: 220–225.

Tanaka, H., Yamashita, T., Aoyagi, H. et al. (1996b). Efficient production of chitinase by *Wasabia japonica* protoplasts immobilized in double-layered gel fibers. *Journal of Fermentation and Bioengineering* 81: 394–399.

Tanimoto, S.Y. and Kinsella, J.E. (2002). Enzymatic modification of proteins: effects of transglutaminase cross-linking on some physical properties of beta-lactoglobulin. *Journal of Agricultural and Food Chemistry* 36: 281–285.

Tavares, C.J., Rebouta, L., Andritschky, M., and Ramos, S. (1999). Mechanical characterisation of TiN/ZrN multi-layered coatings. *Journal of Materials Processing Technology* 93: 177–183.

Truelstrup, H.L., Allan-Wotjas, P.M., Jin, Y.-L., and Paulson, A.T. (2002). Survival of Ca-alginate microencapsulated *Bifidobacterium* ssp. in milk and simulated gastrointestinal condition. *Food Microbiology* 19: 35–45.

Vardi, I. and Tslaf, A. (2013). Method for making a multilayered food product and corresponding product. US Patent Application 0202744 A1.

Wang, W., Xiao, J., Chen, X. et al. (2018). Fabrication and characterization of multilayered kafirin/gelatin film with one-way water barrier property. *Food Hydrocolloids* 81: 159–168.

Willaert, R.G. and Baron, G.V. (1996). Gel entrapment and micro-encapsulation: methods, applications and engineering. *Reviews in Chemical Engineering* 12: 204–205.

Wilson, G.M. and Khomami, B. (1993). An experimental investigation of interfacial instabilities in multilayer flow of viscoelastic fluids. III. Compatible polymer systems. *Journal of Rheology* 37: 341–354.

Wooster, T.J. and Augustin, M.A. (2006). β-Lactoglobulin-dextran Maillard conjugates: their effect on interfacial thickness and emulsion stability. *Journal of Colloid and Interface Science* 303: 564–572.

Xu, J.P., Wang, X.L., Fan, D.Z. et al. (2008). Construction of phospholipid anti-biofouling multilayer on biomedical PET surfaces. *Applied Surface Science* 255: 538–540.

Yang, J., Wu, L., and Liu, J. (2001). Rapid prototyping and fabrication method for 3-D food objects. US Patent 6280785.

Yang, Y., Leser, M.E., Sher, A.A., and McClements, D.J. (2013). Formation and stability of emulsions using a natural small molecule surfactant: quillaja saponin (Q-Naturale (R)). *Food Hydrocolloids* 30: 589–596.

Yu, T.C. and Han, C.D. (1973). Stratified two-phase flow of molten polymers. *Journal of Applied Polymer Science* 17: 1203–1225.

Yuan-Kwan, C. (2008). The Tim Tam test: U.S.A. vs. Australia. *Meniscus*. www.meniscuszine.com/articles/20081021992/the-tim-tam-test-u-s-a-vs-australia/ (accessed March 2015).

Zeeb, B., Gibis, M., Fischer, L., and Weiss, J. (2012). Crosslinking of interfacial layers in multilayered oil-in-water emulsions using laccase: characterization and pH-stability. *Food Hydrocolloids* 27: 126–136.

8

Hydrocolloids to Control the Texture of Three-Dimensional (3D)-Printed Foods

8.1 Introduction

In recent years, the potential applications of three-dimensional (3D) food printing have emphasized this technique's prospects. In general, printable materials have different tastes, nutritional values, and textures. Some of them are sufficiently stable to hold their shape after deposition/extrusion, whereas others may require a post-cooking process. Non-printable, traditional foods require the addition of hydrocolloids to be processed. This chapter includes a brief history of 3D printing and a description of 3D printing options for food. Other topics include cereal-based snacks, and the printability of proteins, carbohydrates, and lipids; infill percentage and pattern; modification of food texture to suit the individual and other requirements for 3D printing technology. This chapter focuses on hydrocolloids in the 3D printing process, with a mention of their use to control the texture of 3D-printed foods. This chapter concludes with some futuristic options, such as 3D printing and laser cooking and a novel application for four-dimensional (4D) food printing.

8.2 A Brief History of 3D Printing

The technology of 3D printing has revolutionized the design and manufacture of novel products. The founder of 3D systems, Charles Hull, is considered the founder of 3D printing (Noorani 2018). Hull received his first patent for 3D printing in 1984, for his stereolithography apparatus, which applied UV light to cure photopolymer resin in a vat to produce prototypes. The rationale for 3D printing is that software can be used to slice a 3D object into layers of a particular thickness, and a machine can stack those layers to form the third dimension (Noorani 2018). In the early 1980s, computer-aided design (CAD) was in its infancy and Hull, with the help of Albert Consulting Group, developed stereolithography, which could be used by any 3D printing machine. Scot Crump developed fused-deposition modeling (FDM) in 1989 and established the company Stratasys. 3D systems and Stratasys are the two most important 3D printing companies in the world (Noorani 2018). Carl Deckard and Joe Beaman advanced the selective laser sintering (SLS) process at the University of Texas in the mid-1980s, independently of the stereolithography and FDM

Use of Hydrocolloids to Control Food Appearance, Flavor, Texture, and Nutrition, First Edition.
Amos Nussinovitch and Madoka Hirashima.
© 2023 John Wiley & Sons Ltd. Published 2023 by John Wiley & Sons Ltd.

processes. Laser sintering is a 3D printing process that uses high-power laser power to sinter powdered materials into solid objects. In the early 1990s, the Massachusetts Institute of Technology (MIT) invented 3D inkjet printing, commonly referred to as 3D printing. The license for 3D printing was given to Z Corporation. In early 2012, 3D systems bought the company, gaining all of the linked patents and licenses (Noorani 2018). Many of 3D systems' and Stratasys' industrial patents expired a few years ago. In particular, expiration of the patent on FDM technology allowed bringing 3D printing technology to the consumer. The 3D printing revolution started in 2005 with the open-source project RepRap. The purpose of the RepRap project was to make a machine that can replicate itself. These 3D printers are inexpensive. Most 3D printers use FDM and stereolithography technology, and can be operated through open-source hardware and software. Some of the most popular consumer 3D printers on the market are MakerBot Replicator and FormsLabs 1+ (Noorani 2018).

The powder-deposition 3D printing method was developed at MIT (Sachs et al. 1992). This method involves the deposition of a thin layer of a specific powder formula, followed by the deposition of an appropriate binder. The latter is indispensable to generate links among the significant powder components. The value of 3D-printed objects obviously depends on a number of variables, including the deposition method, binding criteria, liquid formulation, and printing options (i.e. drops-on-demand or continuous jet), and post-printing uses (Utela et al. 2008; Bose et al. 2013). Much 3D printing expertise has developed, resulting in direct ink writing, FDM, SLS, robot-assisted deposition, laser-assisted bio-printing, and microextrusion (Bose et al. 2013; Murphy and Atala 2014). To date, the most popular studies and technical applications of 3D printing apply to practical scientific uses, such as the manufacture of pharmaceutical dosage forms, metals, polymers (Gburek et al. 2007), and scaffolds for tissues generation (Sobral et al. 2011; Bose et al. 2013; Murphy and Atala 2014).

8.3 3D Printing of Foods

8.3.1 3D Options in Foods

Three techniques are well-suited for 3D printing (Portanguen et al. 2019): (i) extrusion-based printing, the most prevalent technique in food printing (Sun et al. 2018); (ii) inkjet printing (Singh et al. 2010); and (iii) laser-assisted printing (Guillotin et al. 2010). Additive manufacturing (AM) technology has vast potential to construct foods with multiple geometries, unconventional textures, and custom-made nutritional contents (Godoi et al. 2016), and there is also a trend toward culinary applications related to "food design" efforts (Pallottino et al. 2016).

AM, commonly termed "3D printing," now offers huge freedom in design, manufacture, and innovation in various domains. These include mechanical engineering (Chen et al. 2017), aeronautics (Ford et al. 2016), design science (Areir et al. 2017; Lanaro et al. 2017; Takezawa and Kobashi 2017), biomedical engineering (Singh et al. 2017), the pharmaceutical industry (Goole and Amighi 2016; Icten et al. 2017), biotechnology (Krujatz et al. 2017), and foods (Pinna et al. 2017). For the latter, 3D printing technology may offer new prospects for low-cost and improved-value foods (Yang et al. 2015). Furthermore, aside from the type of technology, quite a lot of variables markedly affect the presentation of 3D-printed structures. For example, in the case of FDM, the printer setting should be optimized for extrusion temperature and speed, traveling speed, layer height, fill density, nozzle size, and shell thickness, to name a few (Severini et al. 2016). Interest in 3D printing technology for different food applications is growing (Wegrzyn et al. 2012). Extrusion deposition, FDM, and laser sintering have been used to produce 3D-printed food structures (Severini et al. 2016).

Edible 3D printing is progressively widening its audience from experts to individual private use (https://www.aniwaa.com/buyers-guide/3d-printers/food-3d-printers/). Almost all 3D food printers use extrusion technology, much like regular desktop fused-filament fabrication (FDM) 3D printers (i.e. a model created in a CAD program is imported into a software program specifically designed to work with a direct digital manufacturing (DDM) machine) (Noorani 2018). Instead of using plastic material, however, food 3D printers use paste-type ingredients. In a common 3D printer application, pasty ingredients are pumped from a syringe. Another approach is to use a screw-type syringe for the nozzle (injection port), which is essential for 3D printers, where everything from soft to hard foods can be handled (Figure 8.1). In addition, these machines can have two nozzles. The separate nozzles are filled with hard and soft ingredients, respectively, for the production of long-term care foods. Those materials can be mixed to create a desired "texture" (Figure 8.2).

The most common ingredients for 3D food printing are chocolate (Section 8.4.2), pancake batter (Section 8.4.3), and creams (Section 8.4.3), but there are numerous additional options, such as pizza (Section 8.4.3). These foods are printed layer upon layer, commonly using a syringe-like extruder. Table 8.1 includes a partial list of food 3D printers on the market. Inclusion in this list does not imply a recommendation on our part, or any indication of its relative value. The reader can also check the internet for other food-related applications, including 3D printing of coffee accessories (https://www.youtube.com/

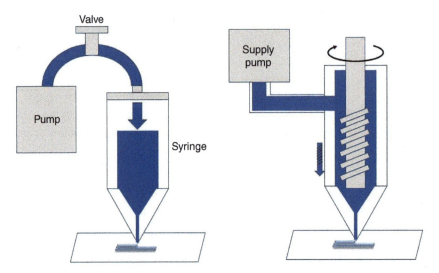

Figure 8.1 3D food printer syringes. *Source:* Ricoh Institute of Sustainability and Business; courtesy of Mr. Akinori Yusa, from Ricoh Institute of Sustainability and Business and Dr. Masaru Kawakami, from Yamagata University; https://blogs.ricoh.co.jp/RISB/environment/post_690.html.

Figure 8.2 Modeling ingredients with a two-nozzle 3D food printer. Photographed by Masaru Kawakami. *Source:* Courtesy of Akinori Yusa.

Table 8.1 Selection of 3D food printers available on the market.

Brand	Product	Build size	Country	Related website
Natural Machines	Foodini	250 × 165 × 120 mm	Spain	https://www.naturalmachines.com/how-it-works
MMuse	Chocolate 3D printer	160 × 120 × 150 mm	China	https://www.3dprintersonlinestore.com/mmuse
PancakeBot	PancakeBot 2.0	445 × 210 × 15 mm	Norway	https://pancakebot.com/
Micromake	Food 3D printer	100 × 100 × 15 mm	China	https://www.facebook.com/Micromake3DPrinter/
Choc Edge	Choc Creator V2.0 Plus	180 × 180 × 40 mm	United Kingdom	http://chocedge.com/
Zmorph	VX	250 × 235 × 165 mm	Poland	https://zmorph3d.com/
byFlow	Focus	208 × 228 × 150 mm	Netherlands	https://www.3dbyflow.com/

Source: Adapted with changes from https://www.aniwaa.com/buyers-guide/3d-printers/food-3d-printers/.

watch?v=X3bEjJ_S2V0), food decorations, and food molds (https://www.youtube.com/watch?v=D16xz-7EdzI).

8.3.2 Special Personalized Foods for the Elderly

3D printing can also be targeted to the development of special personalized foods for elderly consumers (Lipton et al. 2015; Liu et al. 2017). It is presumably possible to define a person's dietary needs and directly custom-print a food that meets their requirements (Lipton 2017). However, there are not many studies on the nutritional value of 3D-printed

234 | *8 Hydrocolloids to Control the Texture of Three-Dimensional (3D)-Printed Foods*

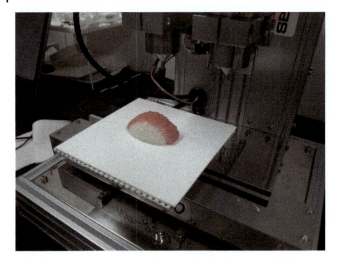

Figure 8.3 Sushi prepared with a 3D food printer. Tuna is "hard," sushi is "soft." Photographed by Masaru Kawakami. *Source:* Courtesy of Akinori Yusa.

foods. An exception deals with the antioxidant activity of 3D-printed fruit-based snacks (Derossi et al. 2018). Long-term care foods for the elderly are becoming a necessity in societies where the percentage of old and very old people is on the rise. Nursing care foods are usually prepared according to individual requirements, taking into consideration nutrients, calorie information, chewing ability, and the like. Generally, this involves pureeing the ingredients in a mixer with a gelling agent to adjust the hardness of the meal, which is extremely laborious. This is where a 3D food printer could help in creating such foods. If and when the concept of 3D-printed foods catches on, nursing care food could be prepared with consistent quality in any quantity, when needed, and according to the circumstances of the person for which it is designed. Therefore, the burden on the caregiver can be dramatically reduced. In addition, combinations of ingredients can be selected to provide texture and umami. The most important feature is that people requiring long-term care will be able to enjoy their "food" (Figure 8.3).

8.4 3D-Printed Food Products

8.4.1 Printed Sugar Products

Sugar is a malleable material that can take on many forms and colors; it is also compatible with 3D printing. Some research groups, such as The CandyFab Project (https://en.wikipedia.org/wiki/CandyFab) and the Sugar Lab from 3D systems, have developed 3D printers dedicated to the printing of sugar and candies with artistic 3D shapes (https://www.3dsystems.com/culinary). Sugar-based powders are the best materials to produce constructions with complex shapes through laser-sintering technology. Classical sugar "cubes" (Figure 8.4) are composed of dry ingredients, including several hydrocolloids, such as maltodextrin and microcrystalline cellulose (https://sugarlab3d.com/products/sugar-cubes-red-ombre-1).

Figure 8.4 Classical geometrical sugar "cubes" in a romantic red ombré featuring a fresh peppermint bouquet. *Source:* Courtesy of The Sugar Lab.

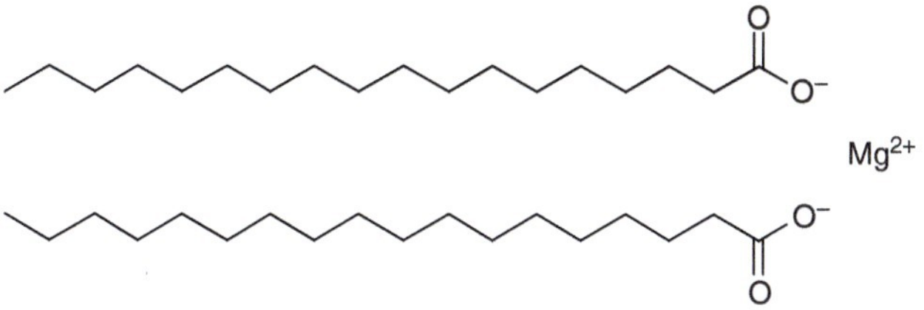

Figure 8.5 Chemical structure of magnesium stearate. *Source:* Edgar181/Wikimedia Commons/Public domain.

8.4.2 Chocolate

Due to its inherent physical properties, chocolate is well-suited to 3D printing. It melts at human body temperature and solidifies upon cooling. It is therefore easy to create tailored chocolate desserts. The company Choc Edge presented two versions of its 3D printer to make all types of chocolate designs. The Choc Creator V2.0 Plus works with a syringe loading system. Users must manually fill the 3D printer's 30-mL syringe with tempered and heated chocolate. The barrel with the syringe can be heated to keep the chocolate warm. The 3Drag printer uses FDM technology, where virtually any chocolate design can be produced (https://www.aniwaa.com/product/3d-printers/choc-edge-choc-creator-2-0-plus/). A FDM-like technique was used to manufacture 3D chocolate objects (Causer 2009; Hao et al. 2010), and several other manuscripts discussed the complex 3D printing of a chocolate object (Lanaro et al. 2017), as well as the design of 3D printers for gastronomical use (Zeleny and Ruzicka 2017). Such printing can produce complex structures from chocolate or sugar, as long as a certain number of important parameters are controlled, including feedstock-vat and extrusion-system temperature, nozzle geometry and height from the forming bed, and rheological behavior, among others (Godoi et al. 2016). To improve the flowability of the chocolate feedstock during deposition, magnesium stearate (Figure 8.5) is added, conferring better "printability" (Mantihal et al. 2017). Linear relationships were found between extrusion speed and both bead diameter and mass of the printed chocolate (Hao et al. 2010). Other studies dealt with confectionery sugar, and with plant- or meat-based purees. In general, product rheology has to be improved by the inclusion of food additives, characteristically xanthan gum or agar–agar for plant foods, or transglutaminase or gelatin for meat products (Lipton et al. 2015).

8.4.3 Pastes, Pizza, Cookies, and Meat

The Netherlands Organisation for Applied Scientific Research offered a 3D printer based on the deposition of pastes produced by mixing several food components (Netherlands Organisation for Applied Scientific Research 2015). Twin-head extrusion and postcooking were utilized to produce a cake shape (Lipton et al. 2010). The previous examples focus on the manufacture of commercial 3D food printer systems geared to consumer interest. The behavior of beef burger preparations, cheese cracker dough, and pizza dough made by 3D printing was studied (Gracia-Julia et al. 2015). The pizza dough and beef preparation were printable using a 4 mm nozzle, whereas the cheese crackers were extruded with smaller nozzles. Better printing of the beef preparation was obtained by controlling the solubilization of the myofibrillar proteins (Gracia-Julia et al. 2015). A formulation of bio-inks based on pectin for 3D-printed edible objects was developed (Vancauwenberghe et al. 2015). Following the printing of the bio-ink formulation, postprocessing was performed, comprised of incubation of the object in a $CaCl_2$ solution to solidify the structure (Vancauwenberghe et al. 2015). Cookies were printed by SLS technique (Aregawi et al. 2015a). Food 3D printing provides the opportunity for customers to create their own made-to-order cookies before their very eyes. This is how Oreo chose to exhibit 3D printers during the South by Southwest (SXSW) 2014 Festival, allowing customers to choose the color of the cream filling in their Oreos (https://www.youtube.com/watch?v=fZjR6pyTJsc). With the best 3D scanners, it may be possible to produce a 3D model of your own head and 3D print it with edible food. Several manufacturers offer the prospect of eating pancakes or waffles in the shape of one's face. In one of their marketing campaigns, the mayonnaise manufacturer Hellmann's even offered customers the option of 3D printing their face on burger buns (https://www.youtube.com/watch?v=_5Zzyq1rKmY). 3D printing allows the customer to join in the creation of a new variety of foodstuffs. Barilla, for instance, organized a competition in 2014 to select the best 3D model design for a new Barilla pasta. Over the course of two months, artists sent in 216 pasta-shape proposals from 20 different countries, and the winning design was sold by Barilla as a new product (https://www.aniwaa.com/guide/3d-printers/food-3d-printing/). 3D printing for the food industry includes, but is not limited to, Foodini 3D printers sold by Natural Machines and Bocusini. These 3D printers can prepare fully certified 3D-printed meals made up of successive layers, similar to the preparation of pizza. For the latter, the dough is cooked throughout the 3D printing process, and tomato sauce (prepared from powder), water, and oil are added, followed by a layer of protein (https://www.naturalmachines.com/how-it-works). Further expanding the limits of 3D printing, some scientists use biomaterials and animal stem cells to create 3D-printed meat. There are many rationales for this idea, including the need to reduce the resources needed for meat production, reducing the carbon and environmental footprint, reducing the slaughter of animals, making the world greener, and the belief that it could serve as a solution to the mounting needs of the world's growing population. Brooklyn-based Modern Meadow has already successfully 3D-printed steak chips made from synthetic animal protein, as well as 3D printing of vegan leather (https://www.forbes.com/sites/simonmainwaring/2021/02/23/purpose-at-work-modern-meadows-path-to-industry-disruption-growth-and-a-sustainable-future/?sh=420d32242136; https://www.modernmeadow.com/#home). Regardless of whether manufacturers want to concentrate on plant-based meat or production lines for cultured meat, numerous companies have tried to take the lead in this field. These include Beyond Meat (https://www.beyondmeat.com/en-US/) and Impossible Foods (https://impossiblefoods.com/), and the Israeli startups Redefine Meat (https://www.redefinemeat.com/) and Aleph Farms (https://www.aleph-farms.com/) which are competing to get their 3D-printed plant-based steaks on our plates. 3D printing may also find a use on long space missions, because a great diversity of products with reduced amounts of raw material can be easily formed. The US National Aeronautics and Space Administration (NASA) has taken an evident interest in 3D food printing and in 2013, it gave a grant to the Texas-based startup systems and Materials Research to create a 3D food printer for astronauts to produce routine meals literally on the fly. NASA was captivated by the potential for this technology in preparing food easily in space, which could be quite handy for astronauts staying in a space station for long periods.

Open Meals is a Japanese startup developing 3D food printing technology. Open Meals was launched in 2016 by Ryosuke Sakaki, an art director at the marketing company Dentsu's Creative Planning Division 3, to revolutionize the way food is prepared. In 2050, according to Sakaki's vision, astronauts working in space will be able to enjoy sushi prepared by famous chefs in Tokyo. The concept – a collaborative project between Yamagata University, Dentsu, automotive manufacturer Denso and film corporation Tohokushinsha – debuted early last year at SXSW in Texas. The platform is research and development of 3D printers and base materials to build food, in an effort to realize such food transfer in the near future (https://asia.nikkei.com/Business/Food-Beverage/Sushi-in-space-Japanese-project-develops-high-tech-food-printer). Other interesting examples include, but are not limited to: food layer manufacture, a new process for constructing solid foods (Wegrzyn et al. 2012); structure design of 3D-printed cookies in relation to texture (Aregawi et al. 2015b); the role of gelation dynamics in food printing (Vallons et al. 2015); and process control in 3D food printing (Klomp et al. 2015).

8.5 Production of Snacks

8.5.1 Cereal-Based 3D Snacks

3D printing technology was employed to produce cereal-based snacks with desired shapes and dimensions (Severini et al. 2016). Commercial type 0 wheat flour was used for the study (Italian Food Regulation 2001). Optimum mixing time and water-absorption capacity of the flour, to reach 500 Brabender Units, was determined using a faringograph test (Severini et al. 2016). Based on the faringograph absorption, the dough was manufactured by mixing 100 g of wheat flour with 54 g of distilled water in a planetary kneader. Ingredients were first mixed for one minute at 60 rpm; the bowl was then scraped down, and mixed again for three minutes at the same speed. The dough was then covered with plastic wrap and left to rest for 30 minutes before use (Severini et al. 2016). The printability of the dough and the value of the cooked samples were studied as a function of infill percentage and layer height. The obtained snacks matched the intended structures, although some dimensional changes were perceived. The increase in layer height produced a reduction in the height of the samples, as well as an increase in their diameter (Severini et al. 2016). This was attributed to the irregular deposition of the dough with increased layer height. On the other hand, the infill level was more important for changes in the solid fraction of both raw and cooked snacks. The breaking strength of the samples was strongly related to the infill level, although there was significant variability in the layer height used during 3D printing (Severini et al. 2016).

8.5.2 Fruit Snacks

3D-printed fruit-based snacks for children were developed using fresh bananas, dried mushrooms, canned white beans, dried non-fat milk and lemon juice mixed with pectin to create the products. The effects of print speed and flow level on the shape, dimensions and microstructure of the snacks were evaluated (Derossi et al. 2018). The hydrocolloid κ-carrageenan blended with an orange concentrate–wheat starch mixture enriched with vitamin D was used for 3D printing of a food mixture (Azam et al. 2018a, 2018b). The rheological data on the 3D printing characteristics of the blends suggested that steam cooking of the orange concentrate–wheat starch mixture produces shear-thinning behavior, which is essential for extrusion-type 3D food printing (Azam et al. 2018a). A combination of 5% wheat starch with orange concentrate significantly increased the yield stress and viscosity. The structure was enhanced by mixing with different gums. The addition of carrageenan gum contributed to better target geometry and good load-bearing capacity. Moreover, collapse over time was prevented, and the best printing conditions were achieved (Azam et al. 2018b).

8.6 Printability of Food Additives

8.6.1 Issues Related to 3D Food Printing

Several issues facing 3D food printing are: satisfactory particle size of the food components and material–material bonding mechanisms (Portanguen et al. 2019). Upstream control of the process steps should suffice to address the first issue. For the second issue, it is recommended that "natural" additives be used, such as unsaturated lipids and blood plasma proteins, to improve solidification on cooling or crosslinking (Portanguen et al. 2019).

8.6.2 Printability of Hydrocolloids

Cellulose (powder) is printable layer-by-layer, on condition that the density, surface tension, and rheological properties of the constructed material be controllable (Holland et al. 2018). Semi-crystalline cellulose was modeled by powder deposition and binding technique. To process the edible objects successfully, semi-crystalline cellulose was ball milled to produce an amorphous powder, and food inks based on xanthan gum were formulated to enable successful jetting with an inkjet printer (Holland et al. 2018). The deposition of only one layer of droplets was not enough to produce a cohesive powder layer that could be lifted cleanly. Three, 5 and 10 ink layers were applied to the structures produced from the powder, which could then be lifted onto glass microscope slides with various degrees of success (Holland et al. 2018). Methylcellulose was used as a reference biomaterial to mimic the printability of numerous types of food applications (Kim et al. 2018). Hydrocolloid concentrations of 9%, 11%, and 13% could scaffold 28-mm diameter cylindrical constructs with heights of 20 mm, 40 mm, and 80 mm, respectively, with no collapse (Kim et al. 2018). The deformation behavior and handling properties of printed foods were classified based on the reference material manufactured at a number of concentrations (5–20%). Shear modulus of all samples was in general agreement with the simulation outcomes based on a dimensional stability test, indicating that the printability of foods can be predicted and classified by comparing their properties to a reference material. The newly established printability classification system consists of grades, along with dimensional stability and degree of handleability. The validity of this classification system was verified by 3D printing tests (Kim et al. 2018).

The feasibility of 3D printing textured variable-microstructure foods was studied (Vancauwenberghe et al. 2017). Pectin-based formulations (i.e. different concentrations of pectin, calcium chloride, bovine serum albumin, and sugar syrup), and different stirring speeds were tested. A coherent and durable 3D structure was attained due to pectin crosslinking by $CaCl_2$. Pectin and sugar syrup concentrations influenced the viscosity of the mixture, and the bovine serum albumin stabilized and aerated it (Vancauwenberghe et al. 2017).

3D printing of low- to high-starch potato purees was studied (Liu et al. 2018). The puree had to include at least 2% starch to be printable. Under those conditions, the material displayed an increase in elastic limit and better extrudability. On the other hand, at 4% starch content, despite the material successfully holding its 3D shape and structure, it had poor extrudability due to overly high viscosity (Liu et al. 2018). Another study demonstrated that 15% (w/w) potato starch in the presence of lemon juice creates printable 3D structures (Yang et al. 2018). The optimal print-process parameters, i.e. nozzle diameter, print-head speed and extrusion rate for fabricating smooth-surfaced constructs with zero deformation were also determined (Yang et al. 2018). Another study showed that a 15% starch solution had better printability when the formulation included semi-skimmed milk powder instead of skim-milk powder (Lille et al. 2018).

8.6.3 Protein Products Applicable for 3D Printing

The AM process focuses on foods with two categories of limitations: (i) thermal treatment as the item for consumption melts, and (ii) shear strain as the product extrudes through a nozzle (Le Tohic et al. 2018). Printability is the set of material properties that give a product with sufficient stability in space to support its own weight (Godoi et al. 2016). Applicability of protein-rich products, such as pureed meat, to 3D printing was reported (Lipton et al. 2015). Other manuscripts dealt with 3D printing of composite calcium phosphate and collagen scaffolds for bone regeneration (Inzana et al. 2014), and the effect of gelatin addition on the fabrication of magnesium phosphate-based scaffolds (Farag and Yun 2014). As a mixture, proteins, carbohydrates, lipids, and water can go through modifications that will affect both fusion and plasticization of the food (Wang et al. 2018). For instance, printability of fish surimi gel was optimal with the inclusion of sodium chloride at 1.5 g/100 g surimi (Wang et al. 2018). Salt addition improved the printability of beef-based preparations once the myofibrillar proteins were solubilized (Gracia-Julia et al. 2015). Electron microscopy demonstrated that the added NaCl leads to myofibrillar protein crosslinking, allowing free amino acids to bind to the proteins, shrinking the void spaces, and changing the structure of the gel into a fine-stranded network (Wang et al. 2018). Novel textures can be formed by the interaction of layers of polysaccharides (such as alginate) with layers of proteins. Promotion of aggregation by application of mechanical stress or changes in temperature, or inclusion of acidic or basic compounds is part of the preparation methodology (Godoi et al. 2016).

Precise and detailed 3D structures are better achieved upon addition of gelable proteins or hydrocolloids in, e.g. structured turkey meat (Yang et al. 2018). Gelatin is derived from collagen obtained from various animal body parts. It is commonly used as a gelling agent in food (Pang et al. 2014). Collagen, elastin, and fibrin can serve as major fabricators of scaffolds (Melchels et al. 2012; Chia and Wu 2015). Gelatin exhibits Newtonian flow in dilute solution except when extended by charged groups, and it is a good candidate as a component of AM bio-inks (Pang et al. 2014). On the one hand, gelatin gels have an exceptional characteristic texture that delivers considerable mouthfeel along with good savor (Godoi et al. 2016). On the other hand, gelatin without additives demonstrates poor self-support features, having higher hardness and fracturability, which give rise to flaws in the samples (Kim et al. 2018). Therefore, the user of gelatin in 3D extrusion processes must take into consideration that charges on the molecule increase its relevant effects on viscosity (Kragh 1961). In addition, one should recall that the major challenge for printing food is to get the desired texture to deliver good mouthfeel and proteins, and significant structural macromolecules are no exception (Portanguen et al. 2019). Fish collagen mixed with vegetable or fruit pastes (avocado, broccoli, carrots, kiwi, and pears) allowed processing of 3D-printed edible objects (Severini et al. 2018). Without the collagen addition, dietary fibers could not be used as the raw materials for 3D printing because most insoluble dietary fiber has no plasticity and poor self-support abilities. The printing did not markedly change the sensorial characteristics of the resultant samples, or their total phenolic content or antioxidant capacity, but the geometrical differences between model and products were significant (Severini et al. 2018).

8.6.4 The Effect of 3D Printing on Lipids

3D printing studies have experimented with high-fat-content foods such as chocolate, and the effect of 3D printing on the structure and textural properties of processed cheese (Le Tohic et al. 2018). In the latter study, 3D printing was studied using a commercially accessible processed cheese as the printing material (Le Tohic et al. 2018). After melting at 75 °C for 12 minutes, the processed cheese was printed by a modified commercial 3D printer at low- or high-extrusion rates. The untreated, melted, and printed cheeses were assessed by texture profile analysis, rheology, colorimetry, and confocal laser-scanning microscopy. Processing (i.e. melting and extrusion) significantly influenced the cheese's properties. Melted and printed cheese samples were significantly less hard, and both exhibited higher degrees of meltability, compared to unprocessed cheese samples. This demonstrates that 3D printing considerably alters the properties of processed cheese, perhaps offering new potential applications for tailoring structures using this novel process (Le Tohic et al. 2018). Another manuscript studied the role of lipids throughout food-printing processes by using milk powder as a source of both proteins and fat (Lille et al. 2018). Two studied formulations included almost the same protein content (21% and 22%, respectively) in solutions of water with skim (0.4% fat) and semi-skimmed (9% fat) milk powder. The skim-milk formulation provided a highly viscous and difficult-to-print paste due to its stickiness. Once the milk powder concentration increased from 50 to 60%, printing was unmanageable. For the semi-skimmed formulation, even at 60% concentration, printability was outstanding, in terms of holding the printed shape and precision. It is possible that the fat acted as a lubricant in the extrusion system and therefore the biomaterial was more fluid. In addition, carbohydrate content differed noticeably between the two formulations (32% and 23%, respectively), and this could also influence the flow rate of the fluid (Lille et al. 2018). Thus, the use of lipids in 3D printing seems promising due to their different melting points and triglyceride compositions, which influence meat tenderness (i.e. textural properties) and flavor. 3D printing methods (especially extrusion and layer-on-layer adhesion) have great potential for fabricating better shaped foods via pre- and postprocessing (Godoi et al. 2016).

8.7 Infill Percentage and Pattern

3D food printing can generate a complex geometry (Mantihal et al. 2019). Several manuscripts regarding food printing have included a description of additive applications (Rapisarda et al. 2018), as well as the effect of this technology on food textural properties (Prakash et al. 2019). In addition, 3D food printing offers the freedom to be inventive in designing structures (Noort et al. 2017). The principal constituent of texture variation in 3D printing lies in the internal structure, comprised of both infill pattern and infill percentage (IP) (RepRap 2016). RepRap Firmware is a comprehensive motion-control firmware intended primarily for controlling 3D printers, but with applications in laser engraving/cutting and computer numerical control (CNC) as well. Unlike most other 3D printer firmwares, it only targets modern 32-bit processors, not outdated 8-bit processors with limited CPU power. It is therefore designed to make good use of the power of modern inexpensive rapid manufacturing processors to implement advanced features (https://www.reprapfirmware.org/). The infill pattern is the type of configuration formed as a design within the printed construction. The infill percentage is the density of the layer-to-layer gaps in the material upon deposition. This structure (pattern and percentage) is designated in the slicing software formally as the printing process (Fernandez-Vicente et al. 2016). The assimilation of infill pattern and percentage in 3D printing generates an exclusive internal microstructure of the printed food. The infill arrangement within a printed food affects the construction's sensorial and textural characteristics. The texture of infill-printed cheese was compared to a cast cheese prepared using a mold. Melted and printed cheese samples were significantly less hard, by up to 49%, and both exhibited higher degrees of meltability, ranging from 14 to 21%, compared to untreated cheese samples (Le Tohic et al. 2018). A cereal-based product printed with 20% infill density required a force of 62 N to break the samples, compared to 26 N for samples printed at 10% infill (Severini et al. 2016). Thus, the infill percentage influences the mechanical strength of the construction, with increasing infill percentage having a tendency to generate a harder construction (Dizon et al. 2018). A lower infill percentage generates a hollow internal structure for the printed food, contributing to its textural variation. The void fraction of chocolate when printed at 60% infill density (with a honeycomb pattern) was $11.6 \pm 2.3\%$, whereas printing at 5% infill resulted in $68.8 \pm 4.4\%$ void (Mantihal et al. 2017).

8.8 Modifying Food Texture to Suit Personal and Other Requirements by 3D Printing Technology

3D printing technology can be used to modify food texture for individual requirements (Vancauwenberghe et al. 2018). Food texture is indispensable to defining the value of food. In addition, it is a vital sensory characteristic affecting an individual's perceptions throughout ingestion. 3D printing addresses the personalization of nutritional requirements and can provide the required texture for elderly people with dysphagia (Lipton et al. 2015). In addition, 3D food printing can offer a texturized and palatable food for patients who would otherwise have to eat a monotonous diet of pureed foods (Noort et al. 2017). Thus, 3D food printing is recognized as a potentially influential technology in tailoring food to individual preferences, principally with respect to texture and shape. The apparent texture is to a large extent linked to the structure and composition of the food (Aguilera and Park 2016; Lupton and Turner 2018). The sensory characteristics (appearance, color, odor, and taste) of 3D-printed fruit and vegetable smoothies were studied (Severini et al. 2018). Printing variables were found to affect the reproducibility of the virtual model. In addition, the 3D samples were better appreciated than non-printed ones; antioxidant levels did not change during storage at 5 °C, and the sanitization of 3D printers was recommended for shelf-life considerations (Severini et al. 2018). However, the perception of 3D printing could pose a barrier for consumer acceptance of printed food items (McCluskey et al. 2016). Although issues such as false impressions, shortage of information, and a negative approach are drawbacks in consumer acceptance of any new food-production technology, preemptive knowledge dissemination and reputable news from dependable sources (i.e. firms, scientists) could overcome consumers' general perception of risks associated with novel food technologies (McCluskey et al. 2016). 3D food printing enables the creation of an extensive assortment of food items with different textures, nutrient contents, shapes, and tastes. To achieve this, a limited number of raw materials/ingredients are used (Mantihal et al. 2019). Texture-tailored foods can be achieved by planning a particular food structure with CAD (Sun et al. 2015). This approach allows combining artistic capabilities with fine dining, as well as providing customization capabilities to the industrial culinary sector (Sun et al. 2015). Furthermore, numerous ingredients (that vary in flavor and nutritional value) can be printed simultaneously using multiple cartridges; therefore, it is a potent tool in personalized nutrition applications (Wegrzyn et al. 2012; Sun et al. 2015; Yang et al. 2015;

Liu et al. 2017). Producing a more wide range of textures and tastes by using suitable hydrocolloids and flavor additives has also been recommended (Cohen et al. 2009).

8.9 Hydrocolloids in 3D Printing

Foods can be either printable or non-printable and may require suitable preprocessing (such as the addition of hydrocolloids) to make them printable. Studies have described extrusion-based 3D printing of lemon juice gel (Yang et al. 2018), vegetable puree (Severini et al. 2018), fish surimi gel (Wang et al. 2018), mashed potatoes (Liu et al. 2018), chocolate (Hao et al. 2010), cheese (Le Tohic et al. 2018), ground meat (Dick et al. 2019), and dairy products (Ross et al. 2019), among numerous others. Alternative food ingredients, such as algae and insect protein, can also be printed (Anukiruthika et al. 2020). The viscoelastic properties of gelatin and different concentrations of κ-carrageenan mixtures for AM applications were compared (Warner et al. 2019). Printability was shown to be a function of the concentration of κ-carrageenan. Furthermore, rapid formation of an elastic network led to highly defined shapes, and design rules were established for thermally gelling hydrocolloids (Warner et al. 2019). The mixtures were printed at two temperatures, just above and much higher than their gelling temperature. Analysis demonstrated that rheological behavior accompanying the coil–helix transition was key to printing the product in a well-defined routine. The printing dependability was related to the magnitude of the storage modulus, which needed to be greater than 23 kPa, along with the rapid construction of an elastic network, recovering no less than 73% of the maximum storage modulus within 200 seconds (Warner et al. 2019).

Another study dealt with room temperature extrusion-based 3D printing using a desktop 3D printer with a syringe extrusion system. Xanthan gum paste, modified starch pastes, and pureed carrot were selected as model inks (Huang 2018). Oscillatory rheology measurements, including strain and frequency sweep, were conducted to study the range of properties suitable for 3D printing. The rheological characterization of the inks provided the upper and lower limits of a printable ink. Power law models were used to analyze the rheology data and the model parameters of the inks were compared to published data on foods to assess their potential suitability as food inks for 3D printing (Huang 2018).

The rheological properties, printability, and 3D-printed geometry and texture of soy protein isolate (SPI) mixtures with sodium alginate and gelatin were investigated (Chen et al. 2019a). SPI and their mixtures demonstrated shear-thinning behavior and could be used as adequate material for 3D printing. The viscosity and elastic modulus for the SPI mixture with gelatin were lower at 35 °C (compared to SPI alone), whereas the rheological index increased rapidly as the temperature decreased to 25 °C, which supported the deposited layers and maintained the designed shape (Chen et al. 2019a). The SPI mixtures with 2, 6, and 10% gelatin printed excellent geometries. The addition of sodium alginate and gelatin to SPI did not cause chemical crosslinking between protein subunits during mixing and 3D printing at 35 °C, while it improved the hardness and chewiness of the 3D-printed geometries. The overall results suggested that a food matrix consisting of SPI, sodium alginate and gelatin is a promising material for 3D food printing (Chen et al. 2019a).

The relationship between rheological properties and printability of three types of starch (potato, rice, and corn starch) for hot-extrusion 3D printing was systematically investigated (Chen et al. 2019b). Each starch sample showed shear-thinning behavior, self-supporting property, and a substantial decrease at higher strains and recovery at lower strains in storage modulus, indicating the suitability of the starch for hot-extrusion 3D printing. Moreover, flow stress, yield stress, and storage modulus increased with a higher starch concentration (Chen et al. 2019b). Starch suspensions with concentrations of 15–25% (w/w) heated to 70–85 °C had better values of flow stress (140–722 Pa), yield stress (32–455 Pa), and storage modulus (1150–6909 Pa) for hot-extrusion 3D printing, which endowed them with excellent extrusion processability and sufficient mechanical integrity to achieve high resolution (0.804–1.024 mm line width). The results provided useful information for the production of customized starch-based food by hot-extrusion 3D printing (Chen et al. 2019b).

The effect of starch from different sources, such as rice, potato, and corn, on extrusion-based 3D printing was also reported by others (Dong et al. 2019; Martinez-Monzo et al. 2019). 3D printing of surimi requires a structural modifier to achieve a stable construct. The effect of adding sweet potato starch (0–10% w/w) on the physical properties (rheological properties, gel strength, water-holding capacity and microstructure) of the surimi gel and the 3D-printed behavior of these gels were studied (Dong et al. 2019). As the starch content increased, the viscosity of the starch–surimi mixture decreased, enabling the surimi to flow out of the printer nozzle. The surimi gel with 8% sweet potato starch showed good gel strength (2021 g·mm), water-holding capacity (82.4%) and microstructure, and less cooking loss (1.9%). The comparison of a traditional method of surimi preparation with 3D printing showed that the latter was softer in terms of gel strength (1398 g·mm) and lower in hardness (945.2 g), although it showed slightly higher cooking loss (6.7%) and lower water-holding capacity (72.7 g) than the conventional product. The results suggested that sweet potato starch can be effectively used as a structural enhancer of 3D-printed complex-shaped surimi (Dong et al. 2019). The temperature and composition of food during the printing process may be key factors affecting rheological properties (Jiang et al. 2019; Martinez-Monzo et al. 2019). The printability of potato puree subjected to different printing variables, such as temperature and product composition, was studied by analyzing its rheological and textural properties (Martinez-Monzo et al. 2019). Viscosity–temperature profiles, flow curves and dynamic oscillation frequency analysis were some of the techniques used in the rheological analysis. Forward extrusion assays of formulated potato puree were used to study the compression force in the 3D printer. Results demonstrated that the formulation with the higher content of dehydrated potato puree at a temperature of 30 °C was the most stable. The printability increased with increasing consistency index and decreasing behavior index and the mean force from the extrusion test was correlated with printability (Martinez-Monzo et al. 2019).

8.10 3D Printing of Hydrocolloid Foods Served in Restaurants

Hydrocolloids can be included in mixtures that are printed and directly consumed in restaurants. For example, "3D Yonezawa beef hot jelly" is prepared from Yonezawa beef consommé soup and agar (Kodama et al. 2017). Yonezawa beef (米沢牛, *Yonezawa gyū*) is *wagyū* (Japanese beef) originating from the Yonezawa region of Yamagata Prefecture,

Figure 8.6 Matsusaka sirloin in Nihonbashi Mitsukoshi food hall, Tokyo, Japan. *Source:* Schellack/Wikipedia Commons/Public Domain.

Japan. Yonezawa is one of the generally recognized three most famous beef brands in Japan (https://en.wikipedia.org/wiki/Yonezawa_beef), along with Kobe beef and Matsusaka beef (Figure 8.6). The 3D product is a rare jelly which is hard to dissolve when warmed, because agar is the gelling agent. Agar made from seaweed usually becomes a jelly at 0.7% (w/w). Jelly made from agar has a high melting temperature (68–84 °C). This helps retain this food's shape in hot soup. The 3D printing mold for the jelly is designed in the motif of a snow lantern that is 4.5 cm in length, 2.4 cm in width, and 1.7 cm in height. The product is served in a local Italian restaurant in Japan as an appetizer or soup garnish (https://www.intechopen.com/chapters/56857).

Another example is 3D carp jelly, which is a jelly in the shape of Yonezawa carp. 3D carp jelly is served in a local carp restaurant as "3D carp hot pot." The jelly is made from soy milk, dried carp flakes, and gelatin. This jelly melts easily and becomes soymilk soup when heated. Since the jelly comprises heart-shaped konjac, the konjac comes out when the jelly melts (Figure 8.7). Gelatin made from collagen, from the skin of cows and pigs, usually forms into a jelly at 2.0% (w/w). Jelly made from gelatin melts at room temperature (20–30 °C). The size of the jelly is 6 cm in length, 2.7 cm in width, and 1.5 cm in height (https://www.intechopen.com/chapters/56857).

Open Meals (Section 8.4.3) is preparing to launch their first 3D-printed sushi restaurant concept – Sushi Singularity – in Tokyo. The restaurant plan is to print sushi, as well as share sushi recipes worldwide through their online platform. Furthermore, the sushi will be hyper-personalized based on individual diners' biometric and genomic data. To achieve this, customers must send in biological samples before they visit the restaurant. The data will then be analyzed by health experts to create the bespoke dishes injected with nutrients that the customer might be lacking (https://www.greenqueen.com.hk/digital-food-revolution-japanese-startup-open-meals-is-3d-printing-sustainable-sushi/). A few examples of the phenomenal ability of Sushi Singularity to produce 3D foods are demonstrated in Figures 8.8 and 8.9. Figure 8.8 presents the printing of Cyber Wagashi. Figure 8.9 shows

Figure 8.7 Color painting of a konjac plant, *Amorphophallus konjac*, with edible corm which is the source of konjac glucomannan. Details of flower, corm and leaves. *Source:* W. Fitch/Wikimedia Commons/Public Domain.

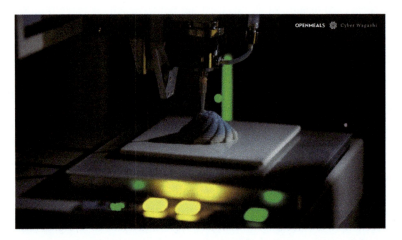

Figure 8.8 Printing of Cyber Wagashi. *Source:* Courtesy of Team Open Meals Japan Patent Pending.

Figure 8.9 The wide and unique selection of sushi that can be 3D printed with advanced technology. *Source:* Courtesy of Team Open Meals Japan Patent Pending.

Figure 8.10 "Gel" that would be suitable as a material to reproduce foods based on acquired food data. *Source:* Courtesy of Team Open Meals Japan Patent Pending.

the wide and unique selection of sushi that can be 3D printed with this extremely advanced technology. Open Meals confirmed that "gel" would be suitable as a material to reproduce foods based on acquired food data (Figure 8.10). Although information is limited, in at least one of the 3D food designs, the presence of the hydrocolloid is denoted. For example, in Figure 8.9, for the Dashi Soup Universe (bottom right), the structures are composed of a membrane-covered "wall" structure and the fabrication methodology is 3D printing plus alginate acid coagulation.

8.11 3D Printing and Laser Cooking

Food layered manufacture (FLM) is an application of AM technology that utilizes food as a material to print 3D food products (Blutinger et al. 2019). In the FLM application, a thin layer of food is deposited or multi-ingredient food products are manufactured (Hertafeld et al. 2018), and therefore precise heat delivery is required to tune heating parameters for each food ingredient (Zoran and Coelho 2011). The ability of laser technology to produce repeatable and targeted energy make it best for use in several areas of food cooking (Blutinger et al. 2018). In addition, to deliver uniform heating, other advantageous features of lasers for food processing include control over energy delivery, consistent 3D assignment of energy, and accurate localized heating (Singh 2013). The use of a blue laser to bake dough was studied. It delivered the required heat to gelatinize starch; however, it lacked the ability to efficiently brown the dough surface (Blutinger et al. 2018). The use of a CO_2 laser cutter to selectively cook the fatty portion of bacon without affecting the meat was reported (Fukuchi et al. 2012). Several patents deal with laser cooking, for example, the use of a CO_2 laser to rapidly cook food (Muchnik 2008), the use of several types of lasers to prepare foods (Singh 2013), and combining lasers and electromagnetic waves in the cooking chamber of a 3D food printer (Gracia and Sepulveda 2015). The performance of a CO_2 mid-infrared laser (operating at 10.6 μm wavelength) during the browning of dough was studied (Blutinger et al. 2019). Dough samples consisting of flour and water were exposed to the infrared laser at different laser power, beam diameter, and sample exposure time. At a laser energy flux of $0.32\,MW/m^2$ (beam diameter of 5.7 mm) and sample exposure time of 180 seconds, a maximum thermal penetration of 0.77 mm and satisfactory dough browning were achieved. These results suggest that a CO_2 laser is ideal for browning thin items as well as for food-layered manufacture (Blutinger et al. 2019). Tandem use of a CO_2 laser and blue laser offered the ability to precisely confine heat, thermally penetrate food, and preserve the shape of the precooked food construct, thus being best for more personalized nutrition in 3D food-printing applications, sheet-dough products, and conventional browning applications in meat manufacture (Blutinger et al. 2019).

8.12 Novel Application for 4D Food Printing

4D printing is a new technologically originating from advances in 3D printing, which demonstrates the desired know-how and additional possible uses (Momeni et al. 2017). Research into 4D printing has garnered a great deal of interest since 2013 when the idea was first introduced by a research group at MIT (Tibbits 2014). Although based on 3D printing technology, it requires supplementary stimulation factors and stimulus-responsive materials (Momeni et al. 2017). Based on certain interaction mechanisms among the stimulation factors and smart materials, along with the suitable design of multimaterial structures from mathematical modeling, 4D-printed structures develop as a function of intelligent behavior and time. Unlike 3D printing, 4D printing is time-dependent, printer-independent, predictable, and aims for functionality/property/and shape evolution. This allows self-assembly, multifunctionality, and self-repair (Momeni et al. 2017). Several manuscripts have described the numerous developments and applications of 4D printing

(Pei 2014; Choi et al. 2015; Xun Khoo et al. 2015; Momeni et al. 2017; Pei et al. 2017; Teng et al. 2021). In one study, two models of 3D-printed food products were created. Model (A) consisted of two parts: lemon juice gel and anthocyanin–potato starch gel. Model (B) also consisted of two parts: potato starch gel without lemon juice and the anthocyanin–potato starch gel. The models were used to illustrate the different color changes of the final product due to their different pH values (Ghazal et al. 2019). The study achieved an evolution in 3D food printing to 4D food printing, where a 4D food product with health-promoting properties was fabricated by printing a 3D food product which changes to a more attractive color with time when exposed to an external stimulus or an internal stimulus, i.e. occurring in another part of the product (Ghazal et al. 2019). The study also revealed that the color intensity of 3D-printed anthocyanin–potato starch gel before and after treatment with pH solutions decreases with increasing potato starch concentration. Conversely, as expected, the color intensity of the 3D-printed anthocyanin–PS gel before and after treatment with pH solutions increased when anthocyanin concentration increased. The information gained from this study could be valuable for the food industry and for AM technology aimed at fabricating a healthy 3D food product with attractive colors that will encourage consumer approval (Ghazal et al. 2019).

References

Aguilera, J.M. and Park, D.J. (2016). Texture-modified foods for the elderly: status, technology and opportunities. *Trends in Food Science and Technology* 57: 156–164.

Anukiruthika, T., Moses, J.A., and Anandharamakrishnan, C. (2020). 3D printing of egg yolk and white with rice flour blends. *Journal of Food Engineering* 265: 109691. https://doi.org/10.1016/j.jfoodeng.2019.109691.

Aregawi, W., Verbonen, P., Vancauwenberghe, V. et al. (2015a). Structural-mechanical analysis of cookies produced by conventional and 3D printing techniques. *Proceedings of the 29th EFFoSt International,* Athens, Greece, vol. 1, pp. 415-419.

Aregawi, W., Verbonen, P., Vancauwenberghe, V. et al. 2015b). Structure design of 3D printed cookies in relation to texture. *Proceedings of the 29th EFFoSt International*, Athens, Greece, 10–12 November 2015, vol. 1, pp. 420–425.

Areir, M., Xu, Y., Harrison, D., and Fyson, J. (2017). 3D printing of highly flexible super capacitor designed for wearable energy storage. *Materials Science and Engineering: B* 226: 29–38.

Azam, R.S.M., Zhang, M., Bhandari, B., and Yang, C. (2018a). Effect of different gums on features of 3D printed object based on vitamin-D enriched orange concentrate. *Food Biophysics* 13: 250–262.

Azam, R.S.M., Zhang, M., Mujumdar, A.S., and Yang, C. (2018b). Study on 3D printing of orange concentrate and material characteristics. *Journal of Food Process Engineering* 41: 12689. https://doi.org/10.1111/jfpe.12689.

Blutinger, J.D., Meijers, Y., Chen, P.Y. et al. (2018). Characterization of dough baked via blue laser. *Journal of Food Engineering* 232: 56–64.

Blutinger, J.D., Meijers, Y., Chen, P.Y. et al. (2019). Characterization of CO_2 laser browning of dough. *Innovative Food Science and Emerging Technologies* 52: 145–157.

Bose, S., Vahabzadeh, S., and Bandyopadhyay, A. (2013). Bone tissue engineering using 3D printing. *Materials Today* 16: 496–504.

Causer, C. (2009). They've got a golden ticket. *IEEE Potentials* 28: 42–44.

Chen, H., Xie, F., Chen, L., and Zheng, B. (2019b). Effect of rheological properties of potato, rice and corn starches on their hot-extrusion 3D printing behaviors. *Journal of Food Engineering* 244: 150–158.

Chen, J., Mu, T., Goffin, D. et al. (2019a). Application of soy protein isolate and hydrocolloids based mixtures as promising food material in 3D food printing. *Journal of Food Engineering* 261: 76–86.

Chen, X., Li, J., Cheng, X. et al. (2017). Microstructure and mechanical properties of the austenitic stainless steel 316L fabricated by gas metal arc additive manufacturing. *Materials Science and Engineering A* 703: 567–577.

Chia, H.N. and Wu, B.M. (2015). Recent advances in 3D printing of biomaterials. *Journal of Biological Engineering* 9: 4. https://doi.org/10.1186/s13036-015-0001-4.

Choi, J., Kwon, O.-C., Jo, W. et al. (2015). 4D printing technology: a review. *3D Printing and Additive Manufacturing* 2: 159–167.

Cohen, D.L., Lipton, J.I., Cutler, M. et al. (2009). Hydrocolloid printing: a novel platform for customized food production. Paper presented at the Proceedings of Solid Freeform Fabrication Symposium (SFF'09). http://utw10945.utweb.utexas.edu/Manuscripts/2009/2009-71-Cohen.pdf (accessed 19 April 2022).

Derossi, A., Caporizzi, R., Azzollini, D., and Severini, C. (2018). Application of 3D printing for customized food. A case on the development of a fruit-based snack for children. *Journal of Food Engineering* 220: 65–75.

Dick, A., Bhandari, B., and Prakash, S. (2019). 3D printing of meat. *Meat Science* 153: 35–44.

Dizon, J.R.C., Espera, A.H., Chen, Q., and Advincula, R.C. (2018). Mechanical characterization of 3D-printed polymers. *Additive Manufacturing* 20: 44–67.

Dong, X., Huang, Y., Pan, Y. et al. (2019). Investigation of sweet potato starch as structural enhancer for 3D printing of *Scomberomorus niphonius* surimi. *Journal of Texture Studies* 50: 316–324.

Farag, M.M. and Yun, H.-S. (2014). Effect of gelatin addition on fabrication of magnesium phosphate-based scaffolds prepared by additive manufacturing system. *Materials Letters* 132: 111–115.

Fernandez-Vicente, M., Calle, W., Ferrandiz, S., and Conejero, A. (2016). Effect of infill parameters on tensile mechanical behavior in desktop 3D printing. *3D Printing and Additive Manufacturing* 3: 183–192.

Ford, S., Mortara, L., and Minshall, T. (2016). The emergence of additive manufacturing: introduction to the special issue. *Technological Forecasting and Social Change* 102: 156–159.

Fukuchi, K., Jo, K., Tomiyama, A., and Takao, S. (2012). Laser cooking. Proceedings of the ACM Multimedia 2012 Workshop on Multimedia for Cooking and Eating Activities – CEA '12, p. 55. Nara, Japan, 2 November 2012. New York: ACM Press.

Gburek, U., Vorndran, E., Muller, F.A., and Barralet, J.E. (2007). Low temperature direct 3D printed bio-ceramics and bio-composites as drug release matrices. *Journal of Controlled Release* 122: 173–180.

Ghazal, A.F.G., Zhang, M., and Liu, Z. (2019). Spontaneous color change of 3D printed healthy food product over time after printing as a novel application for 4D food printing. *Food and Bioprocess Technology* 12: 1627–1645.

Godoi, F.C., Bhesh, S.P., and Bhandari, R. (2016). 3D printing technologies applied for food design: status and prospects. *Journal of Food Engineering* 179: 44–54.

Goole, J. and Amighi, K. (2016). 3D printing in pharmaceutics: a new tool for designing customized drug delivery systems. *International Journal of Pharmaceutics* 499: 376–394.

Gracia, A. and Sepulveda, E. (2015). Apparatus and method for heating and cooking food using laser beams and electromagnetic radiation. US Patent 20170245682A1.

Gracia-Julia, A., Hurtado-Pnol, S., Leung, A., and Capellas, M. (2015). Extrusion behavior of food materials in a 3D food printer. Pectin based bio-ink formulations for 3-D printing of porous foods. *Proceedings of the 29th EFFoSt International*, Athens, Greece, 10–12 November 2015, vol. 2, pp. 1740–1741.

Guillotin, B., Souquet, A., Catros, S. et al. (2010). Laser assisted bioprinting of engineered tissue with high cell density and microscale organization. *Biomaterials* 31: 7250–7256.

Hao, L., Mellor, S., Seaman, O. et al. (2010). Material characterization and process development for chocolate additive layer manufacturing. *Virtual and Physical Prototyping* 5: 57–64.

Hertafeld, E., Zhang, C., Jin, Z. et al. (2018). Multi-material three-dimensional food printing with simultaneous infrared cooking. *3D Printing and Additive Manufacturing* 6: https://www.liebertpub.com/doi/10.1089/3dp.2018.0042.

Holland, S., Foster, T., MacNaughtan, W., and Tuck, C. (2018). Design and characterization of food grade powders and inks for microstructure control using 3D printing. *Journal of Food Engineering* 220: 12–19.

Huang, C.Y. (2018). *Extrusion-Based 3D Printing and Characterization of Edible Materials*. UWSpace http://hdl.handle.net/10012/12899 (accessed 24 April 2022).

Icten, E., Purohit, H.S., Wallace, C. et al. (2017). Dropwise additive manufacturing of pharmaceutical products for amorphous and self-emulsifying drug delivery systems. *International Journal of Pharmaceutics* 524: 424–432.

Inzana, J.A., Olvera, D., Fuller, S.M. et al. (2014). 3D printing of composite calcium phosphate and collagen scaffolds for bone regeneration. *Biomaterials* 35: 4026–4034.

Italian Food Regulation (2001). Decreto del presidente della Repubblica (n. 187) (9 February 2001). https://www.politicheagricole.it/flex/files/2/f/6/D.34ca305e98ded6c87bfc/DPR_187_2001 (accessed 24 April 2022).

Jiang, H., Zheng, L., Zou, Y. et al. (2019). 3D food printing: main components selection by considering rheological properties. *Critical Reviews in Food Science and Nutrition* 59: 2335–2347.

Kim, H.W., Bae, H., and Park, H.J. (2018). Reprint of: Classification of the printability of selected food for 3D printing: development of an assessment method using hydrocolloids as reference material. *Journal of Food Engineering* 220: 28–37.

Klomp, D., Hoppenbrouwers, M., de Kruif, B. et al. (2015). Process control in 3D food printing. *Proceedings of the 29th EFFoSt International*, Athens, Greece, 10–12 November 2015, vol. 1, p. 427.

Kodama, M., Takita, Y., Tamate, H. et al. (2017). Novel soft meals developed by 3D printing. In: *Future Foods* (ed. H. Mikkola) chapter 9. IntechOpen http://dx.doi.org/10.5772/intechopen.70652.

Kragh, A.M. (1961). Viscosity. In: *Determination of the Size and Shape of Protein Molecules* (ed. P.Alexander and R.J. Block), 173–209. Oxford, UK: Pergamon Press.

Krujatz, F., Lode, A., Seidel, J. et al. (2017). Additive biotech-chances, challenges, and recent applications of additive manufacturing technologies in biotechnology. *New Biotech* 39 (Pt B): 222–231.

Lanaro, M., Forrestal, D.P., Scheurer, S. et al. (2017). 3D printing complex chocolate objects: platform design, optimization and evaluation. *Journal of Food Engineering* 215: 13–22.

Le Tohic, C., O'Sullivan, J.J., Drapala, K.P. et al. (2018). Effect of 3D printing on the structure and textural properties of processed cheese. *Journal of Food Engineering* 220: 56–64.

Lille, M., Nurmela, A., Nordlund, E. et al. (2018). Applicability of protein and fiber-rich food materials in extrusion-based 3D printing. *Journal of Food Engineering* 220: 20–27.

Lipton, J.I. (2017). Printable food: the technology and its application in human health. *Current Opinion in Biotechnology* 44: 198–201.

Lipton, J., Arnold, D., Nigl, F. et al. (2010). Multi-material food printing with complex internal structure suitable for conventional post-processing. *2010 Solid Freeform Fabrication Symposium*, pp. 809–815. https://repositories.lib.utexas.edu/bitstream/handle/2152/88304/2010-68-Lipton.pdf?sequence=2&isAllowed=y (accessed 15 April 2022).

Lipton, J.I., Cutler, M., Nigl, F. et al. (2015). Additive manufacturing for the food industry. *Trends in Food Science and Technology* 43: 114–123.

Liu, Z., Zhang, M., Bhandari, B., and Wang, Y. (2017). 3D printing: printing precision and application in food sector. *Trends in Food Science and Technology* 69: 83–94.

Liu, Z., Zhang, M., Bhandari, B., and Yang, C. (2018). Impact of rheological properties of mashed potatoes on 3D printing. *Journal of Food Engineering* 220: 76–82.

Lupton, D. and Turner, B. (2018). "I can't get past the fact that it is printed": consumer attitudes to 3D printed food. *Food, Culture and Society* 21: 402–418.

Mantihal, S., Prakash, S., and Bhandari, B. (2019). Texture-modified 3D printed dark chocolate: sensory evaluation and consumer perception study. *Journal of Texture Studies* 50: 386–399.

Mantihal, S., Prakash, S., Godoi, F.C., and Bhandari, B. (2017). Optimization of chocolate 3D printing by correlating thermal and flow properties with 3D structure modeling. *Innovative Food Science and Emerging Technologies* 44: 21–29.

Martinez-Monzo, J., Cardenas, J., and Garcia-Segovia, P. (2019). Effect of temperature on 3D printing of commercial potato puree. *Food Biophysics* 14: 225–234.

McCluskey, J.J., Kalaitzandonakes, N., and Swinnen, J. (2016). Media coverage, public perceptions, and consumer behavior: insights from new food technologies. *Annual Review of Resource Economics* 8: 467–486.

Melchels, F.P.W., Domingos, M.A.N., Klein, T.J. et al. (2012). Additive manufacturing of tissues and organs. *Progress in Polymer Science* 37: 1079–1104.

Momeni, F., Hassani, N.S.M.M., Liu, X., and Ni, J. (2017). A review of 4D printing. *Materials & Design* 122 (Supplement C): 42–79.

Muchnik, B. (2008). Laser cooking apparatus. US Patent 20080282901A1.

Murphy, S.V. and Atala, A. (2014). 3D bio-printing of tissues and organs. *Nature Biotechnology* 32: 773–785.

Netherlands Organisation for Applied Scientific Research (2015). 3D food printing. https://www.tno.nl/media/2217/3d_food_printing (accessed 30 September 2015). Downloaded 30 September 2015.

Noorani, R. (2018). *3D Printing: Technology, Applications and Selection*. Boca Raton, FL: CRC Press.

Noort, M., Van Bommel, K., and Renzetti, S. (2017). 3D-printed cereal foods. *Cereal Foods World* 62: 272–277.

Pallottino, F., Hakola, L., Costa, C. et al. (2016). Printing on food or food printing: a review. *Food and Bioprocess Technology* 9: 725–733.

Pang, Z., Deeth, H., Sopade, P. et al. (2014). Rheology, texture and microstructure of gelatin gels with and without milk proteins. *Food Hydrocolloids* 35: 484–493.

Pei, E. (2014). 4D printing – revolution or fad? *Assembly Automation* 34: 123–127.

Pei, E., Loh, G.H., Harrison, D. et al. (2017). A study of 4D printing and functionally graded additive manufacturing. *Assembly Automation* 37: 147–153.

Pinna, C., Ramundo, L., Sisco, F.G. et al. (2017). Additive manufacturing applications within food industry: an actual overview and future opportunities. *XXI Summer School "Francesco Turco" – Industrial Systems Engineering*, pp. 18–24. https://re.public.polimi.it/retrieve/handle/11311/1014860/184185/final_35.pdf (accessed 25 April 2022).

Portanguen, S., Tournayre, P., Sicard, J. et al. (2019). Toward the design of functional foods and biobased products by 3D printing: a review. *Trends in Food Science and Technology* 86: 188–198.

Prakash, S., Bhandari, B.R., Godoi, F.C., and Zhang, M. (2019). Future outlook of 3D food printing. In: *Fundamentals of 3D Food Printing and Applications* (ed. F.C. Godoi, B.R. Bhandari, S. Prakash and M. Zhang), chapter 13, 373–381. Oxford, UK: Academic Press.

Rapisarda, M., Valenti, G., Carbone, D.C. et al. (2018). Strength, fracture and compression properties of gelatins by a new 3D printed tool. *Journal of Food Engineering* 220: 38–48.

RepRap (2016). RepRapWiki: G-code. http://reprap.org/wiki/G-code (accessed 27 April 2022).

Ross, M.M., Kelly, A.L., and Crowley, S.V. (2019). Potential applications of dairy products, ingredients and formulations in 3D printing. In: *Fundamentals of 3D Food Printing and Applications*, 175–206. Cambridge, MA: Academic Press.

Sachs, E., Cima, M., Williams, P. et al. (1992). Three dimensional printing: rapid tooling and prototypes directly from a CAD model. *Journal of Manufacturing Science and Engineering* 114: 481–488.

Severini, C., Derossi, A., and Azzollini, D. (2016). Variables affecting the printability of foods: preliminary tests on cereal-based products. *Innovative Food Science and Emerging Technologies* 38: 281–291.

Severini, C., Derossi, A., Ricci, I. et al. (2018). Printing a blend of fruit and vegetables. New advances on critical variables and shelf life of 3D edible objects. *Journal of Food Engineering* 220: 89–100.

Singh, I. (2013). Method and apparatus for plasma assisted laser cooking of food products. US Patent 9107434B2.

Singh, M., Haverinen, H.M., Dhagat, P., and Jabbour, G.E. (2010). Inkjet printing-process and its applications. *Advanced Materials* 22: 673–685.

Singh, S., Ramakrishna, S., and Singh, R. (2017). Material issues in additive manufacturing: a review. *Journal of Manufacturing Processes* 25: 185–200.

Sobral, J., Caridade, S.G., Sousa, R.A. et al. (2011). Three-dimensional plotted scaffolds with controlled pore size gradients: effect of scaffold geometry on mechanical performance and cell seeding efficiency. *Acta Biomaterialia* 7: 1009–1018.

Sun, J., Peng, Z., Zhou, W. et al. (2015). A review on 3D printing for customized food fabrication. *Procedia Manufacturing* 1: 308–319.

Sun, J., Zhou, W., Yan, L. et al. (2018). Extrusion-based food printing for digitalized food design and nutrition control. *Journal of Food Engineering* 220: 1–11.

Takezawa, A. and Kobashi, M. (2017). Design methodology for porous composites with tunable thermal expansion produced by multi-material topology optimization and additive manufacturing. *Composites Part B: Engineering* 131: 21–29.

Teng, X., Zhang, M., and Mujumdar, A.S. (2021). 4D printing: recent advances and proposals in the food sector. *Trends in Food Science & Technology* 110: 349–363.

Tibbits, S. (2014). 4D printing: multi-material shape change. *Architectural Design* 84: 116–121.

Utela, B., Storti, D., Anderson, R., and Ganter, M. (2008). A review of process development steps for new material systems in three dimensional printing (3DP). *Journal of Manufacturing Processes* 10: 96–104.

Vallons, K.J.R., Diaz, J., van Bommel, K., and Noort, M. (2015). The role of gelation dynamics in food printing. *Proceedings of the 29th EFFoSt International*, Athens, Greece, 10–12 November 2015, vol. 1, p. 426.

Vancauwenberghe, V., Delele, M.A., Vanbiervliet, J. et al. (2018). Model-based design and validation of food texture of 3D printed pectin-based food simulants. *Journal of Food Engineering* 231: 72–82.

Vancauwenberghe, V., Katalagarianakis, L., Wang, Z. et al. (2017). Pectin based food-ink formulations for 3-D printing of customizable porous food simulants. *Innovative Food Science and Emerging Technologies* 42: 138–150.

Vancauwenberghe, V., Mbong, V.B.M., Kokalj, T. et al. (2015). Pectin based bio-ink formulations for 3-D printing of porous foods. *Proceedings of the 29th EFFoSt International*, Athens, Greece, 10–12 November 2015, vol. 1, pp. 10–12.

Wang, L., Zhang, M., Bhandari, B., and Yang, C. (2018). Investigation on fish surimi gel as promising food material for 3D printing. *Journal of Food Engineering* 220: 101–108.

Warner, E.L., Norton, I.T., and Mills, T.B. (2019). Comparing the viscoelastic properties of gelatin and different concentrations of kappa-carrageenan mixtures for additive manufacturing applications. *Journal of Food Engineering* 246: 58–66.

Wegrzyn, T.F., Golding, M., and Archer, R.H. (2012). Food layer manufacture: a new process for constructing solid foods. *Trends in Food Science and Technology* 27: 66–72.

Xun Khoo, Z., Ee Mei, T.J., Liu, Y. et al. (2015). 3D printing of smartmaterials: a review on recent progresses in 4D printing. *Virtual and Physical Prototyping* 10: 103–122.

Yang, F., Zhang, M., and Bhandari, B. (2015). Recent development in 3D food printing. *Critical Reviews in Food Science and Nutrition* 57: 3145–3153.

Yang, F., Zhang, M., Bhandari, B., and Liu, Y. (2018). Investigation on lemon juice gel as food material for 3D printing and optimization of printing parameters. *LWT – Food Science and Technology* 87: 67–76.

Zeleny, P. and Ruzicka, V. (2017). The design of the 3d printer for use in gastronomy. *Modern Machinery (MM) Science Journal* 1744–1747. https://www.mmscience.eu/journal/issues/february-2017/articles/the-design-of-the-3d-printer-for-use-in-gastronomy (accessed 20 April 2022).

Zoran, A. and Coelho, M. (2011). Cornucopia: the concept of digital gastronomy. *MIT Press* 44 (5): 425–431. http://muse.jhu.edu/journals/len/summary/v044/44.5.zoran.html (accessed 24 April 2022).

9

Use of Hydrocolloids to Control Food Nutrition

Hydrocolloids have numerous and growing applications in the health realm; e.g. they provide low-calorie dietary fiber, among many other uses. Research into food hydrocolloids has recently been launched, for example, dietary fiber with physiological effects for an elderly society, where the number of people with mastication and swallowing difficulties, and of patients with lifestyle-related diseases is on the rise. Hydrocolloids that have conventionally been used only to provide texture are now being used for their nutritional value; and processing of food fibers to enable their use as a food ingredient with nutritional value and a textural role is being pursued. Furthermore, novel functional food fibers with textural and nutritional benefits are emerging. Hydrocolloids themselves can be exploited for their inherent nutritional value, as they can contain a large proportion of dietary fiber. For example commercially available partially hydrolyzed guar gum (PHGG) is comprised of a high proportion of dietary fiber, and its low viscosity enables its use at sufficiently high concentrations to obtain the benefits of this dietary fiber. The use of functional fibers in baked goods and meat applications, where water binding is key, but also as constituents of beverages will be discussed.

9.1 Nutritional Applications of Natural Hydrocolloids

Hydrocolloids from natural sources have been used in the food industry to a large extent to enhance useful properties, such as quality, safety, stability, and nutritional and health benefits of various food products – baked goods, beverages, confectionary, dressings, sauces, and meat and poultry products (Yemenicioglu et al. 2019). Market pronouncements and estimates predict that the worldwide food hydrocolloid market will increase in value by about 50% in the coming decade. Consequently, extensive effort is needed to increase the production efficiency of the latest hydrocolloid sources, and to assess agroindustrial waste as a new source of hydrocolloids. Traditional applications of hydrocolloids in the food industry rely on their practical properties, mostly their rheological and surface-active ones (Nishinari et al. 2018; Yousefi and Jafari 2019).

Due to their exceptional binding (pasting), coagulation, and emulsification properties, along with outstanding nutritional value, proteins from whole eggs and their

Use of Hydrocolloids to Control Food Appearance, Flavor, Texture, and Nutrition, First Edition.
Amos Nussinovitch and Madoka Hirashima.
© 2023 John Wiley & Sons Ltd. Published 2023 by John Wiley & Sons Ltd.

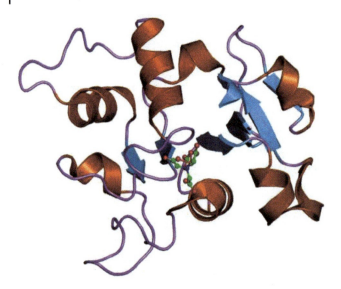

Figure 9.1 Ovotransferrin. *Source:* Jawahar Swaminathan and MSD/EMBL-EBI.

components – egg yolks and whites (albumen) – are widely used by the food industry in dried, fresh, pasteurized, and frozen forms. Diverse proteins or yolk fractions have excellent potential as gelling and emulsifying agents (Laca et al. 2010; Valverde et al. 2016). Purified egg white proteins, such as ovalbumin (excellent source of amino acids), ovotransferrin (Figure 9.1), phosvitin (antioxidant and antimicrobial), and lysozyme (antimicrobial), have additional valuable potential as food additives (Huopalahti et al. 2007). Whey proteins have select applications in nutritional beverages, bars and yogurts. Their major benefits are nutritive value, and their direct physiological effects when consumed in conjunction with physical fitness activities (Kelly 2018). Milk protein concentrates have applications in soap, sauces, yogurt, beverages, pediatric and geriatric foods, and for medical and sports nutrition products. Their major benefits are their nutritive value and their being a source of bioavailable calcium (Patel and Patel 2014).

9.2 Types of Dietary Fibers

The human diet may include numerous fibers as components of foods and/or as additives to improve the nutritional value of foods (Korczak et al. 2017). Acacia gum is a soluble fiber that offers a prebiotic advantage, producing a larger increase in bifidobacteria and lactobacilli compared to the same amount of inulin, with fewer gastrointestinal side effects (Slavin 2013). Arabinoxylan is a major constituent of dietary fiber in whole grains. In wheat, arabinoxylan accounts for approximately 70% of the non-starch polysaccharides in the bran and nearly 90% in the endosperm (Ring and Selvendran 1980). In the gastrointestinal tract, arabinoxylan behaves similarly to soluble fiber, being speedily fermented by the colonic microflora (Lattimer and Haub 2010). The outermost layer of a cereal grain is denoted as "bran." It is composed of the epidermis, seed coat, pericarp, and aleurone layer. Bran is found in a wide range of cereal grains. Oat bran, for example has a β-glucan content of at least 5.5% on a dry-weight basis and total dietary fiber inclusion of at least 16% on a dry-weight basis. About one-third of this total dietary fiber is soluble fiber (Lattimer and Haub 2010). Fructo-oligosaccharides (FOSs) are found in numerous plants, such as artichoke, banana, chicory, garlic, and onion. They are composed of linear chains of fructose units linked by beta bonds (Sabater-Molina et al. 2009). Dietary FOSs are not hydrolyzed by enzymes in the small intestine and thus reach the cecum unaltered. There, FOSs are metabolized by the intestinal microbiota to yield short-chain fatty acids (SCFAs). FOSs are regarded as soluble dietary fiber and can have a prebiotic effect by stimulating the growth of beneficial bacteria (Sabater-Molina et al. 2009). Glucomannan is a soluble fiber from the konjac plant (Horvath et al. 2013). It may have valuable effects on the microbiota in patients with functional gastrointestinal disorders (Horvath et al. 2013). Inulin is a heterogeneous group of fructose polymers linked by beta bonds and terminating with a glucose unit. Inulin is utilized as a functional food ingredient because of its nutritional properties. In the large intestine, it has a prebiotic effect by encouraging the growth of bifidobacterial species, while limiting the growth of pathogenic bacteria such as *Escherichia coli*, *Salmonella*, and *Listeria* (Gibson et al. 1995). To some extent, hydrolyzed guar gum is a water-soluble fiber (WSF) derived from guar gum (Quartarone 2013; Romano et al. 2013). PHGG supplementation reduces symptoms of irritable bowel syndrome in adults and constipation in children (Feldman et al. 1985; Parisi et al. 2005).

Resistant starch is a naturally occurring fiber. However, it can also be produced by the alteration of starch during food processing (Institute of Medicine 2005). Resistant starch behaves like soluble fiber and has been categorized into four basic types (Lattimer and Haub 2010): type 1 starch granules are enclosed in a non-digestible plant matrix; type 2 is its natural form, for example in an uncooked potato; type 3 consists of crystallized starches resulting from distinct cooking and cooling processes; type 4 is chemically modified by esterification or crosslinking and is not found in nature (Korczak et al. 2017).

9.3 Dietary Fiber as a Versatile Food Component

In all Western countries, the consumption of fiber and fiber-containing foods is much lower than optimal levels (Redgwell and Fischer 2005). Although further research is needed to specify the nutritional advantages of dietary fiber, numerous studies have shown that its consumption is inversely related to the risk of a number of chronic diseases (Malkki 2004). Due to its nutritional aspect, dietary fiber can be classified as a functional food, or nutraceutical. Functional foods are those foods that have an explicit health advantage (Redgwell and Fischer 2005). However, it is vital to point out that emphasizing the nutritional benefits of fiber products does not result in their sole use in niche markets (Redgwell and Fischer 2005). A challenge for the industry is to provide dietary fiber for regular consumption in a growing variety of products, without compromising their sensory appeal (Fischer 2004). In this respect, hydrocolloids have revolutionized food fabrication. On the one hand, their capacity to achieve water control allows them to control texture in numerous products, but they are supplemented in insufficient quantities in individual products to enable the claim of dietary fiber content (Redgwell and Fischer 2005). On the other hand, with the rising use of soluble hydrocolloids in prepared foods and growing consumption of such foods, the cumulative content of hydrocolloids in quite a lot of products could make a notable contribution to daily dietary fiber intake. Therefore, the physiological functionality and potential health benefits of hydrocolloids as dietary fibers need more vigorous validation. This will enable the food industry to transmit a much stronger message to the consumer on the beneficial characteristics of these technically important food components (Redgwell and Fischer 2005). Many hydrocolloids are composed of a single type of polysaccharide and while some are insoluble (e.g. microcrystalline cellulose), most of the commonly used ones are water-soluble. They are not equivalent to nor do they have the consumer appeal of natural sources of dietary fiber found in the complex mixture of polysaccharides located in the cell walls of fruit, vegetables, and whole-grain cereals (Redgwell and Fischer 2005). From an industrial perspective, sources of dietary fibers can be separated into three broad classes: (i) hydrocolloids, frequently soluble polysaccharides; (ii) bioactive oligosaccharides; (iii) whole plant cell wall materials derived from cereal grains, fruit, and vegetables. All have cooperative physiological and technological functionalities to varying degrees, which dictate their particular uses (Redgwell and Fischer 2005).

9.4 Food Enriched in β-Glucans

An appropriate diet and healthy lifestyle are related to a lower risk of pathologies, such as type 2 diabetes and obesity, in both children and adolescents. An overweight condition, if not appropriately treated, will persist into adulthood, resulting in further pathologies, such as cardiovascular diseases (Mikkilä et al. 2005). Possible health benefits are extensively linked to the consumption of an appropriate amount of fiber (Buttriss and Stokes 2008). However, outcomes of previous studies have emphasized that total dietary fiber consumption is moderately low among European adolescents compared to the World Health Organization's recommendations. Furthermore, it has been reported that the noteworthy decrease in fiber ingestion in developed countries is associated with a perturbing surge in teenage obesity (Lin et al. 2015). Among dietary fiber compounds, WSF is thought to have numerous positive health effects. WSF can slow bowel absorption, which in turn will decrease cholesterol absorption due to the development of viscous solutions in the gastrointestinal tract (Tan and Seow-Choen 2007). Furthermore, WSF fermentation can yield SCFAs, leading to longer lasting satiety, a diminished acute insulin response and a lower glycemic index for the consumed food (Tan and Seow-Choen 2007; Mathern et al. 2009). Among the WSFs, mushroom β-glucans have attracted interest for their specific health assets (Theuwissen and Mensink 2008; Dalonso et al. 2015). The mushroom *Pleurotus ostreatus* (Figure 9.2) has a high total glucan content: in the range of 14–25% dry weight, with a percentage of β-glucans between 73 and 91.4% (Koutrotsios et al. 2017). *Pleurotus ostreatus* strains with β-glucans up to 32.2% dry weight are also known (Draga and McCleary 2016). A novel flat bread was modified by replacing part of its wheat flour with *P. ostreatus* powder that is rich in β-glucans, which could potentially provide health benefits (Proserpio et al. 2019). Replacing some of the wheat flour in the flat bread was also done to enhance its nutritional profile. Indeed, aside from its high β-glucan content, *P. ostreatus* powder is rich

Figure 9.2 *Pleurotus ostreatus*, the oyster mushroom. *Source:* Roger Griffith/Wikipedia Commons/ Public Domain.

in numerous other nutrients and micronutrients (e.g. high-quality protein and vitamin D), in addition to being poor in fats and salt (Lavelli et al. 2018). The proposed β-glucan-enriched flat bread satisfied both the nutritional and sensory aspects among adolescents. A conclusion of the study was that it is conceivable to develop a β-glucan-enriched product that appeals to adolescents. The developed product might be used to satisfy the daily recommended intake of β-glucans by adolescents (Proserpio et al. 2019).

9.5 Cereal Polysaccharides as the Foundation for Useful Ingredients in the Reformulation of Meat Products

Consumer awareness of meat products is changing with the search for health-related dietary characteristics. Due to meat's fat and cholesterol contents, reformulation of meat products is desired, possibly by combining bioactive constituents or through elimination/ lessening of the detrimental components to modify the perception of health-conscious meat lovers. Consequently, various bioactive components such as antioxidants, dietary fiber, phytochemicals, and vegetable proteins can be used to amend the features of meat products (Kaur and Sharma 2019). As a dietary fiber source, cereal polysaccharides are recognized as health amendments that can prevent emerging diseases. Their antioxidant, antitumor, anti-inflammatory, antimicrobial, and antidiabetic properties have been proven by *in-vitro* and *in-vivo* clinical studies. Therefore, the inclusion of cereal polysaccharides in meat products and their influence on the functional characteristics of the developed novel meat products is of much interest to both producers and consumers (Kaur and Sharma 2019). Bioactive polysaccharides in the cell walls of cereals, in particular arabinoxylan and β-glucan, have been shown to have a helpful physiological effect as a good source of soluble dietary fiber, and a valuable effect on keeping blood glucose, insulin, and cholesterol levels remarkably low (Cui and Wang 2009). β-Glucans from yeast and mushroom sources are beneficial from a nutritional standpoint, and as a natural alternative for the reformulation of meat products (Apostu et al. 2017). Incorporation of barley β-glucan along with other ingredients in Tunisian turkey meat sausage provided a healthier product with lower fat content as well as price. In addition, the replacement of the meat ingredient with barley fiber enhanced the favorable perception of the product (Ktari et al. 2016). Incorporation of 0.15 and 0.30% feruloylated arabinoxylan in frankfurter meat sausages increased phenolic and antioxidant power compared to control samples. Furthermore, physicochemical properties, such as water-holding capacity, pH and titratable acidity, hardness, and diameter, were similarly improved in the sausages sample. This supplementation led to reduced-fat sausages with boosted nutritional and nutraceutical properties (Herrera-Balandrano et al. 2019). In addition, the use of oat emulsion gels and oil-free gels containing β-glucans conferred 30% less calories compared to pork back fat, which is a basic component of meat products. These gels can be included in meat products to improve their nutritional value and at the same time preserve the value and well-being factors of such items for consumption (Pintado et al. 2016).

The words "gelling" and "viscous" are frequently used more or less interchangeably in the nutrition literature to describe soluble fibers. However, the properties of a gel are fairly different from those of a viscous solution, since a gel stretches elastically or breaks under a force of deformation. Even though high-molecular-weight cereal β-glucan for the most part displays viscous flow behavior, gelation may be observed principally in solutions of lower-molecular-weight materials (Doublier and Wood 1995; Bohm and Kulicke 1999). The flow viscosity behavior of oat and barley β-glucans with the same molecular weight does not differ, but their gelation features do; at the same time, the higher the proportion of (1–3)-linked cellotriosyl units in the structure, the more rapid the gelation. Thus, barley β-glucan, at the same concentration and of similar molecular weight as oat β-glucan, will gel more freely. The part of gelation, or gel properties, in governing physiological effects is unknown, even though the second polysaccharide to be allowed a health claim by the FDA, psyllium, is for the most part a gelling polysaccharide mixture (Haque et al. 1993).

It can be concluded that reformulation of meat products through inclusion of cereal polysaccharides should take into consideration the nutritional, organoleptic, safety, and technological profile of the product (Jiménez Colmenero 2000; Grasso et al. 2014). Fat replacers should be used in meat products, which can enhance the texture of reduced-fat meat products, because just decreasing the fat content of such products can result in increased toughness and an associated influence on consumer acceptance of the product (Kaur and Sharma 2019).

Research has confirmed the potential of manufacturing healthier and more attractive meat products for consumers. However, to increase the commercialization of such fortified products, the technology must advance to adapt information about the requirements and adequacy of these polysaccharides as added meat products (Kaur and Sharma 2019). In addition, the meat industry will require exceptional quality raw materials as well as engagement in innovative marketing strategies. The food ingredients industry also needs to collaborate in this endeavor by developing and producing novel healthy ingredients suitable for incorporation in the meat formulations (Kaur and Sharma 2019).

9.6 Health Claims of Hydrocolloids

Food hydrocolloids are obtained from algae, bacteria, fruit, and plant extracts (Phillips and Williams 2009), at contents of between 60 and 90%. These ingredients are well-characterized soluble dietary fibers, although that feature of their functionality is not the main reason for their use in food and beverage preparations (Viebke et al. 2014). The assortment of health benefits that have been associated with consumption of food hydrocolloids are numerous (Brownlee 2011; Kaczmarczyk et al. 2012; Gidley 2013; Mayakrishnan et al. 2013; Xu et al. 2013), including reduction of risk factors in cardiovascular disease, and for immune function, weight management, and colonic health (Slavin 2003; Dettmar et al. 2011). Hydrocolloids offer the opportunity to enhance nutritional value and deliver possible health benefits through control of gastric emptying and ileal brake mechanisms (leading to satiety and possibly preventing obesity), glycemic response (diabetes prevention), plasma cholesterol levels (cardiovascular disease prevention), and carbohydrate fermentation throughout the large intestine (colon cancer prevention). Parallels can be drawn between the functionality of the plant-based foods that the human digestive tract has evolved to digest and the use of extracted hydrocolloids in modern food-structuring technology (Gidley 2013). The constructive properties are a result of the solubility, viscosity (gelling ability), and fermentation ability of the specific hydrocolloid. These features are unique to each hydrocolloid and might be modified within that particular hydrocolloid dependent upon the industrial preparation method or source of the raw material (Phillips and Williams 2009). A number of studies have looked at the effect of consumption of various food hydrocolloids on blood cholesterol levels (Anderson and Chen 1986; Eastwood et al. 1986; Kahlon and Chow 1997; Wolever et al. 2010). Other studies have looked at their effects on the glycemic response (Flammang et al. 2006; Viebke et al. 2014). Moreover, feeding of certain hydrocolloids modifies the composition of the intestinal bacterial flora, stimulating good bacteria – for instance *Bifidobacterium* and *Lactobacillus* (Gibson and Roberfroid 1995; Kolida et al. 2002; Calame et al. 2008). This might lead to an increase in SCFA production (Topping and Clifton 2001) and a decrease in potentially harmful bacteria such as Clostridia. Food hydrocolloids have also been studied in relation to weight management, where consumption of these ingredients might help control energy intake by increasing the feeling of satiety following a meal (Slavin and Green 2007; Willis et al. 2009; Vitaglione et al. 2010; Calame et al. 2011; Guérin-Deremaux et al. 2011; Hess et al. 2011; Monsivais et al. 2011).

In addition to the numerous fields in which food hydrocolloids are used for their valuable physiological effects, other areas have been studied (Viebke et al. 2014), including inhibition of human enamel erosion (Beyer et al. 2010), increase in fecal nitrogen excretion (Bliss et al. 1996), prevention of constipation (Amadio et al. 2009) or diarrhea (Alam et al. 2005), mastication and swallowing difficulties (Funami 2011), absorption of micronutrients (Coudray et al. 2005), reduction of systolic blood pressure (Glover et al. 2009; Phillips and Phillips 2011), and possible prevention of cancer (Pelucchi et al. 2003). Not every specific hydrocolloid will have all of the listed useful physiological effects; nevertheless, as a group of food constituents, they have various beneficial effects (Viebke et al. 2014).

The typical soluble fiber content of some hydrocolloids is as follows: agar 85%, carrageenan 80–90%, gum arabic 85%, guar gum 80–85%, locust bean gum 80–95%, pectin – both for low- and high-methoxyl – 80–95%, starches 60%, inulin 90%, and xanthan gum 8–95%

(Viebke et al. 2014). The hydrocolloids gum acacia and inulin are classified as low-viscosity soluble dietary fibers, which might make them more appropriate for use in beverages and food formulations at higher inclusion levels (Viebke et al. 2014). Submissions of health claims for food hydrocolloids to the European Food Safety Agency (EFSA) (Viebke et al. 2014) are classified into: (i) claims of a recognized beneficial physiological effect according to the EFSA, such as maintenance or achievement of normal blood glucose concentrations, maintenance of normal blood cholesterol concentrations, maintenance or achievement of a normal body weight, reduction of gastrointestinal discomfort; (ii) claims that MAY have a beneficial physiological effect according to the EFSA – changes in bowel function, maintenance of normal bowel function, reduction of postprandial glycemic responses, increase in satiety leading to a reduction in energy intake, increase in calcium and/or magnesium absorption leading to an increase in magnesium and/or calcium retention, maintenance of normal (fasting) blood concentrations of triglycerides; and (iii) claims with NO beneficial physiological effect according to the EFSA – decreasing potentially pathogenic gastrointestinal microorganisms (prebiotic effect), changes in SCFA production and pH in the gastrointestinal tract, maintenance of fecal nitrogen content and/or normal blood urea concentrations (Viebke et al. 2014). One or more such claims were submitted for β-glucan, konjac mannan, alginate, FOS, galacto-oligosaccharide, gum acacia, hydroxypropyl methylcellulose (HPMC), partially hydrolyzed guar gum, pectin, sugar beet fiber, xanthan gum and guar gum in 13 categories (Viebke et al. 2014). Only β-glucan, konjac mannan, HPMC, pectin, and guar gum produced at least one positive outcome, i.e. a positive outcome was achieved in 3 of the 13 health categories. Thus, most health claim submissions were dismissed (Viebke et al. 2014).

Summarizing the approved claims relating to food hydrocolloids, for maintenance of normal blood cholesterol concentrations: β-glucan (at least 3 g/day), konjac mannan/glucomannan (at least 4 g/day), HPMC (at least 5 g/day which should be consumed in two or more servings), pectins (at least 6 g/day), and guar gum (foods should provide at least 10 g/day of guar); for maintenance or achievement of a normal body weight: konjac mannan/glucomannan (at least 3 g of glucomannan should be consumed daily in three doses of at least 1 g each, together with one to two glasses of water before meals, in the context of an energy-restricted diet and for overweight adults); for reduction of postprandial glycemic response, β-glucans from oats or barley (4 g for each 30 g of available carbohydrates per meal), HPMC (at least 4 g per meal), pectins (at least 10 g per meal) (Viebke et al. 2014).

9.7 Miscellaneous Cases of Nutritional and Health Benefits

9.7.1 Health Benefits of Lactic Acid Bacteria (LAB) Exopolysaccharides (EPSs)

Exopolysaccharides (EPS) is a widespread metabolite formed by numerous species of lactic acid bacteria (LAB). EPSs are a diverse group of long-chain, high-molecular-mass polysaccharides that differ in their structure and physiochemical characteristics (Ruas-Madiedo et al. 2009). Screening for EPS production characteristically involves the use of a solid growth medium with added growth factors or nutrients to stimulate EPS production, e.g.

monosaccharides such as fructose or maltose (Schwab et al. 2008; Lynch et al. 2018). Adjustment of processing conditions, such as temperature and pH, may also lead to higher EPS levels (Lynch et al. 2018). Cheese prepared with EPS-producing starters had higher moisture content, with a microstructure similar to full-fat cheese (Hassan and Awad 2005). The EPS-producing LAB that have characteristically been studied and employed in dairy products are strains that are related to dairy fermentation and that typically synthesize heteropolysaccharides (HePSs). It has been suggested that EPS binding and water retention are responsible for the more porous protein network and larger pores of the cheese (Awad et al. 2005).

In yogurt, EPSs can affect the formation of the casein–gel structure by acting as a filler and as nuclei for the formation of serum channels and large pores containing bacterial cells, EPSs, and milk serum (Hassan 2008). The polymer-like behavior of the serum phase in yogurt containing EPSs increases the product's consistency and viscosity (Hassan 2008). LAB EPSs can be exploited for their positive effects in cereal-based products, predominantly through the application of sourdough technology (Moroni et al. 2009).

A study with LAB EPSs in the baking industry was aimed at determining their potential to replace commercial hydrocolloids such as HPMC, guar gum, and xanthan gum. These hydrocolloids are used for their positive impact on product rheology, texture, and shelf life; however, as additives, their inclusion in a food product requires labeling. In contrast, the application of LAB EPSs, when formed *in situ* by starter or adjunct cultures, do not need to be declared on the label, and this is considered advantageous in an industry where consumers frequently demand products with less additives (Waldherr and Vogel 2009). The potentially valuable effects of EPSs in dough and bread are enhanced water absorption by the dough, better dough rheology and machinability, enlarged loaf volume, and decreased bread staling rate (Arendt et al. 2007; Waldherr and Vogel 2009). Due to their water-binding capacity and interaction with structure-forming constituents such as gluten and starch, EPSs can certainly affect baked good and bread quality. The production of dextran throughout wheat sourdough fermentation resulted in an increase in the loaf's volume and enhanced crumb softness compared to sourdough without dextran (Di Cagno et al. 2006).

The technofunctional properties of homopolysaccharides make them a significant tool in the expansion of gluten-free products of value with the desired nutritional advantages. Gluten-free breads were prepared with a sorghum sourdough addition. This dough was created by fermentation with either *Weissella cibaria* or *Lactobacillus reuteri* strains. The strains produced EPSs and oligosaccharides throughout fermentation. Breads fermented with *W. cibaria* were less firm than those fermented with *L. reuteri*, and although the FOSs produced by the latter strain were digested by *Saccharomyces cerevisiae*, glucooligosaccharides produced by *W. cibaria* were detected in the bread, thus possibly enhancing the prebiotic content of the gluten-free product (Schwab et al. 2008). Enrichment of gluten-free pasta with chia seed mucilage improved not only the cooking quality of the end product, but also its nutritional profile, i.e. increased insoluble and soluble fiber, polyphenol and protein content, as well as the in-vitro glycemic response (percentage of slowly digestible starch) (Menga et al. 2017). Incorporation of flax or chia seed mucilage in maltodextrin/protein spray-dried powders carrying probiotic bacteria (*Lactobacillus acidophilus, Lactobacillus plantarum,* and *Bifidobacterium infantis*) retained the biological activity of the cells during the course of the dehydration process and reduced the cells' sublethality in chilling storage and in in-vitro digestion environments (Bernstein et al. 2013; Ricklefs-Johnson et al. 2017). Similarly, there is a growing indication that EPSs from LAB might affect the human hosts' health and that dietary inclusion of such polymers may control immune function or levels of beneficial bacteria in the gastrointestinal tract (Lynch et al. 2018). One of the best-known and broadly commercialized prebiotics is inulin and its associated FOSs (Wong et al. 2006; Delzenne et al. 2011). EPSs with immunogenic properties were shown to be HePSs. This topic has been reviewed in detail (Laino et al. 2016). Early studies demonstrated the activation of macrophages in mice and induction of cytokine production by phosphorylated HePSs produced by *Lactococcus lactis* subsp. *cremoris* KVS20 (Kitazawa et al. 1996). HePSs of certain LAB species have also been confirmed to display antioxidant capacity, i.e. the ability to scavenge hydroxyl and superoxide anion radicals in vitro, and to increase in-vivo levels of antioxidant enzymes (catalase and superoxide dismutase) (Pan and Mei 2010; Liu et al. 2011). An oat-based product, fermented with β-glucan produced by *Pediococcus parvulus* 2.6 (equivalent to 3.5 g β-glucan per day), decreased total cholesterol levels in humans when compared to a control group (Martensson et al. 2005). Thus, the future use of LAB EPSs in the food industry has great potential, mainly because food producers and consumers alike are increasing their demand for clean-label products (Lynch et al. 2018).

9.7.2 Fat Replacers

The role of hydrocolloids as fat replacers in food products has already been mentioned in this chapter. The terms fat substitute and fat replacer have been distinguished (Hsu and Chung 1999; Brennan and Tudorica 2008). A fat replacer is a constituent that fulfills all, or nearly all of the purposes of fat and may or may not offer nutritive value. On the other hand, a fat substitute replaces all of the functions of fat with fundamentally no energy contribution, thus serving the aim of cutting fat and calories in the diet. Hydrocolloid-based fat replacers include inulin, pectin, barley β-glucan, guar gum, okra gum, gum tragacanth, xanthan gum, κ-carrageenan (Figure 9.3), sodium alginate, curdlan, and locust bean gum (Hsu and Chung 1999; Romanchik-Cerpovicz et al. 2002; Aziznia et al. 2008; Brennan and Tudorica 2008). In addition to being fat replacers, some hydrocolloids – when forming hydrogels – intensify the total flavor of the fat-reduced emulsion by extending the diffusion and residence times of non-polar flavor molecules in the system (Chung et al. 2013). One of the most common

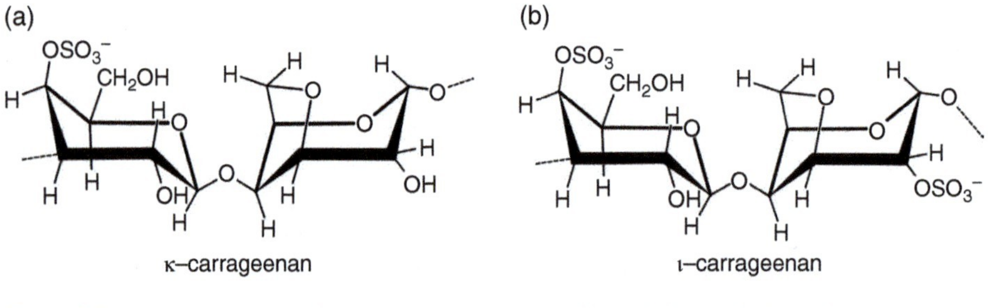

Figure 9.3 (a) κ-carrageenan. (b) ι-carrageenan. *Source:* Roland Mattern/Wikimedia Commons/Public domain.

hydrocolloidal fat-reducing agents is inulin. On top of its beneficial effects on health as a dietary fiber and as a prebiotic ingredient, it serves as a low-calorie sweetener, fat replacer or texture modifier (Bayarri et al. 2010). It was observed that when long-chain inulin was added at a concentration of over 8%, viscosity and creaminess of low-fat milk beverages were similar to the whole-milk beverages (Villegas et al. 2007). In low-fat and whole-milk set yoghurt, inulin addition did not interrupt the casein network and as a result, it could be used in both types of yoghurt with supplementary nutritional benefits (Guggisberg et al. 2009). Inclusion of inulin in low-fat sausages (20% less fat than traditional sausages) conferred sensory characteristics comparable to those of the traditional chorizo (Beriain et al. 2011). The effects of adding inulin and bovine plasma protein as fat replacers on the quality of minced meat were studied. The sensory features of samples with the proper mixture of inulin and protein were found to be comparable to those of the full-fat control samples (Rodriguez Furlán et al. 2014). Locust bean gum and konjac were also studied for their ability to be used in reduced-fat foods (Jiménez-Colmenero et al. 2012, 2013; Ruiz-Capillas et al. 2012; Chung et al. 2013). Fat is of principal importance to product palatability, and it functions as the discontinuous phase of sausage emulsions, affecting sausage tenderness and juiciness. When a gelling agent was also added to the sausages, it improved water-binding ability and the product's heat stability (Li and Nie 2016). Potato starch, locust bean gum and κ-carrageenan were used in low-fat sausages. The κ-carrageenan–locust bean gum interaction improved texture and water retention, as long as both were added in similar proportions. The lowest tested starch proportion could serve as an extender for low-fat meat sauces (García-García and Totosaus 2008).

9.7.3 Benefits of Dietary Fermentable Fibers for Chronic Kidney Disease (CKD)

CKD is a universal health problem that can lead to cardiovascular illness and end-stage renal disease, requiring kidney-replacement therapy (Stenvinkel 2005; McIntyre et al. 2011; Hung and Suzuki 2018). Systemic inflammation plays a major part in CKD progression. Patients with end-stage renal disease have high concentrations of proinflammatory cytokines and high numbers of activated circulating leukocytes in the plasma (Yoon et al. 2007; Kato et al. 2008). There is growing evidence that the pathogenesis of CKD-associated systemic inflammation is affected by modifications in the structure and function of the intestinal microbiome and barrier. Intense alterations in the intestinal microflora composition and hyperpermeability, as well as endotoxemia, are frequently observed in patients with CKD (Vaziri et al. 2016; Jiang et al. 2017). High dietary fiber consumption is linked with a lower risk of inflammation and reduced mortality in patients with CKD (Krishnamurthy et al. 2012). Oral ingestion of oligofructose-enriched inulin diminished serum concentrations of p-cresol and indoxyl sulfate in hemodialysis patients (Meijers et al. 2010). The physicochemical properties of dietary fibers, which influence their physiological effects, diverge, and the clear-cut role of dietary fiber ingestion on the prevention of CKD is not completely understood. The dietary fiber guar gum is found in guar seed. Its key component, galactomannan, is characterized by high fermentability and viscosity. Previous studies have shown that guar gum intake delivers numerous physiological benefits for health, comprising improvement in glucose tolerance (Groop et al. 1993). In addition, PHGG, obtained by controlled partial enzymatic hydrolysis of guar gum, is preferred in food manufacturing due to its lower molecular weight and viscosity. PHGG is thus useful when the valuable physiological effects of guar gum are desired but high viscosity may be a concern (Stewart and Slavin 2006; Hung and Suzuki 2016). Consumption of guar gum and PHGG reduced colonic barrier defect and inflammation in mice with experimental colitis (Hung and Suzuki 2016), possibly by conserving microflora composition and SCFA production. More than a few studies have shown that SCFAs, such as acetic, propionic, and butyric acids, are involved in the regulation of the intestinal tight-junction barrier (Suzuki et al. 2008; Elamin et al. 2013). The effects of fermentable dietary fibers, unmodified guar gum, and PHGG (which differ in their viscosity features) on CKD progression, with a specific emphasis on colonic tight-junction barriers, were studied in mice (Hung and Suzuki 2018). It was concluded that complementary feeding with fermentable dietary fibers such as guar gum and PHGG might prevent or control CKD by reestablishing colonic barrier integrity and microflora composition (Hung and Suzuki 2018).

9.8 Demonstrating the Use of Hydrocolloids in Controlling Nutrition

Nearly all bakery products include sugar and fat in their preparation. In addition to taste, these constituents are vital for the essential texture and shelf life of baked products. Inulin, as a bulk ingredient, demonstrates a variety of functionalities in bakery applications. It can serve as fat replacer, sugar replacer, or texture modifier. Several manuscripts have dealt with the roles of inulin in bakery goods. For example a mixture of maltodextrin and inulin produces cakes with more volume. A combination of native inulin and skim milk can substitute 70% of the fat with no noticeable change in sensory properties (Rodriguez-Garcia et al. 2012). Biscuits have a very wide array of textures, from crispy to soft; nevertheless, consistent texture seems to be a very important quality parameter. If the aim is to reduce sugar and fat, inulin and oligofructose can replace them as bulk ingredients. The most important characteristic of wafers is their crispiness. This quality should be maintained throughout storage. A wafer should also stay crisp when filled with fruit or ice cream, such as sandwich wafers. When inulin is included, the final wafer product becomes crispier. Inulin can bind water, and therefore the product contains less free water and this means that a wafer with inulin can enclose more moisture and stay crisp (Schaller-Povolny et al. 2000).

The ketogenic diet (or keto diet, for short) is a low-carbohydrate, high-fat diet that offers many health benefits. Indeed, numerous studies have shown that this type of diet can help in losing weight and improving health (Paoli 2014). Regular white, wheat, oat, rye, or multigrain breads are very high in carbohydrates and are not allowed on the keto diet (Figure 9.4). Instead, one can consume gluten-free bread, which has high fiber content and lower carbohydrate contents. Such bread is customarily prepared with coconut or almond flour, or a mixture of the two (https://cookinglsl.com/top-10-keto-bread-recipes/). Keto bread rolls (Figure 9.5) were prepared with the inclusion of psyllium husk powder, in addition to inulin. Indeed, psyllium husk has been previously reported as a substitute for gluten in bread (Zandonadi et al. 2009; Cappa et al. 2013). Such a replacement might also affect food odor and texture. In addition, energy and fat are reduced. Replacement of gluten with psyllium in pasta was also reported, expanding the variety of gluten-free foods (Zandonadi et al. 2014). Figure 9.6 shows the preparation of pancakes from a pancake mix that includes

Figure 9.4 A series of pie charts depicting the calorific contributions of carbohydrate, protein and fat in four diets: the typical American diet; the Atkins diet during the induction phase; the classic ketogenic diet in a 4 : 1 ratio of fat to combined protein and carbohydrate (by weight); and the medium-chain triglyceride (MCT) oil ketogenic diet. *Sources:* Colin, vectorized by Fvasconcellos/Wikimedia Commons/Public domain.

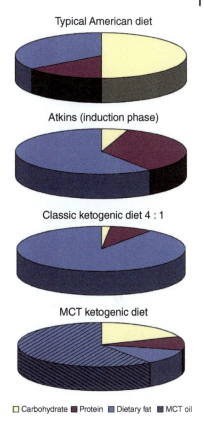

Figure 9.5 Preparation of keto bread rolls with inulin. (a) Inulin powder (left) and psyllium husk powder (right). (b) Dry yeast and inulin mixture after leavening. (c) Dry ingredients in a bowl. (d) Beating egg and egg white, (e) adding softened butter, (f) mixing with half of dry ingredients. (g) Dough. (h) A dough ball. (i) Dough balls in a casserole dish. (j) Expanded dough, and (k) after baking for 15 minutes. (l) Serve. (m) Made in a baking tray.

Figure 9.6 (a) Packages of pancake mix with inulin. (b) Mixing pancake mix, eggs and water or milk, (c) cooking pancake batter in a frying pan. (d) Serve.

barley powder and high levels of inulin. Inulin is added to the pancake recipe to confer prebiotic properties. Inulin is a prebiotic that stimulates the growth of microflora in the intestine, which are beneficial to human health. The effect of inulin on the microbiota of the human intestine has been extensively studied both in vivo and in vitro (Shoaib et al. 2016). It has been concluded that inulin and oligofructose exhibit prebiotic properties when consumed at 5–15 g/day for several weeks (Morris and Morris 2012).

Figure 9.7a shows SOYJOY bars – nutritional cereal bars that are industrially produced with whole soybean flour, fruit, and nuts, and have no wheat. The whole soybean flour provides all of the nutrients of soybeans, including vegetable protein, soy isoflavones, and dietary fiber. These bars, manufactured by Otsuka Pharmaceutical, represent a holistic attitude to people's health and well-being through innovative nutraceutical foodstuffs (https://www.otsuka.co.jp/en/nutraceutical/products/soyjoy/). In general, cereal bars are made of a combination of cereals and accompaniments held together with a binding syrup, commonly based on glucose syrup and sucrose. An improved nutritional outline is provided by the use of inulin and oligofructose in the binder, with no effect on taste or texture. The chain length of the inulin type used and the ratio of inulin to oligofructose govern the texture of the final bar (Anderson-Dekkers et al. 2021). In savory applications, such as the salty snacks presented in Figure 9.7b, inulin can help achieve reduced fat, fiber enrichment, and even improved taste in sodium-reduced products (Anderson-Dekkers et al. 2021). Figure 9.7b also includes a photograph of cookies that contain inulin. The effect of sucrose replacement by erythritol

Figure 9.7 (a) Cereal bars with inulin. (b) Salty snacks and chocolate cookies with inulin.

(Figure 9.8) and inulin was studied in short-dough cookies. Replacement of 25 and 50% of the total sucrose content was tested. Descriptive sensory analysis demonstrated that the replacement of sucrose affects the cookies' visual and textural characteristics. The differences in oral perception were greater than those for manual perception. Overall, sucrose substitutes created a less crispy cookie and lower consumer acceptance, except for the 25% sucrose replacement by inulin (Laguna et al. 2013). Matrix aeration attributes, such as open and crumbly, rated by a trained panel were important properties, and correlated positively with consumer acceptance and negatively with maximum force at break (hardness). The inulin cookies' sensory properties were more similar to the controls than the erythritol cookies. Inulin cookies were softer and erythritol cookies harder than the control cookies; despite this difference, inulin cookies had similar sound characteristics to erythritol cookies (Laguna et al. 2013).

Figure 9.8 Structure of erythritol. *Source:* Su-no-G/Wikimedia Commons/Public domain.

9.8.1 Keto Bread Rolls with Inulin (Figure 9.5)

(Serves 4)

100 g (1 cup) almond flour
13.5 g (1½ Tbsp) skim milk powder
15 g (1½ Tbsp) psyllium husk powder (Figure 9.5a)
4 g (1 tsp) baking powder
3 g (1 tsp) inulin powder (Figure 9.5a)
2.5 g (1 tsp) instant dry yeast
50 g (1) egg
30 g (1) egg white
25 g (2 Tbsp) butter
25 g (a little less than 2 Tbsp) yoghurt (*Hint 1*)
40 mL (⅙ cup) lukewarm water (around 40 °C)
Salt

For preparation, see Figure 9.5.

1) Remove egg and egg white from the refrigerator and bring to room temperature before cooking, and melt or soften butter.
2) Put instant dry yeast and inulin powder into a bowl and add the lukewarm water. Cover the bowl with a plastic wrap and leave in a warm place for 10 minutes (Figure 9.5b) (*Hint 2*).
3) Mix almond flour, skim milk powder, psyllium husk powder, baking powder, and a pinch of salt in another bowl (Figure 9.5c).
4) Put egg and egg white in another bowl and beat it with an electric mixer (Figure 9.5d).
5) Add softened butter (Figure 9.5e), yeast and inulin mixture (2) to the eggs (4) and mix well.
6) Add yoghurt, and half of the powder mixture (3), and mix with a spatula (Figure 9.5f), then add the other half of the powder mixture, and mix dough (Figure 9.5g).
7) Form four dough (6) balls with lightly wetted hands (Figure 9.5h). Place them in a casserole dish (Figure 9.1i) or baking tray lined with baking paper (*Hint 3*).
8) Cover the dish or tray with lightly oiled plastic wrap and leave in a warm place for 60 minutes until dough expands (Figure 9.5j).
9) Preheat the oven to 175 °C, and bake dough rolls for 15 minutes until they brown (Figure 9.5k). Then cover them with aluminum foil and bake them for an additional seven minutes (Figure 9.5l).

Preparation hints:

1) Greek yoghurt is recommended.
2) When you make this bread roll in the cold season, you can put the yeast and inulin mixture in an oven at 35–40 °C and leave it until it froths and thickens.
3) When you use a baking tray, the bread becomes flatter (Figure 9.5m).

References

Alam, N.H., Meier, R., Sarker, S.A. et al. (2005). Partially hydrolyzed guar gum supplemented comminuted chicken diet in persistent diarrhoea: a randomized controlled trial. *Archives of Disease in Childhood* 90: 195–199.

Amadio, L., Stocco, E., and Dodi, G. (2009). The prebiotic effects of a new mixture of soluble fermentable fibres in the treatment of chronic constipation. *Pelviperineology* 28: 55–58.

Anderson, J.W. and Chen, W.L. (1986). Cholesterol-lowering properties of oat products. In: *Oats: Chemistry and Technology* (ed. F.H. Webster), 309–333. American Association of Cereal Chemists: St. Paul, MN.

Anderson-Dekkers, I., Nouwens-Roest, M., Peters, B., and Vaughan, E. (2021). Inulin. In: *Handbook of Hydrocolloids*, 3e (ed. G.O. Phillips and P.A. Williams), chapter 17, 537–559. Cambridge, UK: Woodhead Publishing.

Apostu, P.M., Mihociu, T.E., and Nicolau, A.I. (2017). Technological and sensorial role of yeast β-glucan in meat batter reformulations. *Journal of Food Science and Technology* 54: 2653–2660.

Arendt, E.K., Ryan, L.A., and Dal Bello, F. (2007). Impact of sourdough on the texture of bread. *Food Microbiology* 24: 165–174.

Awad, S., Hassan, A., and Muthukumarappan, K. (2005). Application of exopolysaccharide-producing cultures in reduced-fat Cheddar cheese: texture and melting properties. *Journal of Dairy Science* 88: 4204–4213.

Aziznia, S., Khosrowshahi, A., Madadlou, A., and Rahimi, J. (2008). Whey protein concentrate and gum tragacanth as fat replacers in nonfat yogurt: chemical, physical, and microstructural properties. *Journal of Dairy Science* 91: 2545–2552.

Bayarri, S., Chulia, I., and Costell, E. (2010). Comparing λ-carrageenan and an inulin blend as fat replacers in carboxymethyl cellulose dairy desserts. Rheological and sensory aspects. *Food Hydrocolloids* 24: 578–587.

Beriain, M.J., Gomez, I., Petri, E. et al. (2011). The effects of olive oil emulsified alginate on the physico-chemical, sensory, microbial, and fatty acid profiles of low-salt, inulin-enriched sausages. *Meat Science* 88: 189–197.

Bernstein, A.M., Titgemeier, B., Kirkpatrick, K. et al. (2013). Major cereal grain fibers and psyllium in relation to cardiovascular health. *Nutrients* 5: 1471–1487.

Beyer, M., Reichert, J., Heurich, E. et al. (2010). Pectin, alginate and gum Arabic polymers reduce citric acid erosion effects on human enamel. *Dental Materials* 26: 831–839.

Bliss, D.Z., Stein, T.P., Schleifer, C.R., and Settle, R.G. (1996). Supplementation with gum Arabic fiber increases fecal nitrogen excretion and lowers serum urea nitrogen concentration in chronic renal failure patients consuming a low-protein diet. *American Journal of Clinical Nutrition* 63: 392–398.

Bohm, N. and Kulicke, W.-M. (1999). Rheological studies of barley $(1\rightarrow 3)(1\rightarrow 4)$-β-glucan in concentrated solution, mechanistic and kinetic investigation of the gel formation. *Carbohydrate Research* 315: 302–311.

Brennan, C.S. and Tudorica, C.M. (2008). Carbohydrate-based fat replacers in the modification of the rheological, textural and sensory quality of yoghurt: comparative study of the utilization of barley beta-glucan, guar gum and inulin. *International Journal of Food Science and Technology* 43: 824–833.

Brownlee, I.A. (2011). The physiological roles of dietary fibre. *Food Hydrocolloids* 25: 238–250.

Buttriss, J.L. and Stokes, C.S. (2008). Dietary fibre and health: an overview. *Nutrition Bulletin* 33: 186–200.

Calame, W., Weseler, A.R., Viebke, C. et al. (2008). Gum Arabic establishes prebiotic functionality in healthy human volunteers in a dose-dependent manner. *British Journal of Nutrition* 100: 1269–1275.

Calame, W., Thomassen, F., Hull, S. et al. (2011). Evaluation of satiety enhancement, including compensation, by blends of gum Arabic. A methodological approach. *Appetite* 57: 358–364.

Cappa, C., Lucisano, M., and Mariotti, M. (2013). Influence of psyllium, sugar beet fibre and water on gluten-free dough properties and bread quality. *Carbohydrate Polymers* 98: 1657–1666.

Chung, C., Degner, B., and McClements, D.J. (2013). Designing reduced-fat food emulsions: locust bean gum–fat droplet interactions. *Food Hydrocolloids* 32: 263–270.

Coudray, C., Rambeau, M., Feillet-Coudray, C. et al. (2005). Dietary inulin intake and age can significantly affect intestinal absorption of calcium and magnesium in rats: a stable isotope approach. *Nutrition Journal* 4: 29–36.

Cui, S.W. and Wang, Q. (2009). Cell wall polysaccharides in cereals: chemical structures and functional properties. *Structural Chemistry* 20: 291–297.

Dalonso, N., Goldman, G.H., and Gern, R.M.M. (2015). β-(1→3),(1→6)-glucans: medicinal activities, characterization, biosynthesis and new horizons. *Applied Microbiology and Biotechnology* 99: 7893–7906.

Delzenne, N.M., Neyrinck, A.M., and Cani, P.D. (2011). Modulation of the gut microbiota by nutrients with prebiotic properties: consequences for host health in the context of obesity and metabolic syndrome. *Microbial Cell Factories* 10 (Suppl. 1): S10. https://doi.org/10.1186/1475-2859-10-S1-S10.

Dettmar, P.W., Strugala, V., and Richardson, J.C. (2011). The key role alginates play in health. *Food Hydrocolloids* 25: 263–266.

Di Cagno, R., De Angelis, M., Limitone, A. et al. (2006). Glucan and fructan production by sourdough *Weissella cibaria* and *Lactobacillus plantarum*. *Journal of Agricultural and Food Chemistry* 54: 9873–9881.

Doublier, J.-L. and Wood, P.J. (1995). Rheological properties of aqueous solutions of (1→3)(1→4)-β-D glucan from oats (*Avena sativa* L.). *Cereal Chemistry* 72: 335–340.

Draga, A. and McCleary, B.V. (2016). Measurement of β-glucan in mushrooms and mycelial products. *Journal of AOAC International* 99: 364–373.

Eastwood, M.A., Brydon, W.G., Path, M.R.C., and Anderson, D.M.W. (1986). The effect of the polysaccharide composition and structure of dietary fibers on cecal fermentation and fecal excretion. *American Journal of Clinical Nutrition* 44: 51–55.

Elamin, E.E., Masclee, A.A., Dekker, J. et al. (2013). Short-chain fatty acids activate AMP-activated protein kinase and ameliorate ethanol-induced intestinal barrier dysfunction in Caco-2 cell monolayers. *Journal of Nutrition* 143: 1872–1881.

Feldman, W., McGrath, P., Hodgson, C. et al. (1985). The use of dietary fiber in the management of simple, childhood, idiopathic, recurrent, abdominal pain. Results in a prospective, double blind, randomized, controlled trial. *American Journal of Diseases of Children* 139: 1216–1218.

Fischer, M. (2004). Dietary fibers – new challenges for research. In: *Dietary Fibre: Bioactive Carbohydrates for Food and Feed* (ed. J.W. van der Kamp, N.G. Asp, J. Miller Jones and G. Schaafsma), 59–66. Wageningen: Wageningen Academic Publishers.

Flammang, A.M., Kendall, D.M., Baumgartner, C.J. et al. (2006). Effect of a viscous fiber bar on postprandial glycemia in subjects with type 2 diabetes. *Journal of the American College of Nutrition* 25: 409–414.

Funami, T. (2011). Next target for food hydrocolloid studies: texture design of foods using hydrocolloid technology. *Food Hydrocolloids* 25: 1904–1914.

García-García, E. and Totosaus, A. (2008). Low-fat sodium-reduced sausages: effect of the interaction between locust bean gum, potato starch and k-carrageenan by a mixture design approach. *Meat Science* 78: 406–413.

Gibson, G.R. and Roberfroid, M.B. (1995). Dietary modulation of the human colonic microbiota: introducing the concept of prebiotics. *Journal of Nutrition* 125: 1401–1412.

Gibson, G.R., Beatty, E.R., Wang, X., and Cummings, J.H. (1995). Selective stimulation of bifidobacteria in the human colon by oligofructose and inulin. *Gastroenterology* 108: 975–982.

Gidley, M.J. (2013). Hydrocolloids in the digestive tract and related health implications. *Current Opinion in Colloid and Interface Science* 18: 371–378.

Glover, D.A., Ushida, K., Phillips, A.O., and Riley, S.G. (2009). Acacia(sen) SUPERGUM™ (Gum arabic): an evaluation of potential health benefits in human subjects. *Food Hydrocolloids* 23: 2410–2415.

Grasso, S., Brunton, N.P., Lyng, J.G. et al. (2014). Healthy processed meat products – regulatory, reformulation and consumer challenges. *Trends in Food Science & Technology* 39: 4–17.

Groop, P.H., Aro, A., Stenman, S., and Groop, L. (1993). Long-term effects of guar gum in subjects with non-insulin-dependent diabetes mellitus. *American Journal of Clinical Nutrition* 58: 513–518.

Guérin-Deremaux, L., Pochat, M., Reifer, C. et al. (2011). The soluble fiber NUTRIOSE induces a dose-dependent beneficial impact on satiety over time in humans. *Nutrition Research* 31: 665–672.

Guggisberg, D., Cuthbert-Steven, J., Piccinali, P. et al. (2009). Rheological, microstructural and sensory characterization of low-fat and whole milk set yoghurt as influenced by inulin addition. *In ternational Dairy Journal* 19: 107–115.

Haque, A., Richardson, R.K., and Morris, E.R. (1993). Xanthan-like 'weak gel' rheology from dispersions of ispaghula seed husk. *Carbohydrate Polymers* 22: 223–232.

Hassan, A.N. (2008). ADSA Foundation Scholar Award: possibilities and challenges of exopolysaccharide-producing lactic cultures in dairy foods. *Journal of Dairy Science* 91: 1282–1298.

Hassan, A.N. and Awad, S. (2005). Application of exopolysaccharide-producing cultures in reduced-fat Cheddar cheese: cryo-scanning electron microscopy observations. *Journal of Dairy Science* 88: 4214–4220.

Herrera-Balandrano, D.D., Báez-González, J.G., Carvajal-Millán, E. et al. (2019). Feruloylated arabinoxylans from nixtamalized maize bran byproduct: a functional ingredient in frankfurter sausages. *Molecules* 24: 2056. https://doi.org/10.3390/molecules24112056.

Hess, J., Birkett, A., Thomas, W., and Slavin, J. (2011). Effects of short-chain fructo oligosaccharides on satiety responses in healthy men and women. *Appetite* 56: 128–134.

Horvath, A., Dziechciarz, P., and Szajewska, H. (2013). Glucomannan for abdominal pain-related functional gastrointestinal disorders in children: a randomized trial. *World Journal of Gastroenterology* 19: 3062–3068.

Hsu, S.Y. and Chung, H.Y. (1999). Comparisons of 13 edible gum-hydrate fat substitutes for low fat Kung-wan (an emulsified meatball). *Journal of Food Engineering* 40: 279–285.

Hung, T.V. and Suzuki, T. (2016). Dietary fermentable fiber reduces intestinal barrier defects and inflammation in colitic mice. *Journal of Nutrition* 146: 1970–1976.

Hung, T.V. and Suzuki, T. (2018). Dietary fermentable fibers attenuate chronic kidney disease in mice by protecting the intestinal barrier. *Journal of Nutrition* 148: 552–561.

Huopalahti, R., López-Fandiño, R., Anton, M. et al. (2007). *Bioactive Egg Compounds*. New York, NY: Springer.

Huttenlocher, P.R., Wilbourn, A.J., and Signore, J.M. (1971). Medium-chain triglycerides as a therapy for intractable childhood epilepsy. *Neurology* 21: 1097–1103.

Institute of Medicine (2005). *Dietary Reference Intakes for Energy, Carbohydrate, Fiber, Fat, Fatty Acids, Cholesterol, Protein, and Amino Acids*. Washington, DC: The National Academies Press https://doi.org/10.17226/10490.

Jiang, S., Xie, S., Dan, L.V. et al. (2017). Alteration of the gut microbiota in Chinese population with chronic kidney disease. *Scientific Reports* 7: 2870. https://doi.org/10.1038/s41598-017-02989-2.

Jiménez Colmenero, F. (2000). Relevant factors in strategies for fat reduction in meat products. *Trends in Food Science and Technology* 11: 56–66.

Jiménez-Colmenero, F., Cofrades, S., Herrero, A.M. et al. (2012). Konjac gel fat analogue for use in meat products: comparison with pork fats. *Food Hydrocolloids* 26: 63–72.

Jiménez-Colmenero, F., Cofrades, S., Herrero, A.M. et al. (2013). Konjac gel for use as potential fat analogue for healthier meat product development: effect of chilled and frozen storage. *Food Hydrocolloids* 30: 351–357.

Kaczmarczyk, M.M., Miller, M.J., and Freund, G.G. (2012). The health benefits of dietary fiber: beyond the usual suspects of type 2 diabetes mellitus, cardiovascular disease and colon cancer. *Metabolism* 61: 1058–1066.

Kahlon, T.S. and Chow, F.L. (1997). Hypocholesterolemic effects of oat, rice and barely dietary fibers and fractions. *Cereal Foods World* 42: 86–92.

Kato, S., Chmielewski, M., Honda, H. et al. (2008). Aspects of immune dysfunction in end-stage renal disease. *Clinical Journal of the American Society of Nephrology* 3: 1526–1533.

Kaur, R. and Sharma, M. (2019). Cereal polysaccharides as sources of functional ingredient for reformulation of meat products: a review. *Journal of Functional Foods* 62: 103527. https://doi.org/10.1016/j.jff.2019.103527.

Kelly, P. (2018). Whey protein ingredient applications. In: *Whey Proteins: From Milk to Medicine* (ed. H.C. Deeth and N. Bansal), 335–375. London, UK: Academic Press.

Kitazawa, H., Itoh, T., Tomioka, Y. et al. (1996). Induction of IFN-γ and IL-1α production in macrophages stimulated with phosphopolysaccharide produced by *Lactococcus lactis* ssp. *cremoris*. *International Journal of Food Microbiology* 31: 99–106.

Kolida, S., Tuohy, K., and Gibson, G.R. (2002). Prebiotic effects of inulin and oligofructose. *British Journal of Nutrition* 87: S193–S197.

Korczak, R., Kamil, A., Fleige, L. et al. (2017). Dietary fiber and digestive health in children. *Nutrition Reviews* 75: 241–259.

Koutrotsios, G., Kalogeropoulos, N., Stathopoulos, P. et al. (2017). Bioactive compounds and antioxidant activity exhibit high intraspecific variability in *Pleurotus ostreatus* mushrooms and correlate well with cultivation performance parameters. *World Journal of Microbiology and Biotechnology* 33: 98. https://doi.org/10.1007/s11274-017-2262-1.

Krishnamurthy, V.M.R., Wei, G., Baird, B.C. et al. (2012). High dietary fiber intake is associated with decreased inflammation and all-mortality in patients with chronic kidney disease. *Kidney International* 81: 300–306.

Ktari, N., Trabelsi, I., Bkhairia, I. et al. (2016). Using barley β-glucan, citrus, and carrot fibers as a meat substitute in turkey meat sausages and their effects on sensory characteristics and properties. *Journal of Food Process Technology* 7: 620. https://doi.org/10.4172/2157-7110.1000620.

Laca, A., Paredes, B., and Diaz, M. (2010). A method of egg yolk fractionation. Characterization of fractions. *Food Hydrocolloids* 24: 434–443.

Laguna, L., Primo-Martín, C., Salvador, A., and Sanz, T. (2013). Inulin and erythritol as sucrose replacers in short-dough cookies: sensory, fracture, and acoustic properties. *Journal of Food Science* 78: S777–S784.

Laino, J., Villena, J., Kanmani, P., and Kitazawa, H. (2016). Immunoregulatory effects triggered by lactic acid bacteria exopolysaccharides: new insights into molecular interactions with host cells. *Microorganisms* 4 (3): 27. https://doi.org/10.3390/microorganisms4030027.

Last, A.R. and Wilson, S.A. (2006). Low-carbohydrate diets. *American Family Physician* 73: 1942–1948.

Lattimer, J.M. and Haub, M.D. (2010). Effects of dietary fiber and its components on metabolic health. *Nutrients* 2: 1266–1289.

Lavelli, V., Proserpio, C., Gallotti, F. et al. (2018). Circular reuse of bio-resources: the role of *Pleurotus* spp. in the development of functional foods. *Food and Function* 9: 1353–1372.

Li, J.-M. and Nie, S.-P. (2016). The functional and nutritional aspects of hydrocolloids in foods. *Food Hydrocolloids* 53: 46–61.

Lin, Y., Huybrechts, I., Vereecken, C. et al. (2015). Dietary fiber intake and its association with indicators of adiposity and serum biomarkers in European adolescents: the HELENA study. *European Journal of Nutrition* 54: 771–782.

Liu, C.F., Tseng, K.C., Chiang, S.S. et al. (2011). Immunomodulatory and antioxidant potential of *Lactobacillus* exopolysaccharides. *Journal of the Science of Food and Agriculture* 91: 2284–2291.

Lynch, K.M., Zannini, E., Coffey, A., and Arendt, E.K. (2018). Lactic acid bacteria exopolysaccharides in foods and beverages: isolation, properties, characterization, and health benefits. *Annual Review of Food Science and Technology* 9: 155–176.

Malkki, Y. (2004). Trends in dietary fiber research and development: a review. *Acta Alimentaria* 33: 39–62.

Martensson, O., Biorklund, M., Lambo, A.M. et al. (2005). Fermented, ropy, oat-based products reduce cholesterol levels and stimulate the bifidobacteria flora in humans. *Nutrition Research* 25: 429–442.

Mathern, J.R., Raatz, S.K., Thomas, W., and Slavin, J.L. (2009). Effect of fenugreek fiber on satiety, blood glucose and insulin response and energy intake in obese subjects. *Phytotherapy Research* 23: 1543–1548.

Mayakrishnan, V., Kannappan, P., Abdullah, N., and Ahmed, A.B.A. (2013). Cardio protective activity of polysaccharides derived from marine algae: an overview. *Trends in Food Science & Technology* 30: 98–104.

McIntyre, C.W., Harrison, L.E., Eldehni, M.T. et al. (2011). Circulating end toxemia: a novel factor in systemic inflammation and cardiovascular disease in chronic kidney disease. *Clinical Journal of the American Society of Nephrology* 6: 133–141.

Meijers, B.K., De Preter, V., Verbeke, K. et al. (2010). p-Cresyl sulfate serum concentrations in haemodialysis patients are reduced by the prebiotic oligofructose-enriched inulin. *Nephrology Dialysis Transplantation* 25: 219–224.

Menga, V., Amato, M., Phillips, T.D. et al. (2017). Gluten-free pasta incorporating chia (*Salvia hispanica* L.) as thickening agent: an approach to naturally improve the nutritional profile and the in vitro carbohydrate digestibility. *Food Chemistry* 221: 1954–1961.

Mikkilä, V., Räsänen, L., Raitakari, O. et al. (2005). Consistent dietary patterns identified from childhood to adulthood: the cardiovascular risk in Young Finns Study. *British Journal of Nutrition* 93: 923–931.

Monsivais, P., Carter, B., Christiansen, M. et al. (2011). Soluble fiber dextrin enhances the satiating power of beverages. *Appetite* 56: 9–14.

Moroni, A.V., Dal Bello, F., and Arendt, E.K. (2009). Sourdough in gluten-free bread-making: an ancient technology to solve a novel issue? *Food Microbiology* 26: 676–684.

Morris, C. and Morris, G.A. (2012). The effect of inulin and fructo-oligosaccharide supplementation on the textural, rheological and sensory properties of bread and their role in weight management: a review. *Food Chemistry* 133: 237–248.

Nishinari, K., Fang, Y., Yang, N. et al. (2018). Gels, emulsions and application of hydrocolloids at Phillips Hydrocolloids Research Centre. *Food Hydrocolloids* 78: 36–46.

Pan, D. and Mei, X. (2010). Antioxidant activity of an exopolysaccharide purified from *Lactococcus lactis* subsp. *Lactis* 12. *Carbohydrate Polymers* 80: 908–914.

Paoli, A. (2014). Ketogenic diet for obesity: friend or foe? *International Journal of Environmental Research and Public Health* 11: 2092–2107.

Parisi, G., Bottona, E., Carrara, M., and Cardin, F. (2005). Treatment effects of partially hydrolyzed guar gum on symptoms and quality of life of patients with irritable bowel syndrome. A multicenter randomized open trial. *Digestive Diseases and Sciences* 50: 1107–1112.

Patel, H. and Patel, S. (2014). Milk Protein Concentrates: Manufacturing and Applications. US Dairy Export Council, Technical Report. https://www.thinkusadairy.org/resources-and-insights/resources-and-insights/application-and-technical-materials/milk-protein-concentrates-manufacturing-and-applications (accessed 15 May, 2022).

Pelucchi, C., Talamini, R., Galeone, C. et al. (2003). Fibre intake and prostate cancer risk. *International Journal of Cancer* 109: 278–280.

Phillips, A.O. and Phillips, G.O. (2011). Biofunctional behavior and health benefits of a specific gum arabic. *Food Hydrocolloids* 25: 165–169.

Phillips, G.O. and Williams, P.A. (2009). *Handbook of Hydrocolloids*, 2e. Cambridge, UK: Woodhead Publishing.

Pintado, T., Herrero, A.M., Jiménez-Colmenero, F., and Ruiz-Capillas, C. (2016). Emulsion gels as potential fat replacers delivering β-glucan and healthy lipid content for food applications. *Journal of Food Science and Technology* 53: 4336–4347.

Proserpio, C., Pagliarini, E., Laureati, M. et al. (2019). Acceptance of a new food enriched in β-glucans among adolescents: effects of food technology neophobia and healthy food habits. *Foods* 8: 433. https://doi.org/10.3390/foods8100433.

Quartarone, G. (2013). Role of PHGG as a dietary fiber: a review article. *Minerva Gastroenterologica e Dietologica* 59: 329–340.

Redgwell, R.J. and Fischer, M. (2005). Dietary fiber as a versatile food component: an industrial perspective. *Molecular Nutrition & Food Research* 49: 421–535.

Ricklefs-Johnson, K., Johnston, C.S., and Sweazea, K.L. (2017). Ground flaxseed increased nitric oxide levels in adults with type 2 diabetes: a randomized comparative effectiveness study of supplemental flaxseed and psyllium fiber. *Obesity Medicine* 5: 16–24.

Ring, S. and Selvendran, R.R. (1980). Isolation and analysis of cell-wall material from beeswing wheat bran (*Triticum aestivum*). *Phytochemistry* 19: 1723–1730.

Rodriguez Furlán, L.T., Padilla, A.P., and Campderros, M.E. (2014). Development of reduced fat minced meats using inulin and bovine plasma proteins as fat replacers. *Meat Science* 96: 762–768.

Rodriguez-Garcia, J., Laguna, L., Puig, A. et al. (2012). Effect of fat replacement by inulin on textural and structural properties of short dough biscuits. *Food and Bioprocess Technology* 6: 2739–2750.

Romanchik-Cerpovicz, J.E., Tilmon, R.W., and Baldree, K.A. (2002). Moisture retention and consumer acceptability of chocolate bar cookies prepared with okra gum as a fat ingredient substitute. *Journal of the American Dietetic Association* 102: 1301–1303.

Romano, C., Comito, D., Famiani, A. et al. (2013). Partially hydrolyzed guar gum in pediatric functional abdominal pain. *World Journal of Gastroenterology* 19: 235–240.

Ruas-Madiedo, P., Salazar, N., and de los Reyes-Gavilán, C.G. (2009). Biosynthesis and chemical composition of exopolysaccharides. In: *Bacterial Polysaccharides: Current Innovations and Future Trends* (ed. M. Ullrich), 279–310. Norfolk, UK: Caister Academic Press.

Ruiz-Capillas, C., Triki, M., Herrero, A.M. et al. (2012). Konjac gel as pork backfat replacer in dry fermented sausages: processing and quality characteristics. *Meat Science* 92: 144–150.

Sabater-Molina, M., Larque, E., Torrella, F., and Zamora, S. (2009). Dietary fructooligosaccharides and potential benefits on health. *Journal of Physiology and Biochemistry* 65: 315–328.

Schaller-Povolny, L.A., Smith, D.E., and Labuza, T.P. (2000). Effect of water content and molecular weight on the moisture isotherms and glass transition properties of inulin. *International Journal of Food Properties* 3: 173–192.

Schwab, C., Mastrangelo, M., Corsetti, A., and Ganzle, M. (2008). Formation of oligosaccharides and polysaccharides by *Lactobacillus reuteri* LTH5448 and *Weissella cibaria* 10M in sorghum sourdoughs. *Cereal Chemistry* 85: 679–684.

Shoaib, M., Shehzad, A., Omar, M. et al. (2016). Inulin: properties, health benefits and food applications. *Carbohydrate Polymers* 147: 444–454.

Slavin, J. (2003). Why whole grains are protective: biological mechanisms. *Proceedings of the Nutrition Society* 62: 129–134.

Slavin, J. (2013). Fiber and prebiotics: mechanisms and health benefits. *Nutrients* 5: 1417–1435.

Slavin, J. and Green, H. (2007). Dietary fibre and satiety. *Nutrition Bulletin* 32: 32–42.

Stenvinkel, P. (2005). Inflammation in end-stage renal disease: the hidden enemy. *Nephrology* 11: 36–41.

Stewart, M.L. and Slavin, J.L. (2006). Molecular weight of guar gum affects short chain fatty acid profile in model intestinal fermentation. *Molecular Nutrition and Food Research* 50: 971–976.

Suzuki, T., Yoshida, S., and Hara, H. (2008). Physiological concentrations of short chain fatty acids immediately suppress colonic epithelial permeability. *British Journal of Nutrition* 100: 297–305.

Tan, K.Y. and Seow-Choen, F. (2007). Fiber and colorectal diseases: separating fact from fiction. *World Journal of Gastroenterology* 13: 4161–4167.

Theuwissen, E. and Mensink, R.P. (2008). Water-soluble dietary fibers and cardiovascular disease. *Physiology and Behavior* 94: 285–292.

Topping, D.L. and Clifton, P.M. (2001). Short chain fatty acids and human colonic function: roles of resistant starch and non-starch polysaccharides. *Physiological Reviews* 81: 1031–1064.

Valverde, D., Laca, A., Estrada, L.N. et al. (2016). Egg yolk and egg yolk fractions as key ingredient for the development of a new type of gels. *International Journal of Gastronomy and Food Science* 3: 30–37.

Vaziri, N.D., Zhao, Y.Y., and Pahl, M.V. (2016). Altered intestinal microbial flora and impaired epithelial barrier structure and function in CKD: the nature, mechanisms, consequences and potential treatment. *Nephrology Dialysis Transplantation* 31: 737–746.

Viebke, C., Al-Assaf, S., and Phillips, G.O. (2014). Food hydrocolloids and health claims. *Bioactive Carbohydrates and Dietary Fiber* 4: 101–114.

Villegas, B., Carbonell, I., and Costell, E. (2007). Inulin milk beverages: sensory differences in thickness and creaminess using R-index analysis of the ranking data. *Journal of Sensory Studies* 22: 377–393.

Vitaglione, P., Lumaga, R., Montagnese, C. et al. (2010). Satiating effect of barley beta-glucan-enriched snack. *Journal of the American College of Nutrition* 29: 113–121.

Waldherr, F.W. and Vogel, R.F. (2009). Commercial exploitation of homo-exopolysaccharides in non-dairy food. In: *Bacterial Polysaccharides: Current Innovations and Future Trends* (ed. M. Ullrich), 313–329. Norfolk, UK: Caister Academic Press.

Willis, H.J., Eldridge, A.L., Beiseigel, J. et al. (2009). Greater satiety response with resistant starch and corn bran in human subjects. *Nutrition Research* 29: 100–105.

Wolever, T.M.S., Tosh, S.M., Gibbs, A.L. et al. (2010). Physicochemical properties of oat β-glucan influence its ability to reduce serum LDL cholesterol in humans: a randomized clinical trial. *American Journal of Clinical Nutrition* 92: 723–732.

Wong, J.M., De Souza, R., Kendall, C.W. et al. (2006). Colonic health: fermentation and short chain fatty acids. *Journal of Clinical Gastroenterology* 40: 235–243.

Xu, X., Xu, P., Ma, C. et al. (2013). Gut microbiota, host health, and polysaccharides. *Biotechnology Advances* 31: 318–337.

Yemenicioglu, A., Farris, S., Turkyilmaz, M., and Sukru Gulec, S. (2019). A review of current and future food applications of natural hydrocolloids. *International Journal of Food Science and Technology* 55 (4): https://doi.org/10.1111/ijfs.14363.

Yoon, J.W., Pahl, M.V., and Vaziri, N.D. (2007). Spontaneous leukocyte activation and oxygen-free radical generation in end-stage renal disease. *Kidney International* 71: 167–172.

Yousefi, M. and Jafari, S.M. (2019). Recent advances in application of different hydrocolloids in dairy products to improve their techno-functional properties. *Trends in Food Science and Technology* 88: 468–483.

Zandonadi, R.P., Assuncao Botelho, R.B., and Coelho Araujo, W.M. (2009). Psyllium as a substitute for gluten in bread. *Journal of the American Dietetic Association* 109: 1781–1784.

Zandonadi, R.P., Botelho, R.B.A., and Araujo, W.M.C. (2014). Psyllium as a substitute for gluten in pastas. *Journal of Culinary Science and Technology* 12: 181–190.

Index

a

Abelmoschus esculentus 176
abrasion 215
acacia gum 222, 256
acetaldehyde 82, 86–7
Acetobacter hansenii 121
Acetobacter xylinum 121
acetone 82, 97
acetophenone 86–7
acetyl 91–2, 176, 186
achiote 212–13
adhesion 12, 49, 93, 148, 150, 192, 194, 196, 205, 217, 221, 226, 241
Adrià, Ferran 23–4
adsorption 83, 89, 98, 129, 134, 172, 198, 200, 210, 215
adzuki bean paste 11
agar 17, 24–6, 29, 31, 34–5, 56, 118, 125, 160–1, 178–83, 186, 188, 194, 196, 201–2, 244–5, 261
agar–agar 1, 21, 29, 31, 35, 125, 194–6, 236
aggregation 88, 91, 120, 123, 125, 132, 209, 240
agriculture 93, 131, 157, 165, 184–5, 206, 222, 275
agro industrial 61, 255
air
  bubbles 154
  conduction 141, 144
Albania 2

alcohol 17, 55, 59, 61, 75, 84–5, 151, 215, 228
alcoholic beverages 215
al dente 33
Aleph Farms 237
alginate 11, 15, 18–19, 21–4, 30, 32–4, 37–8, 53, 56–7, 60, 73, 80, 86–7, 92, 98, 100, 103–4, 106, 118–20, 127–8, 149, 161–2, 174, 198–200, 215, 220–1, 223–4, 226–9, 240, 243, 247, 262, 264, 271–2
almond(s) 61, 140, 143, 164, 266, 269–70
aluminum foil 10, 270
amino acids 61–2, 77, 80, 209, 221, 240, 256, 273
Amorphophallus konjac 246
amorphous 193–4, 239, 251
amylose 83, 97, 99, 104, 119, 122, 134, 148, 150–1, 177, 185, 190
angle 3, 5, 45–7, 49, 50, 57, 198
anionic 200, 209
annatto 212–13, 225, 228
anthocyanin 59, 60, 66, 68, 249
antibacterial 71, 103, 175
antimicrobial 92, 104, 198, 256, 259
antioxidant 67, 71, 92, 176, 198, 227, 234, 240, 242, 256, 259–60, 264, 274–6
apparent viscosity 28, 87, 103, 112, 114, 119, 121–3
appetizer 2, 245
apple 41, 52, 60, 72, 78, 103–4, 128–9, 138, 160, 187, 201, 217, 219–20
arabinogalactan 118–19

Use of Hydrocolloids to Control Food Appearance, Flavor, Texture, and Nutrition, First Edition.
Amos Nussinovitch and Madoka Hirashima.
© 2023 John Wiley & Sons Ltd. Published 2023 by John Wiley & Sons Ltd.

arabinoxylans   118, 131, 176, 186, 273
Archimedes   5, 20
Armenia   26
Arnott's Biscuits   194
aroma   26, 54, 61, 74, 77–8, 83–6, 88–91, 97–101, 103–4, 106–7, 126, 140, 215
artificial salmon eggs   1
Artocarpus heterophyllus   52
atmospheric pressure   90, 105
atomic force microscopy   55
atomization   7, 23, 58, 65, 227
average
  dimension   14
  DP   149
  Feret diameter   145
  final product size   9
  human response   45
  initial feed size   9
  molecular mass   193, 213
  person   45
  piece size   207
  projected area   14–15
  size   7, 200
  yield growth   171

b

bacteria   59, 79, 101, 121, 171, 186, 215, 257, 261–3, 275
bagels   5
baked goods   10–11, 140, 150, 169, 178, 194, 228, 255
bakery   11, 16, 79, 94, 98, 120, 149, 157, 160, 169, 194, 223, 266
baking
  molds   16
  soda   4
balls   5, 10, 30, 141, 267
banana   56–7, 71, 136, 138, 146, 149–50, 162–3, 201, 257
Barilla   236
batter   4, 74, 80, 92–6, 148–9, 157, 160, 163, 169, 232, 268, 270
bean
  flour   158
  paste   11, 218
  proteins   157

beef   3, 4, 38, 189, 236, 240, 244–5
beer   112, 192, 215, 219–20
beeswax   99, 193
beverages   42, 67, 68, 72, 74, 81–2, 86–8, 102–3, 106, 112, 120, 184, 191, 193, 199, 210, 215, 255–6, 262, 265, 275, 278
Bifidobacterium infantis   263
biodegradable   61, 68, 131, 197, 225, 227
bio-inks   236, 240
biomedical   71, 222, 229, 231
biopolymers   61, 125, 132, 178, 189–90, 208–11, 224
biscuits   11, 136, 140, 143–4, 149, 158, 194, 266, 276
bite   7, 11, 96, 138, 142, 166–7, 175, 188, 192, 194
bitterness   86, 88
Bixa orellana   213, 228
bixin   215
black pepper   128
blood   41, 114, 238, 259, 261–2, 275
boiled
  buckwheat konjac noodles   181
  chicken   181
  minced carrots   27
boiling point   87
bolus   26–7, 131
Bond   9, 33
  work-index   10
bone
  conductance   141
  conducted sounds   143
  conduction   141–2, 158, 163
  marrow   35
  mediated sounds   141
  regeneration   240, 251
  tissue engineering   250
borek   2
bovine
  gelatin   225
  liver   215, 227
  plasma protein   265, 276
  serum albumin (BSA)   82, 97–8, 123, 187, 199, 239
Braunschweiger   5

bread
  baked   95, 168, 169, 170, 184
  crumb   170
  dough   170, 185
  extruded   143
  flat   154, 258
  frozen   170
  β-glucan-enriched   259
  gluten free   36, 74, 92, 94, 96, 104, 266
  gluten in   266
  manufacturing   169
  rice   188
  rolls   266, 267, 269, 270
  sorghum sourdough   79
  sorghum-wheat composite   101
  texture   168, 185, 270
  volume   170
  wheat   79
  wheat-faba bean composite   106
  white   185
  whole grain pearl millet   106
  whole grain sorghum   79, 106
breading   93, 139, 161, 163
breakfast cereals   138, 152, 162–3
brik   2
brisket   4
brittleness   140, 151, 178, 196
broth   216
browning   54, 57–9, 146, 198, 227, 248–9
buffalo milk   78, 100, 102, 104, 171–2
burek   2
burger   236
butanol   21
butter paper (BP)   54
buttery   85, 215
butyric acid   86–7

c

cakes   5, 11, 16, 34, 67, 103, 145, 157, 187, 266
calcium   30, 125, 127, 171, 256, 262, 271
  alginate   22, 38, 215, 221
  carbonate   22, 206
  chloride   22, 24, 30, 56, 211, 239
  ion   21, 145
  lactate   30, 32

pectinate   22, 33
phosphate   239, 251
salts   77
setting bath   30
calorific contributions   267
camel   78, 171
candy   16, 67, 116, 178–80, 193–4, 220
capillary jet   18, 33
capsaicin   214
capsanthin   214
capsorubin   214
carbohydrates   72, 84, 99, 122, 130, 174, 196, 230, 240, 262, 266, 272, 278
carboxymethylcellulose (CMC)   11, 53, 80, 84, 86, 88, 103–4, 115, 172
β-carotene   53, 66, 212
carotenoid   212
carp jelly   245
carrageenan   21–4, 34, 38, 56, 80, 118, 174–5, 177, 180, 184, 186–7, 190, 196, 201, 238, 261
  beads   37
  gel   37, 38, 98
  jelly   180, 183
  powder   181, 183
ι-carrageenan   11, 100, 103, 177, 190, 223, 264
κ-carrageenan   11, 21–2, 37, 88, 91, 105, 117, 119, 129, 130, 133, 169–70, 173–4, 176–8, 184–6, 188, 190, 209, 227, 238, 243, 254, 264–5, 272
λ-carrageenan   82, 86, 89, 90, 97, 177, 271
carriers   36, 58, 66, 70, 200, 203
carrot chips   147, 159
carrots   27, 39, 68, 70, 240
casein   82–3, 99, 100, 105, 133, 153, 162, 165, 172, 184, 210, 228, 263, 265
β-casein   209
caseinate   82–3, 209, 184, 199, 209, 227
casing   10
cassava
  root   206
  starch   36, 151
Casson equation   114, 129
casting   25
catalase   215, 264

cellular solids   202–5, 223, 227
cellulose   54, 70, 93, 98, 121, 193, 238
  backbone   93
  -based edible film   70
  derivative   36, 54, 70, 162, 177, 190
  ether   92, 94, 99
  gum   116
  hydroxyl groups   170
ceramics   202
cereal   72, 77, 94, 130, 138, 149, 153, 157, 161, 164–5, 184–8, 191, 228, 253, 270, 272, 274, 277
  bars   138, 268–9
  -based product   65, 242, 253, 263
  -based snacks   153, 230, 237
  -based 3D snacks   237
  -based tubes   206
  clusters   149
  crops   176
  flour   94, 153
  β-glucans   170, 190, 260
  grain(s)   256–8, 271
  ingredients   192
  layers   206
  polysaccharides   259–60, 274
  product   157, 161
  sources   206
  tube   206
charged   23–4, 120, 199, 200, 208–9, 213, 215, 240
cheese   3, 5, 11, 26, 53, 66–7, 78, 99, 172–3, 183–6, 187, 188, 242–3, 263
  balls   5, 30, 141
  -based sauce   2
  cast   241
  Cheddar   74, 78–9, 102, 104, 184, 271, 273
  coatings for   55
  cracker   236
  Egyptian kariesh   172
  full-fat   263
  infill-printed   241
  layer   214
  low-fat   101, 103, 172, 190
  melted   2, 241
  moisture   172
  powder   31
  processed   225, 240–1, 252
  soft   172
  wheels of   5
chewiness   167, 172, 175–9, 187, 189, 243
chewy foods   166–7
chicken   1, 64, 95, 96, 161, 181, 270
  balls   5
  battered   96
  buffalo   5
  coated   93
  fried   5, 74, 92–3, 95
  marinated   93–4, 96
  nuggets   143
  teriyaki   61
chili
  pepper paste   2
  powder   156
Chipos and Pringles   156
chitosan   21, 38, 80, 175, 187, 191, 198–9, 221–3, 226–7
Choc Edge   233
chocolate   4, 25, 37, 114, 129, 212, 223, 232–3, 235, 240, 242–3, 251–2, 277
  bar   4, 194
  choco   195
  chunks   4
  cookies   269
  covering   194
  designs   235
  desserts   235
  fat-based   194
  flavored   56, 212
  milk   56
  paste   10, 103
  printed   236
  spread   81, 97, 99
  syrup   56, 112
chroma   42–3, 46
  meter   43
chronic kidney disease (CKD)   265, 273–5
CIE (Commission Internationale de l'Eclairage)   43–7, 49
circle   3–5, 12, 14, 21

circular shape   4, 5
coacervation   23
cob loaf dip   5
cocoa   56, 71, 114, 212, 228
coextrusion   165, 194, 205–7, 223, 226–8
coffee   87–8, 112
  accessories   232
  bitterness of   88
  disposable cups   202
  instant   87
  manufacturing   6
cold gels   24
color   1, 7, 26, 40, 41–6, 48, 50–1, 53–60, 62,
    65–9, 72–3, 76–7, 79, 101, 105, 107,
    116, 126, 146, 151, 155, 161, 168, 198,
    207, 212–13, 216–17, 219, 224, 236,
    242, 246, 249, 251
colorant   42, 60, 66, 211, 213–14, 225
colored products   192, 211
computer-aided design (CAD)   230
confectionery   10, 25, 51, 192–4, 206–7,
    216–17, 221, 224, 226, 236
consistency   10–11, 27, 56, 88, 113, 116, 119,
    129, 182, 194–5, 199, 244, 263
cookies   5, 61, 145, 150, 194–5, 236–7, 249,
    268–9, 274, 277
cooling tunnel   10, 207
corn
  fiber gum   117, 135
  starch   134, 151–2, 160, 191, 244, 250
crackers   4, 138, 151, 161, 173, 236
crackly   136, 142–4, 158–9
crepes   5
crispy   56, 70, 93, 136–7, 140–4, 146,
    148–9, 153, 158, 161–2, 164, 194, 206,
    266, 269
Cross equation   115
crunchy   93, 136–8, 140–4, 146, 154–5,
    157–9, 161–4, 179–80, 185
  products   136, 154
crust   2, 12, 56, 69, 142, 150, 155, 162,
    168, 180
cupcakes   5
curdlan   156, 176, 186, 195–6, 201, 224,
    226, 264

cutting   7, 12, 15, 26, 38, 70, 179, 207, 241
  device   206
  mechanical   23
  method   36
  techniques   1
Cyber Wagashi   245–6

d

3D
  printed foods   230, 234, 239
  printers   231–7, 241–2
  printing   208, 222–5, 230–45, 247–54
    technology   242, 248
4D printing   248, 250, 252–4
dairy foods   171, 188, 191, 273
deacetylation   176
γ-decalactone   82
deep fat frying   56, 65, 69, 71, 149,
    158, 161–2
Dentsu   237
depositor   10
desserts   11, 122, 129, 211, 216, 235, 271
dextran   79, 80, 84, 106, 116, 120, 226,
    229, 263
dextrin   65, 146–7, 275
diameter   7, 11, 13, 14, 18, 19, 20, 21, 22, 31,
    36, 83, 145, 156, 208, 236, 238, 239,
    248, 260
die   10, 152, 206, 207
dietary fibers   184, 240, 256, 258, 261–2,
    265–6, 272, 274, 277
diffused light   48
dill   126
dilute region   115
dispersion   20–1, 28, 30, 55, 98, 117–18,
    121–2, 135, 219, 225
disulfide   83, 85
double-layered   133, 196, 201, 204, 210–11,
    214–16, 226, 229
dough   2, 4, 10, 15, 94–6, 149, 150, 151, 154,
    156–7, 159, 169, 170–1, 185
  adhesion   12
  balls   5
  bread   170, 185
  composition   94, 99, 150, 161, 164

dough (cont'd)
  crust 12
  layers 2
  leavened 3
  pasta 171
  potato-based 156
  product 11
  rheology 33
  rice 11
  scraper 16
  volume 169
  wheat 131
doughnuts 5, 11
dressing
  French 107, 126–8, 132
  Italian 126–8
drying 51, 53, 57, 59, 65, 66, 67, 69, 72, 98, 119, 120, 151, 153, 155, 163, 179, 180, 188, 197, 206
  cabinet 53
  freeze 144
  hot-air 7
  microwave 69
  spray 58, 66, 67, 68, 69, 70, 71, 72, 73, 224
  sun 52
dry yeast 94–6, 267, 269–70
dull 40, 48
durometer 167
durum wheat 29, 171, 192
dynamic viscosity 111, 133, 178
dysphagia 26, 28, 33, 36, 38, 120, 127–8, 131–4, 242

e

edible films 54–5, 61, 66–8, 70, 104, 198
Edo Period 179
efflux-type viscometer 107
egg(s) 68, 69, 210, 256, 267, 269, 270, 273
  albumin 82, 83
  coating 55
  hard-boiled 5
  white 57, 256, 267, 269, 270
  yolk 25, 249, 274, 277
elderly 26–7, 35, 233–4, 242, 249, 255
electrostatic 23, 33–4, 36, 198, 208–10, 220, 223–4, 228

ellipsoid 13–14, 21, 23
elongation ratio (ER) 22
emulsification 7, 17, 23, 35, 36, 117, 126, 132, 188, 210, 211, 216, 226, 255
emulsifiers 103, 126, 130, 209–10, 227
encrusting discs 10
energy 7, 8, 9, 10, 17, 45, 125, 140, 141, 145, 152, 182, 194, 202, 206, 248, 249, 261, 262, 264, 266, 273, 275
Escherichia coli 257
esters 75, 78, 80, 82, 84, 89, 103, 127, 130, 151, 193
ethyl ferulate 61
ethyl phenylacetate 61
ethyl vanillin 78
Euclidean 5, 33
eugenol 82–3, 92, 103
exopolysaccharides 104, 262, 275, 277–8
extruded snack 151–3, 158–9, 162
extrusion 10, 80, 102, 151–3, 157, 159, 160, 192, 194, 205–8, 220, 222–3, 225–8, 230–2, 235–6, 238–41, 243–4, 250–2, 254

f

fabricated 155–7, 164, 209, 249, 250
Farmer Producer Organization (FPO) 53–4, 69
fat replacers 79, 101, 172, 186–7, 260, 264–5, 271, 276
FDA 16, 55, 67, 260
features 7, 11, 115, 123, 140, 164, 198, 209, 240, 241, 248, 249, 259, 260–1, 266
  flavor 80, 83
  sensory 7, 265
  textural 142, 166, 193
fennel 126
fenugreek 116–17, 130, 275
filled pastry 2
filling(s) 1, 2, 10, 11, 12, 48, 76, 120, 143, 193, 194, 206, 223, 228, 236
filter(s) 6, 45, 49
fines 6
firmness 6, 24, 136, 144, 151, 166, 167, 178, 195

fish
  fingers 80, 105, 184, 189
  products 11, 80, 100, 166, 174–5
flavorings 75, 77, 126–7, 211
flavor(s) 1, 51, 55, 61, 77–8, 84,
    85, 98, 102–3, 105, 126, 193,
    195, 224
  compounds 74, 80–6, 88–90, 97–100,
    102–4, 106, 132, 215
  release 11, 26, 74, 83–5, 88–92, 98–102,
    106, 126, 132, 173, 185
flour 4, 79, 93–6, 102, 148, 150–1, 153–4,
    158, 160–3, 165, 169, 170, 181, 183,
    188–9, 191–2, 206, 237, 248–9, 258,
    266, 268, 269–70
flow
  equations 107, 112
  rate 18, 22, 241
fluid gels 107, 124–5, 130–2, 134
fluidity 111
Foodini 233, 236
food(s)
  center-filled 10
  commodities 42
  layered 208
  liquid 7, 27, 129
  nutrition 255
  powder 6, 7, 58, 65
  semi-liquid 7, 129
  spherical 5, 30
forming 10
four-layered 201, 203
fragrance 77, 100, 101, 104
freeze-drying 66, 144
fresh market 7
fried chicken 5, 74, 92–3, 95, 161
from a bottle 107
Froot Loops 77, 101
fruit 7, 11–12, 14, 16, 26, 41–2, 48, 51–5, 59,
    63, 65–6, 67, 69–72, 77, 88, 92, 98,
    101–2, 118, 129, 137, 138, 158, 160–1,
    168, 198, 200–1, 203, 206, 211–13,
    221–2, 234, 238, 240, 242, 250, 253,
    258, 261, 266, 268
  leathers 51, 72
  snacks 238

frying 56–7, 64–5, 69, 71, 93–6, 136, 139,
    146–51, 157–63, 165, 268
fused-deposition modeling (FDM) 230

g

galactomannans 116–17, 135, 188, 201
garlic 85, 93, 95–6, 103, 126, 128, 156, 257
gel
  beads 17, 22, 36, 38, 215
  films 25
gelatin 24, 36, 61, 82, 89–91, 93, 97, 99, 105,
    119–20, 135, 176, 192–4, 196–8, 200,
    209–10, 216–20, 222, 224–6, 228–9,
    236, 240, 243, 245, 250, 253, 254
gellan 21, 24, 27, 32, 39, 85, 89, 91, 98, 125,
    130–1, 133–4, 177–8, 186, 193
gelling agent 1, 24, 29, 63, 85, 91, 120–1,
    125, 179, 214, 234, 240, 245, 265
generally recognized as safe (GRAS) 55, 178
geometrical 1, 4, 6, 20, 174, 235, 240
glassy-state 83
gloss 1, 40, 42, 44, 46, 48–50, 51–2, 54, 55–6,
    58, 60–2, 64, 66–8, 70, 72
  range 50
  units (GU) 49
glossmeter 49–50, 54–5
glucomannan 22, 116–17, 172, 176–8, 182,
    186, 190–1, 246, 257, 262, 273
L-glutamic acid 76, 104
gravity-separation processes 8
green beans 48, 147
grinding 7–10, 35, 137
ground product 6
guanosine monophosphate (GMP) 77
guar gum 27–8, 53, 59, 60, 62–4, 79, 84–6, 101,
    103, 115–17, 123, 127, 129–31, 134–5,
    144–6, 156, 158, 172, 174, 178, 187–9,
    255, 257, 261, 262–6, 270–1, 273, 276–7
gum
  acacia 53, 262
  arabic 58, 92, 116, 120, 130–1, 187, 193–4,
    222, 261, 271, 273, 276
  karaya 116
  tragacanth 57, 59, 60, 97, 103, 116,
    264, 271
gummy worms 25

h

half-moon 1
ham consommé 23–4
hardness 9, 79, 98, 140, 143, 145, 148, 150, 155, 158, 165–7, 170, 172–9, 181, 186, 188–9, 197, 205, 234, 240, 243–4, 260, 269
harissa 2
hazelnuts 140, 163
headspace 85, 88–91, 97–8, 101–3, 106
health claims 184, 261–2, 278
heart and soul 95
heat transfer 10, 15, 114
Helmholtz, Hermann von 43
Herschel–Bulkley 113, 134
hot
   extrusion 3D printing 244, 250
   sugar mass 10
hue 30, 42–3, 46–7, 57, 59, 195, 213
Hull 230
hydrocolloid 24, 27–8, 40, 53, 55–8, 60, 63, 65, 75, 78, 82, 84–6, 88, 90–2, 97–8, 106–7, 115–16, 118–19, 121, 124–5, 129, 134, 145, 149, 150, 172, 174, 178–9, 181, 184–5, 189, 191, 195, 200–3, 208, 221, 226, 238–9, 247, 250, 255, 261, 264, 272
   applications 24, 36, 70, 132, 226
   beads 14–15, 17, 23, 36
   films 40, 55
   foods 23, 244
   gelling 180
   gels 90–1, 101, 125, 179, 192, 201
   gums 24, 102
   products 17, 192
   thickeners 86, 115–16
hydroxymethylfurfural 58–9
hydroxypropylcellulose (HPC) 54
hydroxypropyl methylcellulose (HPMC) 11, 21, 23, 54–5, 61, 74, 80–1, 86, 92, 93–4, 96, 115–16, 146, 156–7, 169–70, 262–3

i

ice cream 37, 56, 105, 116, 120, 172–4, 184–90, 266
Ikeda, Kikunae 75–6
index 6, 11, 37, 62, 113, 278
   absorption 150
   Bond-work 10
   browning 54, 57
   color 58
   consistency 113, 244
   expansion 152
   lightness 57
   lower glycemic 258
   packing 6
   refractive 49
   rheological 243
infill percentage 230, 238, 241–2
injection molding 207
inosinic acid 75–77
interfacial layers 209–10, 229
inulin 11, 136, 149, 159, 163, 175, 182, 184, 186, 188, 191, 256–7, 261–2, 264–78
β-ionone 82
isoamyl acetate 82, 86, 92, 98
Italy 3

j

jackfruit 52–4, 66, 151, 161
jet diameter 18–19
juice 127, 168
   apple 103, 217, 219, 220
   cactus pear 71
   citrus 219
   fruit 59, 66, 222
   ginger 62, 64
   lemon 64, 126, 216, 219, 238–9, 243, 249, 254
   lime 156, 219
   melon 23
   milk beverage 88
   onion 127
   orange 47, 67, 101
   pineapple 58, 65
   pomegranate 73
   raspberry 88, 97
   spray-dried 58
   tomato 7, 34, 87–8, 123, 134
juiciness 166–8, 190, 265

k

kamaboko   175, 187, 191, 197
ketchup   120, 127
keto bread rolls   266–7, 269
Kick, Friedrich   9, 34
Kick's law   9
kinematic viscosity   111
knot foie   24
kohakuto   179
Kubelka–Munk   46–7, 54–5

l

L*a*b* color space   43, 45–7
Lactobacillus plantarum   200, 223, 263, 272
Lactobacillus reuteri   263, 277
laminar flow   108, 110, 112
laser cooking   230, 248, 250, 252–3
length   7, 18, 22, 31, 121, 245, 268
lentils   1, 17
Leucaena   117
lightness   43, 45–7, 53, 57, 59, 62
lignin   121
lipoxygenase   7
liquid
　core   30
　jet   18
Listeria   257
locust bean gum (LBG)   27, 59, 60, 91, 115, 134, 144, 163, 178, 201, 261, 264–5, 271–2
low methoxy pectin (LMP)   21, 63–4

m

magnesium stearate   235
maize   122–3, 131, 153–4, 163–4, 176, 186, 206, 273
maltodextrins   58, 151
mango   52–3, 66–71, 147, 161, 198
　chips   147, 161
　snacks   147
Manilkara zapota   52
marinade   94, 96
marshmallow   138
Massachusetts Institute of Technology (MIT)   231
mastication   1, 7, 26–7, 34, 38–9, 75, 136, 140, 142, 158, 160, 167, 255, 261
material(s)   7–12, 23–5, 35, 38–9, 50, 57–8, 68, 70, 75, 80, 84, 99, 100, 102, 110, 115, 118, 121, 124–5, 141, 144–5, 150, 156, 162, 166–8, 176, 193–4, 197–9, 203, 205–8, 209, 212, 214–16, 221, 222–6, 229–32, 234, 237–40, 241–3, 247–54, 258, 260–1, 271, 276
Maxwell, James Clerk   44–5
mayonnaise   24, 112, 116, 126, 128, 131, 236
meat   2, 4, 11, 16, 26, 38, 68, 74, 75, 80–1, 93, 99, 100, 102–3, 112, 120, 148, 168, 174–6, 182, 188–91, 206, 236–7, 239, 240–1, 243, 248, 250, 255, 259–60, 265, 270–4, 276–7
　analogs   80–1, 99
meatballs   5, 182, 186
melon   23–4
Mentos   193
mesh size   6, 7
mesquite gum   122, 124, 130–1, 134
methyl siloxanes   16, 32
Mexican   2, 5, 154, 189
microbeads   22, 38
Micromake   233
microstructure   33, 54–5, 66, 72, 80–1, 91, 99, 100, 104, 122, 125, 129–30, 135, 146–7, 189, 205, 224, 238–9, 241, 244, 250–1, 253, 263
milled foods   26
minced foods   27, 38–9
mints   193
mirin   61–2, 64
mixed gels   176–8
mixing   7, 10, 29, 44, 61, 85, 95–6, 116, 118, 127, 189, 217, 236, 237, 238, 243
mizu-Shingen mocha   183
MMuse   233
　molds   1, 10, 15–17, 25, 34, 183, 233
molecular weight   28, 79, 84, 93, 115, 117–20, 122–3, 127, 129, 150, 163, 170, 188, 210, 260, 265, 277
mozzarella   5, 100, 214
multilayer adsorption   210

multilayered
  emulsions   208–9, 227
  films   192, 197–8
  food products   192, 194–5, 205
  gelatin jelly   192, 216, 219–20
  gels   196, 200, 202–3, 215, 221
  liposomes   199
  particles   199, 200, 226
  sweets   218
  texturized fruit   200–1
Munsell   42–3, 69

n

nachos   2, 3
nanometer   23
nano-multilayer coatings   198–9
Naples   3
nappage neutre   40, 63–4
NASA   237
natural
  hydrocolloids   79, 255, 278
  machines   233
neutral mirror glaze   40, 63–5
Newtonian
  behavior   115, 122
  flow   112–13, 120, 240
  fluid   107, 110, 112–14, 121, 129
  liquid   18
  plateau   115
noodle   1, 15, 29, 181
nursing care foods   234

o

oat
  β-glucan   170–1, 187–8, 260, 278
  gum   84, 101
oblong   20
odor   70–1, 74–8, 83–8, 90, 92, 100, 103, 105,
    132, 242, 266
oil absorption   56, 71, 139, 146, 149, 151,
    157, 161–3
oil-in-water emulsions   116
oil–water interface(s)   83, 92–3, 97–8,
    209, 210
Okonomiyaki   5

onion   2, 126–8, 257
Open Meals   237, 245–7
optical   207
  categorization   205
  characterization   225
  experience   40
  properties   41, 47, 54–6
  sensor   53
  society   43, 68
oral cavity   7, 26, 28
oregano   126
Oreo   5, 194–5, 236
organic acids   61, 77, 171
orientation   6, 15
oropharynx   26
ovate   20
oven   3, 16, 34, 56, 65, 96, 156–7, 171,
    206, 270
ovotransferrin   256
oxygen   7, 16, 36, 54–6, 76, 93, 149, 156,
    197–9, 216, 278

p

PancakeBot   233
pancake(s)   5, 236, 266
  mix   266, 268
papain   21, 37
parallelogram   4
parsley   2
particle
  shape   6, 7
  size   6, 7, 10, 16–17, 23, 35, 37, 55, 125,
      157, 175, 238
pasta   11, 17, 28–9, 150–1, 171, 181, 188, 236,
    263, 266, 275
paste
  hot chili pepper   2
pastry   143, 205, 214
  baked   1
  bars   17
  filled   2
  layer   214
  prebaked   214
  triangular-shaped   2
  unbaked   214

peach   42
peanut butter   5, 112, 114, 129
pea(s)   1, 18, 42, 147–8, 165
   starch   178–9, 187, 189
pectin   11, 21–2, 53–7, 60, 63–4, 66, 80, 84, 88, 90, 92, 97, 100–1, 105, 117–18, 120, 131, 144–5, 149–50, 162–3, 172, 175, 187, 198–9, 209–10, 222–3, 226, 228, 236, 238–9, 251, 254, 261–2, 264, 271
pekmez   59, 60, 65, 71, 73
permeability   6, 54–5, 66–8, 70, 146, 277
peroxidase   1
personalized foods   233
phenolic compounds   84
pie
   casing   10
   charts   266
   filling   11, 48
   molds   10
pizza   2, 3, 5, 34, 136, 149, 214, 223, 232, 236
Plantago major   117
Pleurotus ostreatus   258–9, 274
Poiseuille   107, 111, 133
polygon   1
polypropylene
   bags   170
   films   198, 222
   pouches   54
polyvinyl acetate (PVA)   55
popcorn   138–9
   balls   5, 30
pork   5, 38, 143, 186, 260, 274, 277
porous   6, 35, 56, 160, 251, 254, 263
potato   5, 11, 42, 67, 69, 71, 124, 138–9, 146, 151, 155–7, 159, 162–4, 176, 186, 239, 244, 249–50, 252, 257, 265, 272
   chips   138–9, 146, 155–7, 159, 164
powder color index   58
power equation   113–14
printed sugar products   234
projected area   12, 14–15
propylene glycol alginate (PGA)   73, 92, 127–8

pseudoplasticity   112, 115, 126
psychological impact   48, 51, 72
pudding   211–12
pullulan   40, 60–5, 67, 70–1, 73
pureed foods   242

q

quantum satis   121
quasi-monodispersed   18
quillaja saponin   227, 229
quince (*Cydonia oblonga*)
   fruit   118
   gum   117

r

raspberry   88, 97
Rayleigh   18–20, 36
Rayleigh–Ritz method   205
ready to eat   11, 80, 155, 162, 196, 228
Redefine Meat   237
reflective   40, 49, 206
refractive index   49
relative viscosity   111
relaxation   19
retinol   82
Reynolds   108–10, 133
   number   110
RGB color space (model)   43, 44
rheological parameters   7
rheopexy   123, 132, 135
rice   4, 5, 7, 11, 48, 61–3, 94, 96, 148, 152, 161, 168, 170, 176–8, 181, 186–8, 190, 193, 206, 244, 249–50, 274
   starch pastes   176–7
Rittinger, Peter von   8
   equation   37
roughness   1, 159
round   1–3, 12–13, 17, 20, 51, 154
roundness   1, 12–13, 15, 21, 33
rum   5

s

sake   61–2, 64, 95–6
salmon   5, 30–2, 226
Salmonella   257

salt   4, 24, 31, 62, 68, 77, 84, 91, 93–6, 100, 103, 114, 118, 120–1, 126–8, 151, 154, 156, 169, 174–5, 186, 189, 206, 211, 240, 259, 268–71
saltine   4
samosa   1, 104
sandwich   4, 149, 173, 194, 266
sapodilla   52
sardine   62–4, 185
sausage casings   55
scones   5
sediments   6
Seiryu   179–82
shear
  rate   28, 110–16, 119–21, 123–4
  stress   86, 98, 112–14, 116, 121–3, 175, 208
  thickening   122–4, 133
  thinning   28, 115–16, 120, 122–3, 174, 184, 238, 243–4
shellac   55, 69, 81, 103
shrimp balls   5
Sideritis stricta   58
sieve   6, 7, 10, 23
silicone   15, 16
  elastomer   16
  mat   25
  molds   16, 17
  oil   36
  spatula   18
  tube   1, 29, 31
siloxanes   16, 35
sirloin   245
size
  control   7
  enlargement   10
  reduction   7–9, 33, 157
sliminess   121
slit screens   7
snack foods   10, 136, 151, 153, 155, 159, 162
soda cracker   4
sodium chloride   22, 86, 122, 240
soy protein   82, 97–9, 122, 134, 153, 220, 227, 243, 250
  isolate (SPI)   122, 220, 227, 243, 250
spaghetti   1, 17, 24, 28–9, 31, 33

sphere   4, 5, 11, 13–15, 20–2, 134
sphericity   1, 12–16, 21–3, 28, 36, 38
spherification   1, 30
spheroid   13, 20, 22–3, 30
spirulina   153, 160
spreads   74, 81, 97, 99, 103
staling inhibitors   79
standard observer   45–6
starches   58, 83, 115, 119, 122, 124, 134, 136, 148, 150–1, 158–62, 165, 177, 178, 186, 187, 190, 250, 257, 261
Stokes   84, 108, 111, 133
strawberry   5, 63, 78, 90, 97, 100, 201
String-type agar   179, 182–183
Sub-Saharan Africa countries   57
surface tension   19, 23, 92, 238
surfactant   17, 54–6, 92, 95, 158, 229
surimi   175, 184, 187–9, 197, 240, 243, 244, 250, 254
sushi   5, 30, 32, 104, 174, 234, 237, 245, 247
suspension   23, 114–15, 121, 125–6, 156, 195–6
syringe   17, 22, 24, 29, 31, 232, 235, 243

t

tacos   5
tara gum   116
tart   5, 63
tartan   45
taste   26, 34, 51, 54, 70–1, 74–5, 77, 79–82, 84, 86, 88–9, 91–4, 98–9, 102–3, 107, 126, 137, 140, 151, 208, 218, 242, 266, 268
tetradecanoic acid   82
textural sensation   1
thickening   36–7, 86, 93, 104, 107, 115–16, 118–22, 130, 132, 134, 188, 190, 226
  agent(s)   27, 38, 116, 120, 122, 127, 128, 275
  properties   116–18
  purpose   28, 117–18
thin-walled pores   145
thixotropic   27, 117, 121–3, 135
Tikhonov   124
Tim Tams   195

Toblerone   4
tomato   5, 7, 29–31, 34, 37, 54, 65, 67, 87, 88, 112, 123, 126, 134, 198, 214, 236
tongue   26, 74, 122
topography   40
topology   254
topping   3, 56, 212
tortilla   2, 154, 155, 156
  chips   2, 136, 155, 156
totopos   2
transparent   25, 49, 110, 180–1, 198
tuna   2, 224, 234
Tunisia   2
turbulent flow   17, 108, 110, 122

u

umami   62, 75, 77, 99, 101, 234
unit operations   7, 8, 33

v

vacuum frying   136, 146–7, 157–9, 161, 165
vegetable   11–12, 53–4, 65, 67, 70, 72, 126, 137, 138, 146, 152, 158, 168, 176, 181, 191, 206–7, 212, 240, 242–3, 253, 258–9, 268
  puree   243
vibrating
  capillary apparatus   34
  nozzle   18
videofluorographic (VF)   28, 36
Vienna   9
viscosity formers   27, 84, 107, 116
viscous liquid   19, 27, 196
visual appeal   1, 15–16
vitamin(s)   29, 171
  C   53–4, 199
volatile flavor compounds   74, 84, 88, 97, 100, 103

w

waffle   4, 11
wagashi   179, 218, 245–6
water activity   70, 138, 141, 145, 149, 158, 160, 162–3, 170
watercore   41, 67
wavelength   18, 43, 45, 47, 248
Weber   19, 20, 38, 123, 134
Weissella cibaria   263
wet-classification   8
whey proteins   83, 100, 134, 200, 222, 223, 256, 274
wholegrain   79, 80, 106
Wich   3

x

xanthan   11, 25, 27–8, 38–9, 56–7, 68, 73, 79, 80–1, 84–6, 115–16, 120, 123, 126–8, 131–4, 144–5, 158, 174, 178, 185, 188–9, 201, 236, 239, 243, 261–4, 273
x-ray diffraction   76, 83, 205, 224
x-rays   205

y

yam   57, 65, 181, 206
yeast   4, 34, 61, 94–6, 169, 188, 208, 215, 228, 259, 267, 269, 270
yield stress   112–14, 116, 130, 134, 208, 238, 244
yoghurt   59, 68, 97, 129, 131, 265, 269–71, 273
Young, Thomas   43

z

zein   146, 197
zero shear viscosity   86, 115, 134
zeta potential   123
zinc   21
Zmorph by Flow   233